THE LURE OF THE SEA

The Lure of the Sea

The Discovery of the Seaside in the Western World 1750–1840

Alain Corbin

Translated by Jocelyn Phelps

University of California Press
Berkeley and Los Angeles

University of California Press
Berkeley and Los Angeles, California

First University of California Press edition published 1994

This English translation © 1994 by Polity Press.
Published with the assistance of the French Ministry of Culture.
First published in France as *Le Territoire du vide* © 1988 by Aubier, Paris.

ISBN: 0–520–06638–3

Library of Congress Catalog Card Number: 93–60742

Typeset in 10½ on 12 pt Sabon
by Apex Products Ltd
Printed in Great Britain by Biddles Ltd, Guildford and Kings Lynn

This book is printed on acid-free paper.

Contents

Foreword

Specialists of cultural history now know how to examine institutions, objects and practices. They will not, however, let themselves tackle emotionally charged systems. Yet, only by understanding such systems can these specialists give meaning to their patient and fruitful work.

The trouble is that in this field the issues of quality of the documents and validity of evidence pose especially thorny problems. If one undertakes a case-study, one finds oneself running the risk of not being representative, of reaching hasty conclusions, or of vainly assembling an insignificant collection of examples. Even if one is careful not to neglect elements that were imported from elsewhere or the social mobility of emotions, a limitation of one's approach to an inventive elite appears to reduce the historian's realm to superficial issues. Should one make a naïve interpretation, watch for the meaning of the slightest little document, break away from established preconceptions, or refuse to adopt a god's-eye view, then one may find oneself drifting far from the major interpretations of the course of history, edging away from hermeneutics. This is all the more significant when groping along an unfamiliar path.

At the end of the day, however, the most serious error in my eyes is psychological anachronism. The worst is the satisfied, abusive, and blind certainty of having understood the past. What is indispensable is to map the boundaries of what the mind can imagine, identify the mechanisms driving new emotions, trace the origins of desires and the way in which suffering or pleasure was experienced at a given time, describe habitus, and reconstitute the logic behind systems of vision and evaluation. There is no other means for understanding people from the past than attempting to see through their eyes and live with their feelings. Only by subjecting ourselves to this process can we re-create the desire for the shore that swelled and spread between 1750 and 1840.

List of Illustrations

Acknowledgements

Now that this work has reached its conclusion, I would like to thank those who helped me in various ways to complete it: André Bordeaux, professor of English literature, who translated the quotations from major authors who have not yet been published in French; Professor Jean Lafond, who shared with me his vast knowledge of seventeenth-century French literature; Robert Beck, who helped me collect the documents concerning Germany; Elisabeth Deniaux, who graciously reread certain allusions to antiquity; Josette Pontet, who knows the Biarritz area so well; and Dominique Raoul-Duval, who was responsible for the pictorial research.

I would also like to thank the following for permission to publish the various plates: the Bibliothèque nationale for plates 1, 2, 6, 7–14, 22–4; the National Maritime Museum, Greenwich, for plates 3, 26, and 27; Giraudon, for plates 4 and 21; the Altonaer Museum, Hamburg, for plate 7; the Bury Art Gallery and Museum, for plate 15; the Yale Center for British Art, for plate 16; Bulloz, for plates 17 and 20; Museum Stiftung Oskar Reinhart, for plate 18; Jörg P. Anders, Berlin, for plate 19; and the French National Museums, for plate 24.

Part I

Unconsciousness and the
First Premises of Desire

1

The Roots of Fear and Repulsion

With few exceptions,[1] the classical period knew nothing of the attraction of seaside beaches, the emotion of a bather plunging into the waves, or the pleasures of a stay at the seaside. A veil of repulsive images prevented the seaside from exercising its appeal. Blindness, as well as horror, was built into an overall system of appreciation of natural landscapes, meteorological phenomena, and coenaesthetic[2] impressions (those which created a sense of existence on the basis of a collection of bodily sensations), whose outline had gradually taken shape since the Renaissance.[3] To understand properly the origins of the way in which coastal landscapes were 'read' and the new practices that emerge surrounding them around 1750, it is first necessary to grasp how coherent the cluster of representations were that formed the basis of the repulsion.[4]

THE ABYSS CONTAINING THE DEBRIS FROM THE GREAT FLOOD

The interpretation of the Bible, especially the book of Genesis, the Psalms, and the book of Job, exerted a deep influence on the ways in which the sea was portrayed.[5] Both the story of the Creation and that of the Flood coloured the world of collective imagination with their own specific features. Genesis imposed the vision of the 'great abyss', a place of unfathomable mysteries,[6] an uncharted liquid mass, the image of the infinite and the unimaginable over which the Spirit of God moved at the dawn of Creation.[7] This quivering expanse, which symbolized, and actually was, the unknowable, was frightful in itself. There is no sea in the Garden of Eden. There is no place within the enclosed landscape of Paradise for the watery horizon whose surface extends as far as the eye can see. To attempt to fathom the mysteries of the ocean bordered on sacrilege, like an attempt to penetrate the impenetrable nature of God, as St Augustine, St Ambrose, and St Basil repeatedly pointed out.[8]

This unconquerable element was evidence that Creation remained unfinished. The ocean was the remnant of that undifferentiated primordial substance on which form had to be imposed so that it might become part of Creation. This realm of the unfinished, a vibrating, vague extension of chaos, symbolized the disorder that preceded civilization. A firm belief began to appear which held that already in antediluvian times, it was only with difficulty that the raging ocean could be contained within its bounds.[9] Consequently, the ocean inspired a deep sense of repulsion; for the classical age was apparently unaware of the desire to return to the womb of the Creator and the longing to be swallowed up that were to haunt the Romantics.

Since Creation was organized around the appearance of man, who was both its goal and its focus,[10] this vestige without form remained alien to him. A creature fashioned in the image of God would never make his abode outside the garden or the city.[11] Besides, the Mosaic text mentions only the creatures of the air and the field; marine species, hidden in the mysterious darkness of the deep, could not be named by man, and consequently were outside his dominion.

Even more telling is the story of the Flood. According to the authors, the ocean was an instrument of punishment, and in its actual configuration was the remnant of the disaster. According to Mosaic cosmogony, there are two great expanses of water: that which fills the oceans and that which is held within the vaults of heaven. In separating them, the Creator drew two dividing lines: the coast, which delineates the respective domains of the sea and the earth, and the clouds, that shifting *limes*[12] established between the waters of heaven and the atmosphere breathed by man. Opinions were divided as to which of the two watery bodies had submerged the earth during the Flood.[13]

In any case, the ocean spoke to pious souls. Its roaring, its moaning, its sudden bursts of anger were perceived as so many reminders of the sins of the first humans, doomed to be engulfed by the waves; its sound alone was a permanent appeal to repent and an incitement to follow the straight and narrow path.

The Flood marked a temporary return to chaos; this return of the boundless waves haunted the cultured minds of the Renaissance. The invasion of the waters was a major pictorial theme, whose evolution can be traced from the ceiling of the Sistine Chapel to Nicolas Poussin's evocation of the ocean in winter.[14] French poets at the end of the sixteenth century, especially du Bartas in his *Sepmaine*, were fond of dwelling at length on the story of this disaster.[15] A hundred years later, the Flood stood at the heart of the debate provoked by the great theories regarding the origins of the earth. After all, without the Flood, the history of our globe and its relief seemed completely incomprehensible.

It is worth dwelling for a while on these cosmogonical theories,[16] which have been analysed up to now primarily from the sole point of view of the history of science; they create a vivid picture of the link that developed between the scholarly evocation of the great catastrophe of the past and the contemporary assessment of landscapes. In this respect, Thomas Burnet's *Theory of the Earth* assumes a special significance. This book, to which constant reference was made throughout the eighteenth century, is both backward-looking and premonitory. It was contemporary with the rise of natural theology, which was to transform images of the sea and its shores; moreover, it foreshadowed the great changes in the field of aesthetics which were eventually to lead people to sample this hideous beauty.

According to the British theorist, there was no sea in Paradise, nor later in the antediluvian earth inhabited by Adam and his descendants after the Fall; all men lived on the same continent. The surface of this primitive globe was as gentle as a beach. 'The face of the Earth before the Deluge was smooth, regular and uniform, without mountains, and without a Sea ... it had the beauty of Youth and blooming Nature, fresh and fruitful, and not a wrinkle, scar or fracture in all its body; no rocks or mountains, no hollow Caves, nor gaping Channels ... The Air was calm and serene.'[17] There were no hurricanes on the antediluvian earth. An eternal spring prevailed, as in the golden age evoked by Virgil.

The Flood occurred when God opened up the great abyss of the waters; universal chaos then spread a second time over the darkness and the mists of the earth. Even the stormy sea cannot provide an adequate image of this cosmic turmoil; the recession of the waters, ordained by God, went on for a long time; for a long time the flooding ocean continued to sweep into subterranean caverns. The present sea is merely this great abyss once again restrained by God; its basins, its shores, and the mountains that fix its boundaries date from the Flood; they comprise 'the most frightful sight that Nature can offer'. [18]

As a result, it seemed very likely that the bottom of the sea was terribly chaotic, as one might suspect from the anarchic distribution of islands. If that horrible, monstrous ground ever became visible, men would see the most misshapen cavity on earth stretch out before their eyes. 'So deep, and hollow, and vast; so broken and confus'd, so every way deform'd and monstrous. This would effectually waken our imagination, and make us enquire and wonder how such a thing came in Nature.' [19]

In fact, the coastline was nothing but ruins. This accounted for its irregularity and the incomprehensible pattern of the reefs which border it; there was no order to be found there. Because they were so unaesthetic, the sea and its shores could not, according to sound theology, date from Creation; they could not be the result of Nature's original work. The ocean was nothing but an abyss full of debris; at best, it might be conceded that it formed the least unsightly landscape which could have resulted from the temporary return of chaos. [20]

William Whiston's *New Theory of the Earth* also aroused considerable interest. It relates to a system of appreciation that is fairly close to Burnet's, although the interpretation of the unfolding of the planet's history is quite different from one book to the other. According to Whiston, the primitive Earth looked very much like the Earth today; it too included an ocean that was salty and whose tides were gentle; however, this ocean did not keep men apart, as they were all gathered together on a single continent. Its shape was different from, and its volume less than, that of the present seas; moreover, it was not afflicted by storms.

In the sixteenth or seventeenth century of Creation, the fountains of heaven opened up and caused a world-wide flood that completely upset the structure of the globe. Whiston, however, presents a calmer picture of the disaster than Burnet. Throughout the forty days, the waters that submerged the land remained fairly calm, in order to avoid overwhelming the Ark. As the waters of the Flood receded, the coastlines of continents, which from then on were separated from one another, took on their complex outline. The waters, deeper in their centre than before, continued to be rocked by terrible storms. For Whiston, as for Burnet, the oceans are indeed remnants of the Flood; but according to the former, the catastrophe merely modified the shape, appearance, and outline of the primitive ocean.[21]

Until about 1840, marine disasters remained central to the natural history of the earth, and then to geology; we shall return to this. Even very late on, a number of scholars endeavoured to vindicate the Genesis narrative; as late as 1768, Alexander Cattcott in his *Treatise on the Deluge* gives a step-by-step commentary on the Old Testament story, which he considers perfectly satisfactory.[22] Like most defenders of the biblical text writing in the eighteenth century, he backs up his argument by making use of accounts of the Flood recently collected from ancient peoples such as the Assyrians, the Persians, the Babylonians, the Egyptians, the Greeks, and the Latins. He even refers to Indian and Chinese traditions. In his opinion, the sand on the sea-shore, the erratic boulders to be found on certain beaches, and natural chasms cannot be explained without reference to the Flood.

In the aftermath of the French Revolution, when the theories of Burnet, Woodward, and Whiston were discarded as obsolete, a new generation of 'catastrophist' scholars appeared: with different arguments and in a different scientific context they continued to uphold the accuracy of the sacred text.[23] For example, according to Richard Kirwan,[24] the abruptness of the coasts of Ireland, Scotland, and the islands on their fringes was a result of the onslaught of the great southern ocean, whose incursion caused the inundation. In his opinion, the air that poisoned the Earth was also a by-product of the Flood; it was the evil-smelling vestige of the mephitism[25] that prevailed as the waters receded, when the Earth remained covered with the dead and rotting flesh of drowned animals. In order to escape from these emanations, Kirwan asserted, men went on dwelling in the mountains for a long time. This interesting belief arose out of the obsession with infection which was part and parcel of the neo-Hippocratic tradition, and it reinforced the repulsive image of the seaside.[26]

For our purpose it is fundamental to grasp the huge importance accorded the Flood by scholars writing in the watershed period of the seventeenth and eighteenth centuries. All of them placed the disaster at the centre of their cosmogonies; all reasoned within the framework of a restricted temporality; and all believed that the histories of mankind and the Earth were one and the same, composed as they were of simultaneous episodes.

It is clear how the ocean, a threatening remnant of the Flood, came to inspire horror, as did the mountains,[27] that other chaotic vestige of the disaster, which were 'pudenda of Nature',[28] ugly, aggressive warts that grew on the surface of the new continents. This repulsive interpretation was in keeping with the certainty that the world was in decline. No matter how zealously they worked, men would never be able to re-create the antediluvian Earth, on whose surface the traces of earthly paradise could still be seen.

The endless movement of the seas suggested the possibility of a new flood;[29] it was part of the vague threat that hung heavily over the sanctuaries of happiness. Of course, caution is necessary on this point. The description of the apocalypse in the book of Revelation makes it clear that the final 'conflagration' will not come from water, that vestige of the chaotic, diluvial past, but from fire sent by God. The universal blaze was to ensure the victory of the purifying agent.[30] When Christ returns, the sea will have disappeared.

However, the ocean's anger was expected to play a part at the beginning of the series of cataclysms. Among the fifteen signs heralding 'the advent of Our Lord', the *artes moriendi*, widely circulated from the fifteenth century, assigned water a devastating role.[31] The sea would submerge the mountains before tumbling into the bowels of the earth; the fish and the ocean monsters would rise to the surface, uttering their cries; and the waters would howl at the fire coming from heaven.

This religious cosmogony, summarized all too briefly here, sets forth certain stereotypes for understanding the sea and the creatures dwelling in it, and endows them with a rich symbolic significance. In the figure of Leviathan, 'the dragon that is in the sea',[32] the Bible established the monstrous nature of fish, which also follows logically from the Creation story. The dragon that is rent asunder by the archangel St Michael first rose out of the sea.[33] The peregrinations of the Irish monks in the Middle Ages, especially of St Brendan,[34] later reinforced this interpretation. According to Benedeit's account, it took all the hero's saintliness to appease the horrible beasts that crawled out of the depths of the abyss. Beowulf had to dive into the sombre loch to kill the nameless female who begat the monster Grendel in another legend that testifies to the terror inspired by the marine creatures that appeared on the shores of the northern ocean. In the sixteenth century, the Swedish bishop Olaus Magnus believed firmly in sea monsters. Even in 1751, following a thorough investigation conducted among sailors, Erich Pontoppidan devoted a long chapter of his *Natural History of Norway*[35] to the sea serpent that fishermen called the Kraken. The poets of the seventeenth century also turned their attention to the theme of horrifying viscous contact with these nightmarish creatures born out of black waters[36] and risen from the chaotic world of gloomy caverns. Spenser, who settled in Ireland, tells how the holy pilgrim who accompanied Sir Guyon on his way to the Bower of Bliss was able, by touching the waves with his rod, to subdue the threatening beasts and compel them to return to the depths of the ocean.[37] Milton, in a striking image, makes the sea monsters huddle together and copulate in palaces submerged by the waters of the Flood.[38]

The ocean, that watery monsters' den, was a damned world in whose darkness the accursed creatures devoured one another. Gaston Bachelard and Gilbert Durand have emphasized the fascination that a child experiences when seeing for the first time a little fish swallowed up by a bigger one.[39] The cruel world of this food-chain in which the swallower is promptly swallowed, represents the realm of Satan and the powers of Hell. Consequently, there was nothing fortuitous about great storms; sailors saw in them the Devil's hand, or they imagined the roughness of the sea to be caused by the souls of the damned who haunted the atmosphere's intermediate zone.[40] The same image reappears in erudite culture: the description of the first circle of Hell in the *Divine Comedy* combines the ancient pattern of repulsion towards the infernal rivers' black waters with the unleashing of the diabolical storm. According to Françoise Joukovsky,[41] the image of the Satanic sea becomes more evocative in France at the end of the sixteenth and the beginning of the seventeenth century. It then gradually loses its force until it becomes a mere device intended to renew the hackneyed stereotypes of the Virgilian tempest.[42]

The demonic nature of the angry sea provided a justification for exorcism.[43] Sixteenth-century Portuguese and Spanish sailors occasionally immersed relics in the waves. These sailors were convinced that storms would not abate by themselves, but that the intervention of the Holy Virgin or St Nicholas was required. This gives an added dimension to the figure of Christ stilling the waves on the lake of Tiberias and chiding his frightened apostles for their little faith.[44]

The chaotic ocean, that unruly dark side of the world which was an abode of monsters stirred up by diabolical powers, emerges as one of the persistent figures of madness; the unforeseeable

violence of its winter storms was proof of its insanity. Jean
Delumeau stresses the frequent association established between
the sea and madness; in this connection he evokes the image of
Tristan abandoned by the mariners on the coast of Cornwall
and that of the Ship of Fools, a floating body used to seclude
lunatics, who were thus committed to the element that was in
keeping with their unpredictable temperament. [45]

The shifting immensity of the sea is also a bearer of misfortune. In both Shakespeare's early and later plays, the tempest
at sea, wild beasts, comets, illness, and vice all weave a network
of associations, evoking a world in conflict, dominated by disorder.
The grey winter ocean, dismal and cold, generates various kinds
of fear; it fosters the haunting dread of being caught by sudden
death and deprived of extreme unction, far from the home fires; of
being delivered, body and soul, unburied, to these endless waves
that know no rest. [46] The desire to ward off the brutal advent
of death explains why the practice of propitiatory rites persisted
here and there.

Pious literature had long devoted considerable space to the
symbolism of the sea and its shores; a sermon by pseudo-Ambrose
and, even more surely, a long excerpt from St Augustine's *De
Beata Vita* could in this respect be regarded as foundation texts.
In the opinion of the Fathers of the Church, the immensity of
the waters represented both the source of life and a vision of
death; [47] the Mediterranean was both angelic and diabolical, of
theological as well as geographical importance; for in spite of
the violence of its storms, it made possible Paul's missionary
travels, thereby facilitating the spread of the divine Word and
the establishment of the Christian diaspora. Life, which was
perceived as a passage and a journey strewn with perils, unfolds
in the midst of a world as unstable as the sea, the intangible
domain of vanity, in the midst of which loved ones and objects
are rolled along in a moving space without a 'petrified shell'. [48]
The evocation of this 'very bitter sea' [49] becomes a stereotype in
French poetry during the last thirty years of the sixteenth century.
This is how poets, often Huguenots fond of hyperbole and violent
images, discovered the ocean, which had been almost totally
absent from the smiling landscapes of the Renaissance. One Sieur
de Valagre perceived the world as an edifice sinking into 'the
waves of the sea' on which it was erected and as an 'ocean
of desires, appetites, jealousies, designs and projects'. Siméon
de la Rocque saw it as a 'seething, deep Sea/Which has neither
shore nor rest'. The world was also modelled on the image of

the maelstrom, the spiral abyss that fascinated Leonardo da Vinci,[50] and whose depths threaten to suck in the soul.

Flemish and later Dutch seascape painting was established on this symbolism;[51] the waves represent the fragility of life and the precariousness of human institutions. They demonstrate why it is necessary to have faith in God. Roman painting of the seventeenth century, especially that of Claude Lorrain, was also very open to the religious symbolism of the sea.[52]

The Church came to be considered as a ship, with the Holy Spirit at the helm guiding it to the eternal haven which was the object of Christian longing; sin, by contrast, made mortals stray far from the path to salvation.[53]

The sea was also sometimes interpreted as a symbol of Purgatory,[54] like a crossing that provides the sinner assailed by the punitive storm with an opportunity to repent and return to the straight and narrow path. What was beginning to appear here was the image of the redeeming sea that fosters the sailor's faith. For the ageing author of *Pericles*, *The Winter's Tale*, and *The Tempest*, passionate beings who have been caught up in the chaos of the world experience a deep moral crisis in the course of a sea voyage and shipwreck. It is through disaster or apparent loss and separation that the heroes recover their sense of meaning and the possibility returns of a world full of music and harmony among human beings;[55] but this concept has made us stray from the negative images. We shall return to it.

The sea-shores and the people dwelling there were constituents of the repulsive images evoked above. The line of contact between the world's constituent elements was also a front along which they oppose one another and reveal their madness; this was where the precarious balance established among them was in danger of being destroyed. This boundary was where the Flood would mark its return and the chain of cataclysms be triggered. It was on this shore, more than anywhere else, that Christians could come and contemplate the traces of the Flood, meditate upon that ancient punishment, and experience the signs of divine wrath. Only the harbour, that scene of desire, nostalgia, and collective rejoicing, escapes this repulsive pattern.

The burning sands of the desert and the beach, together with the swamp and the craggy mountain, form one of the images of Gehenna; they line the third circle of Dante's *Inferno*; it would be interesting to reflect on what the sight of the strand and the 'horrible' desolation of the sea bottom laid bare by the ebbing tide suggested to the men of that time.

ANTIQUITY'S CODIFICATION OF THE SEA'S ANGER

The reading of ancient texts reinterpreted by the humanists, like the quest for and contemplation of ancient art, produced other images of the sea and its shores. These combined with images derived from Judaeo-Christian tradition.[56]

The sixteenth- and seventeenth-century authors whose works constitute our sources turn only very rarely to the ancients in describing the spectacle offered by the waves and the beach;[57] they seem indifferent to the emotion that emerges from the seascapes of the *Georgics* and to the subtlety of the Alexandrian poets. Thus French Renaissance poets are almost entirely ignorant of the stillness of the sea, which gets barely a passing mention from them; the few pieces about the sea that can be found in their works are nothing but processions of mythological deities, usually inspired by texts celebrating Venus or Neptune in book 5 of the *Aeneid*. On the other hand, because their emotional reactions are so strong, as Lucien Febvre and Robert Mandrou have emphasized, these poets are very sensitive to everything in ancient texts that evokes fear and horror.

In the sixteenth century, the wrath of the sea in the *Aeneid* already stood out as a stereotype intended to give new impetus to the descriptions of storms that marked medieval accounts of travels to the Holy Land. These accounts were haunted by the fear of the void and the presumed proximity of monsters.[58] Virgilian stereotypes were borrowed partly from Homer, and were adapted by Ennius and Pacuvius and somewhat enriched by Ovid, and Seneca and then Lucian,[59] who were also widely read. They inspired novels, epics, and lyrical poems, as well as travel accounts. This model shapes the way the storm is depicted in Rabelais's *Fourth Book* of Pantagruel's adventures, as well as in *The Lusiads*. It continued to command the attention of tragic writers throughout the eighteenth century,[60] and left its mark on Thomson's vision of the tempest.[61] Monique Brosse shows how deep an impact it was to make on western Romantic literature about the sea.[62]

The sea storm in antiquity, as codified and described by second-century rhetoricians inspired by the stories in the *Aeneid* and Ovid's five descriptions, comprises a series of specific stereotypes that endow it with the appearance, albeit inaccurate, of a cyclone. At first, winds come rushing from the four corners of the horizon and fight noisily with one another; the sailors' cries, the whistling in the ropes, the din of the waves, and thunder comprise the sounds that help set the scene. The waters, laden with sand, silt, and foam, rise up like mountains, laying bare the earth at the bottom of the abyss. The clash of the waves makes the flanks of the boat tremble; at the heart of the darkness streaked with lightning, blinding rains make the sky appear to collapse. With the tenth wave, the most terrible of all, death is unavoidable, unless the sailor at prayer is saved by divine intervention.

As this well-known model was reiterated, it reinforced the image of the terrible sea: a road without a road, on which man drifts in the hands of the gods, under the permanent threat of hostile water, which is a symbol of hatred that extinguishes the passion of love as it does fire.

Horace, like Tibullus, Propertius, Ovid, and later Seneca,[63] detests the 'unsociable' ocean that keeps men apart. He condemns navigation, which he sees as a challenge to divine forces. The Adriatic terrifies him; the sea for Horace is stormy and hungry for ships in distress. A scene of bloody battles, it abounds with beasts and traps. In this respect the poet is not representative of the majority of his contemporaries, but this is not important here; the main point is that he remained widely read as long as classical culture maintained its ascendancy.[64]

Ancient literature presents the sea – and later the Atlantic Ocean – as a supreme enigma. It makes of them the primary source of scholars' grief; the seventeenth century repeatedly asserts, albeit without great conviction, that Aristotle committed suicide because he failed to elucidate the complexity of the Euripus currents.[65] Of course, oceanographic science had evolved profoundly since antiquity; however, it remained troubled by three major problems, which had already been raised by the Greeks. The first concerns the distribution and the outline of the lands and the seas; their locations seem without order, contrary to the natural arrangement of the elements according to Aristotle's *Physics*;[66] and water should in fact cover the whole of the earth.[67]

The way that water circulated over the globe also remained a partial enigma. Admittedly, even before its accuracy was demonstrated by Halley, people were quite familiar with the theory of the cycle of evaporation and precipitation which is thought to date from Aristotle;[68] but this idea of an exchange of water between the sea, the atmosphere, and the Earth was considered insufficient. Plato's model, which suggested that water passed through the centre of the earth, continued to be widely accepted. This belief in a subterranean connection between the earth and the ocean lent credence to the idea of the existence of horrible caverns at the bottom of the seas.[69] Cultural imagery was haunted by the idea of waters moving deep below, and this inspired Father Kircher's *Mundus subterraneus*, explains the appeal of the *Submarine Voyage*,[70] and supports the earth science theories developed by Burnet and Woodward. Observers were certain that there was beneath the surface of the earth a vast reservoir with a complex network of canals that swelled the tides and distributed water to the rivers and the sea. This made the shoreline seem less important, and drove the quest for another voyage, an in-depth one, so to speak.

This fascinating subterranean connection was often mentioned when attempting to account for currents and tides. On the issue of tides there were a great number of opposing theories. Of course, the role of the moon had been detected since Pytheas; and in 1687, Newton was to provide a definitive explanation of the phenomenon. Before him, Galileo and Descartes had also put forward dazzling explanations, the former attributing a major role to the rotation of the earth, the latter suggesting that the moon exerted pressure on the atmosphere. However, credence continued to be given to other theses that relied on astrology and an animal vision of the sea. It is important to bear them in mind in order to recreate the mental universe of Descartes' contemporaries in all its complexity.

Ancient culture, heedful of the images of boundaries,[71] focuses more strongly on appreciating coastlines than on the sea itself. This is not surprising. Paul Pedech shows how the experience of navigation for the Greeks granted a paramount role to the profile of coasts in geographical consciousness. 'It is mainly the sea', writes Strabo, 'that gives the earth its outline and its shape, fashioning gulfs, the high seas, straits, and equally isthmuses, peninsulas and capes.'[72] Greek travellers describe their coastal peregrinations and itineraries.[73] The first ambition of Avienus's poetic geography in his *Ora Maritima* was to provide a continuous description of the shoreline, to map out a trail of sandy or barren beaches, of ponds, sea-shores, and rocky headlands.[74] Nowhere does Homer say that Ulysses really loves the sea; it is, symbolically, the longing for the shores of Ithaca that leads him to set sail. This same feeling is what prompts the hero of Fénelon's *Télémaque* to climb the cliff in order to gaze at the sight of the sea.[75] In ancient epics, the shore keeps alive the dream of a fixed abode prescribed by the gods or provides the focus for the hope of return.

At the same time, there is no lack of episodes in classical mythology and literature to strengthen the negative vision of the shore. The site of hope and success can also become a cold land of exile and a sojourn of misfortune. On the beach of Naxos, as she is pursuing Theseus, Ariadne plunges into the sea and lets her tears mingle with the seething water of the waves; Racine's *Phèdre*, ignoring Dionysus, asks, 'Wounded by what love did you die deserted on a barren shore?'[76] It is with blackness in his soul that Ovid in his loneliness paces up and down the gloomy shore of Tomes. In *Télémaque*, which is nothing but a succession of seaside scenes, the beach – site of escape, shipwrecks, and nostalgic tears – also becomes the privileged backdrop to farewells and heart-rending complaints.[77]

On the sea-shore monsters lurk: Scylla surrounded by her barking dogs and the sly Charybdis who gulps down and vomits her victims. When the Greek Poseidon or the Etruscan Nethuns, originally chthonic forces, gods of earthquakes and tidal waves, became the gods of the sea, they inherited sovereignty over the monsters with which the Aegean world had filled the waters. Poseidon's sons were mostly maleficent giants, such as Polyphemus the Cyclops or the bandit Sciro.[78] Every late eighteenth-century tourist dreamed of visiting the straits of Sicily and confronting the terrifying Homeric creatures to be found there. For the neoclassical traveller, discovering the abyss would soon become a compulsory stage in a voyage experienced as an initiatory rite. Once on site, visitors laughed at how slight the danger actually was, while at the same time relishing their memories of child-like fright.

The shore in antiquity was also the receptacle of the sea's rejections; it is along the beach that the ocean purges itself and throws up its monsters. Seneca illustrates this: 'It is in the nature of the sea to cast back on its shore every secretion and every impurity ... and this purging occurs not only when the storm is stirring the waves, but when the deepest calm prevails.'[79]

Strabo also speaks of 'the repulsive' or 'purgative movement of the sea'.[80] According to Pliny the Elder,[81] the Fortunate Islands, situated off the African continent, were 'infected with the putrefaction of the animals that the sea continually throws back on their coasts'. In the seventeenth century, amber was still considered the richest and most spectacular product of these marine excretions. According to Fournier and Bouhours, coastal people still regarded the stinking matter that was cast back on the shores of Venice and Messina as excreta of the sea;[82] they believed that

the salty foam was the ocean's sweat. The Venetians called the
rhythm of the tide 'il viva dell'aqua'. This same perspective
explains why the tides were thought of as so many bouts of
fever affecting the sea. In 1712, the English poet Diaper used
these very terms to describe the pollution on nauseating shores
where dolphins chose to die in order not to contaminate the
purity of the high sea's air and the clarity of its waters.[83]

In Greek literature, every boundary zone is a dangerous area
in which the activities of deities, human beings, and animals,
living in confused, dangerous proximity, threaten to interfere
with one another.[84] The sea-shore of antiquity, as imagined in
the classical period, remains haunted by the possibility of a
monster bursting forth or of the sudden incursion of foreigners,
who are comparable to monsters; as a natural setting for unexpected violence, it is the privileged scene for abductions. It
would take too long to mention all the episodes, continually rehashed in painting and literature, that reflect this perspective
and confirm the link established by antiquity between landscapes
and the way in which wars unfold.[85] The abduction of Europa
and the installation of the Danaeans' camp on the banks of the
Tiber and their landing, weapons in hand, to face Turnus's
companions are the most obvious examples. The monster spring-
ing up out of the sea ready to devour its prey in Corneille's
Andromède[86] and the Theramenes narrative [in Racine's *Phèdre*]
that relates the unhappy fate of Hippolytus are also part of this
long chain of stereotypes.

The outline of the ancient shore, when evoked in the modern
period, was often imagined as a hesitant boundary threatened
with the possibility of being broken down. Such a rupture would
in turn violate the harmonious peace of laborious, hard-working
lives. This image was strengthened by recollections of the numerous
scourges that have come from the sea since the early Middle
Ages.[87] The traces left by the Norman and Saracen invasions,
the spreading of the Black Death via the sea, and the misdeeds
of pirates, wreckers, smugglers, and port bandits, all left a
damaging mark on the image of the seaside; later, the great naval
battles at the ends of the seventeenth and eighteenth centuries
would make the Channel coasts bristle with parallel walls of
stone. For the eighteenth-century traveller, the key to appreciating a shoreline, a harbour, or a port lay in evaluating its defences.

For outsiders, the sea-shore was also the place where they
made the disturbing discovery of the folk who inhabited it; it
was the dangerous setting in which a choice had to be made

between the joys of savouring hospitality and the fear of encountering monsters, between the vision of Nausicaa and an invasion by Polyphemus.

At the dawn of the eighteenth century, Daniel Defoe summarized and reworked these evil images of the sea-shore. Robinson Crusoe's island features all the characteristics of the Garden of Eden after the Fall: there is a prospect of serene happiness provided man does not spare his sweat, organizes his time, and carefully orders his labour. Throughout its pages, the novel [*Robinson Crusoe*] adopts a widely recognized Promethean perspective, and symbolically retraces the stages of civilization: food gathering and fishing, agriculture, and stock raising. But this Garden of Eden is situated inland, in the midst of meadows and groves. The solitary individual has at his disposal a series of interconnected havens that run into the underworld, that ultimate safeguard against threats to his privacy.

Here the beach is the scene only of the disasters whose marks it still bears: the ship ripped apart and smashed against the coastal reefs. The shore is where its *useful* remains were cast up. Above all, the sand bears the imprint of the wild threatening forces that symbolize desire. This is where cannibals indulge in their orgies, as Robinson watches, fascinated, a voyeur threatened by the animality of collective rejoicing. It is from the shore that danger menaces the womb-like shelter which the hero has lovingly built for himself; it is also from the sea that mutineers disembark. Robinson does not linger on the sand that bears the marks of these incursions and on which he observes the savages in their nakedness; he does not play or bathe; his only bold act is to take Friday away from the group of savages, thereby endowing him with an identity that turns him into a companion and inaugurates a relationship that borders on homosexuality.[88]

This novel is the forefather of all castaway stories, on which the negative appreciation of the sea-shore long continued to exercise its influence. It contains an echo of the tales told by sailors who, since the end of the fifteenth century, expanded mankind's knowledge of this planet. The image of the savage is just one more element in the age-old catalogue of the threats coming from the sea and its shores.

The images of the ocean and its shores were obviously influenced by the experiences of modern navigation; but this impact should not be exaggerated. Until at least around 1770, the recollections drawn from ancient literature and the reading of the Bible exerted a greater influence on the collective imagination than did the accounts of exotic travels. To understand why, it is enough to bear in mind the number of hours that a cultured person devoted to edifying reading and to Greek and especially Latin works; their number far exceeds the hours that he could spend on travel accounts; and in fact, the cultural contribution of the latter was integrated within these more ancient, deeply rooted patterns. Paradoxically, the history of seafarers and navigation, however prestigious it might be, is not the best way to understand and analyse images of the sea and its shores; it nevertheless remains indispensable.

The appalling fate of sailors in modern times gave rise to an abundant scientific and medical literature which reinforced the negative images of the ocean. Elsewhere I have described at some length[89] the way in which the ship comes to be seen as a pre-eminently maleficent place. Between its damp wooden flanks, sources of fermentation and putrefaction are at work; at the bottom of the black, stinking void which is its hold, all possible miasmas are concentrated in the bilge. It was from the ships, so people believed, that infection often spread and epidemics started. The ship in the harbour threatened the health of the city. At sea, it sapped sailors' strength. The crossing provoked scurvy, a disease that carried the symbolism of rotting the flesh of its victims. The decay of the food taken aboard and the discovery of exotic diseases provided conclusive evidence that the ship was a place of putrefaction.

The sea itself seemed to be rotting. One of the most firmly held beliefs of neo-Hippocratic medicine in the seventeenth and eighteenth centuries was that its emanations were unhealthy. Salt, which in great quantities retards decay, hastens it in small doses. The mephitic vapours that rose from the sea made the coasts stink. This smell of the shores, comprising emanations that eighteenth-century chemists later endeavoured to analyse, was thought to be caused by decaying matter deposited by the sea. Seaweed, excreta, and organic debris tossed up on the beaches played a part, so it was thought, in producing the bad air that often prevailed on the sea-shores. Paradoxically, as we shall see, this disgust with the unhealthy, overheated beach was increasing at a time when people were beginning to admire the salubrity of the open, wind-blown northern strands.

Furthermore, well before Coleridge's famous text[90] – indeed, as early as Pytheas – the fantasy existed of a coagulated ocean and a thick, rotting sea teeming with creatures born out of decay. This mass would delay the ship's advance. The imagery for this fantasy drew a wealth of material from the descriptions of the Sargasso Sea.

In view of this, it comes as no surprise that seasickness was such an acutely awful experience. This scourge seemed to blight all those who chose to travel by sea, with the exception of the captain and the sailors. It is impossible to analyse the images of the sea and its shores without taking into account the horror of this malady. Dizziness and the smell of vomit added to the distress of sensitive tourists who were already disgusted by the physical surroundings and the appearance of the crew. The system of appreciation was not merely the result of an individual's point of view and cultural baggage; it began with coenaesthetic experiences, especially when they made themselves felt with as much force as the nausea provoked by the ship's pitching and rolling.

It should be noted here in passing, once and for all, that the horrors of seasickness, described as early as the Middle Ages by pilgrims on their way to the Holy Land,[91] seem to have increased in the eighteenth century, particularly among women. This raises the problem of the historical reality of these coenaesthetic experiences. The 'tourists'[92] who set down their travel recollections did not have the sturdy, resistant endowment that characterized the indefatigable sailors of the past. A range of factors – the vogue of the sensitive soul, erudite discourses on the role of the diaphragm, vapours and psychological troubles affecting women and amply described by doctors, a heightened fear of the miasmatic effluvia, and an increasing awareness of the health risks posed by the proximity of this putrescence, which in turn quickened voyagers' sensitivity to smells, and also no doubt the misguided dietary recommendations made to these victims – all contribute to explain the growing anxiety at the thought of setting sail, as well as the increasing disgust provoked by the sight of others vomiting repeatedly.

Montesquieu complained of a 'frightful bout of seasickness' which he suffered in 1726 between Genoa and Porto Venere.[93] In 1739, the President de Brosses left Antibes aboard a felucca bound for Genoa. 'In my opinion', he later wrote, 'the least painful discomfort caused by the sea is vomiting; what is hardest to bear is the mental dejection, so great that one would not bother to turn one's head to save one's life, and the horrible smell that the sea wafts to one's nose.'[94] Once he debarked at Speretti, a tiny village, he shunned the sea-shore: 'I had developed such a loathing of the sea that I could not even think about it.'

This did not stop him from gazing at it with pleasure a few days later, in calm weather, before describing it once again as 'impertinent' and 'a malignant beast'.[95]

Travellers at the beginning of the following century also enjoyed recalling this torture that took on something of an initiatory import. Adolphe Blanqui[96] [a French economist and brother of the socialist and revolutionary Louis-Auguste Blanqui] expressed surprise that seasickness was not included by the ancients in their repertory of human afflictions. This suggested to him that human sensibilities had changed:

> It is a serious [disability] for us, who are not as simple as our forefathers, and it certainly has an important place in the history of the traveller's tribulations. As soon as land is out of sight, joy and movement vanish from the ship; all conversations are suddenly broken off, and the rosiest faces abruptly lose colour and take on a ghastly greenish complexion. One often sees women stretched out on the deck in a state of total dejection, unaware of everything that goes on around them ... Everyone seems to withdraw into himself.[97]

It should be added that in this period, when steamship navigation first began (1824), the smoke from coal fires further increased passengers' discomfort

For sensitive Romantics, the experience could become quite dramatic. Storm-tossed by the tempestuous waves beating against the Scottish shores, the Marquis de Custine was convinced that his final hour had arrived; in spite of his very earnest wish to visit the Hebrides, he had to give up and return by a land route.[98]

This brings to a close to this too brief catalogue of repulsive images of the sea and its coasts. These images are rooted in a framework of perception that precedes a growing sensitivity to the lure of the seaside. Yet the changing attitude that made a new outlook possible began as early as the seventeenth century. Between 1660 and 1675, the mysteries of the ocean began to fade with the progress made in England by oceanography.[99] In the same period, Satan began to disappear from the Western intellect.[100] Above all, after the ephemeral attention paid by a group of baroque poets to marine enchantment, three phenomena began to pave the way for a transformation of the system of appreciation: the idyllic vision of the prophets of natural theology; an exaltation of the fruitful shores of Holland, a land blessed by God; and the fashion for the classical voyage along the luminous shores of the Bay of Naples.

2

The First Steps towards Admiration

THE ENCHANTMENT OF THE WATERY MIRROR AND THE SOURCE OF PROFOUND CERTAINTY

It would be wrong to imagine that total blindness and a general insensitivity to nature preceded the genesis of the system of appreciation that developed during the Age of the Enlightenment. But the ways of reading the landscape and the types of desire and enjoyment to which it gave rise before 1720 conform to a discourse and a configuration of feelings that were in keeping with classical epistemology.

At the dawn of the seventeenth century, a group of French poets often described as baroque spoke of the joy they experienced in the presence of the sea-shore. Théophile, Tristan, and above all Saint-Amant,[1] who had been accustomed since childhood to walking along the coastline of the Caux region of France, proclaimed the pleasure of standing on a cliff, strolling along the strands, and gazing at the ever changing sea. Not only does this expanse provide the poets with a metaphor to describe man's fate as he confronts the dark forces that surround him[2] or the initiatory trials through which lovers must go,[3] but within the network of ancient stereotypes governing the evocation of the Virgilian tempest, the procession of deities, or sea gods' revels, specific ways of enjoying a given setting were beginning to take shape. As Tristan, a lover of the beaches near La Rochelle, writes:

Nul plaisir ne me peut toucher
Fors celuy de m'aller coucher
Sur le gazon d'une falaise,
Où mon deuil se laissant charmer
Me laisse rêver à mon aise
Sur la majesté de la mer.[4]

[I can experience no pleasure
like that of lying down
on the green of a cliff,
where my sorrow, come under a spell,
lets me dream as I wish
about the majesty of the sea.]

This enjoyment of the seaside is in keeping with the baroque poets' taste for motion; it is further nourished by their longing for surprise. For them, the unending agitation of the waters and the glistening sunlight created a magical realm, a 'perpetual source of imaginary creations'.[5] The continuous metamorphosis, the magical highlights, and the reflection of the sky in this watery mirror, suggesting that the universe was reversible,[6] fulfilled the expectations of people who could intimate from the spectacle of nature that the world was a game of illusions.

In 1628, Saint-Amant sampled the charms of seclusion. Abandoning the pleasant countryside and the deep woods, he chose the wild coast of Belle-Île as his 'desert'. The experiences he evokes in *Le Contemplateur* are part of the ritual of meditation; his knowledge of the Scriptures guides his emotions.[7] Saint-Amant went down to the beach to gaze at the amazing boundaries within which God had decided to imprison the abyss. He went there dreaming of 'the sad effects of the Flood', and imagined with terror the sea of the apocalypse that 'burns like spirits'.[8] The great spectacle of the sun rising out of the sea, which he watched for early in the morning, recalled the Resurrection, and foreshadowed the day of judgement and the ascension of the just.

Memories of antiquity combined with biblical images; he [Saint-Amant] too experienced the Virgilian vision of the 'liquid plain'; the sea gods revelled in Thetis's bosom. In Saint-Amant's opinion, the alternating ebb and flow symbolized the unknowable that drove Aristotle to suicide and allowed the Fathers of the Church to illuminate the mystery of Creation. The vision of the storm that he watched from the sea-shore was strongly influenced by Lucretius. Like the English, who were fond of rural sports,

Saint-Amant claimed that he enjoyed trolling for fish and hunting rabbits along the shore. He admitted that he would remain seated for hours atop a cliff, gazing at the marine horizon, listening to the sea-gulls taunting the empty spaces with their strange cries. Then he would go down to the beach, and pace up and down for a long time; his walks became a stimulus for his meditations, and created an opportunity to collect shells. The mirror of calm waters, the illusions it created, and the versatility of the ocean fascinated him.

Saint-Amant shows no trace of the horror aroused by the spectacle of infinity. Obviously he was quite happy to contemplate the limitless expanse of waters that, in most cases, offended his contemporaries. Already, he feels the attraction of the feudal ruins bordering the sea-shore; his melancholy gleefully conjures up the skeletons that must haunt their underground corridors. This makes it easier to understand why his work, translated into English in 1716, contributed on the other side of the Channel to the emergence of the aesthetics of the sublime.

But the emotions we are dealing with here are rarely expressed. Like all sensual landscapes in the eighteenth century, the sea viewed from the shore is, as Jacques Thuillier points out, 'only expressed in literature in short excerpts; one must cut and prune in order to quote, and one inevitably comes back to the same authors and the same stanzas'.[9]

There are, in the second third of the century, a few scattered testimonies that outline a somewhat different system of evaluation. The exploration of the sea-shore became part of a way of experiencing nature that conformed with the life-style of a social elite seeking renewal and refreshment. For people who were morally demanding, seclusion was not an abdication, but rather 'a melancholy, lucid decision by the soul no longer to claim it can impose order on anything if not itself, far away from a world in which experience has taught it to recognize the invincible presence of evil'.[10] This seclusion takes place at the intersection between a stoical conception of moral life and the Christian goal that encourages solitary meditation as the earthly form of heavenly bliss.

The self-imposed seclusion on the island of Jersey of the Norman Henri de Campion, pursued in 1644 by Mazarin's henchmen, fits with this model. In 1654, in his *Mémoires*, he evokes the great certitudes he acquired on those shores. Campion was given a very civil welcome by the governor Carteret and his wife, and he devoted one-third of his time to reading, another third to social life, and the rest to 'walking along the sea-shore, or on the solitary rocks surrounding my abode, and as the latter faced the sea, it offered me a view of this vast, changeable element.

I could watch at my ease the storm as well as the lull. In that place I found plenty of food for thought about the fragility of human affairs, of which I was still learning some of the consequences ... I took so much strength in that sentiment (of responsibility for his own happiness), that I spent seven months in that wild place, without any anxiety or impatient desire to leave it.'[11]

This type of meditation was not purely eremitic. The seaside also lent itself to the pleasures of conversation. It entailed a delicate balance between solitary retreat and the madding crowd, and it implied selecting 'a few specific individuals with whom one communicates in order to avoid the boredom of solitude and the overwhelming pressure of the multitude'.[12] Moisant de Brieux succeeded in achieving the same mixture of meditation and selective friendship. This man of letters from Caen, accustomed to worldly salon life, also enjoyed living in his home in Bernières.[13] The house looked directly on to the sea. This surprised his friends, who would have better understood windows opening on to hedge-lined fields. But Brieux took care not to allow any shrubbery to rob him of his view of the open sea. Like Campion, he enjoyed dreaming 'in front of that vast, changing element' and taking long walks on the beach.

This experience is further evidence that the marvelling at the richness of the sea which the adepts of natural theology would describe at the end of the century was preceded by another mode of appreciation that was distinct from a previous model based on links between the sea and dreaming. There was a range of ways to enjoy experiencing place: being attracted by the rest procured by this retreat, engaging in meditation and conversation, indulging in reverie fostered by the environment,[14] taking certain forms of bodily exercise, or being fascinated by the light shimmering off the aquatic mirror. Yet those who enjoyed these pleasures did not go so far as to depict the spectacle of nature as the authors of loco-descriptive poetry were to do at the beginning of the next century.

A RECEPTACLE FOR DIVINE WONDERS

At this stage a brief parenthesis must be opened. Between 1690 and 1730, what since the seventeenth century had been called natural theology in France and physico-theology in England unfolded across the Western world. This marked the beginning of an unforeseen split between the popular modes of appreciation of nature and the concepts of pious scholars, who were starting to view the external world in a new light.

Natural theology[15] in fact implements a transition. It marks the twilight of the vision of a living, harmonious world, as outlined in the *Timaeus*, systematized by Aristotle and the Alexandrines, described by Raymond de Sebonde at the end of the fifteenth century, and popularized by the Neoplatonists during the Renaissance. This system entailed a belief in mysterious correspondences between the physical world and the spiritual one, between the human and the divine, between man – the microcosm – and the universe, the macrocosm. The external world, composed of a network of analogies, did not yet appear primarily as an enigma to be resolved through observation or as a cluster of forces to be mastered through scientific knowledge.

In its modernity, natural theology ceases to analyse man and the universe in terms of analogies; it considers the external world as a spectacle. Yet, despite the blows dealt to this conception by progress in astronomy and the hypothesis that there are many inhabited worlds, by the revelation of the infinitely small and by the discovery of vast deserts, physico-theologians remained faithful to their anthropocentric view of the universe.

These pious scholars suggested that the spectacle of nature had a meaning; and, by the same token, they forbade anyone to be indifferent toward this spectacle. They saw the external world as an image given by God to his most perfect creature. This explains the importance attributed at the time to the theme of Paradise lost, that fascinating primal stage on which the divine design could unfold in all its perfection.

The beauty of nature bore witness to the power and bounty of the Creator. God regulated the spectacle, through both the laws that he had established in his infinite wisdom and his immediate, providential interventions. Descartes' God-the-clockmaker or Newton's passive creator of nature intervenes directly through miracles whenever he sees fit.[16]

Since the Flood, the earth has enjoyed great stability.[17] Every creature on the planet fulfils God's design, and every object has its function. According to William Derham, the present earth is indeed the most beautiful, the most pleasant, the most healthy that could be imagined.[18] Far from having left nothing but a chaotic heap of ruins, as Burnet believed, the Flood seemed to Woodward to be a disaster that was necessary for human happiness. After dissolving the primitive earth under the waves of the diluvian ocean, God refashioned the earth and adapted it to the new human frailty.[19]

Physico-theologians rejected the idea of a world in decline, whose progressive dilapidation supposedly revealed the corruption initiated by the sin of the first humans. Woodward, for one, wrote that 'the earth, the sea and the whole of nature will for ever remain in the state they now enjoy, without ageing or decaying, without disturbance, without any disorder, without anyone encroaching upon anyone else, without revolutions and the order of things being upset or changed'.[20] Only the final conflagration, discussed at length by Whiston,[21] would eventually disfigure the globe remodelled for Noah.

The present Earth therefore appeared to be a book written by the Creator and intended for man. God's purpose, Derham asserts, is to let his works be admired 'by reasonable Creatures'.[22] 'Providence has made the air invisible', the Abbé Pluche affirmed in turn, 'in order to allow us to witness the spectacle of nature.'[23]

The goal of natural theology is edification. Man must become a pious reader of God's book. His Creator endowed him with five senses in order that he might magnify divine power and bounty. The physico-theologians extol empirical observation; indeed, the very existence of God ensures that his work is intelligible. The Creator appreciates the work of the scholar, who devotes himself to discerning the religious meaning of the workings of nature.

Such a vision of the world gave rise to a scientific approach whose goal was to inventory all of creation; it was the force behind attempts at classification which reached their zenith with Linnaeus, and systematics actually revealed the plan of Creation.[24] In this manner, a close link developed between the collector's patience, the scholar's curiosity, and the Christian's piety.

This brief overview highlights one of the deep motivations for touristic travel: henceforth, members of society's elite would seek in it the opportunity to experience this new relationship with nature. They discovered the previously unknown pleasure to be found in an environment transformed into a spectacle. Natural theology, after all, involved educating the eye.[25] It also compelled individuals to make their observation of the world into a hymn praising divine greatness and goodness. Every one of God's creatures, in its own way, proclaims his glory, and it is man's duty to gather up this multitude of praises and lay it at the feet of his Creator.

Across the Channel, physico-theology was attuned to the rites of
the Church of England, which it powerfully reinforced.[26] Morning service included a psalm and hymns praising God. Sanctuaries
echoed with the sounds of *Te Deum laudamus* and *Benedicite
omnia opera Domini*, which praises the sun and the moon, the
mountains and the hills, the dew and the frost, and magnifies the
beauty of the seas and rivers and even of sea monsters. Religious
poetry became broadly penetrated by the Creation theme,[27] and
the lyricism of sacred texts singing of the wonders of nature inspired
profane poetry. The art of the hymn underwent a profound rebirth
as physico-theology pursued its expansion.

This new sensibility touched the West as a whole. In England
its way was paved between 1640 and 1660 by the aristocratic
vogue for retiring to the countryside. This was a compensatory
dream for landowners who felt their power threatened and so set
out in quest of images of Paradise before the Fall or of the golden
age described by Virgil.[28]

This religious sensitivity to the spectacle of nature that touched
England was no less great in the United Provinces. In 1714,
Nieuwentijdt wrote an impressive *summa* of natural theology.
His aim was to refute Spinoza and 'prove the divinity of the Holy
Scriptures by the objects of Nature'.[29] This enterprise was also
in keeping with the sensibilities of Lutheran circles in northern
Germany. Between 1715 and 1720, Brockes produced his *Earthly
Pleasure in God*; this work, of which the first volume was
published in 1721, numbers no less than 100,000 lines of verse.
In it, the author lauds the plants, birds, sky, and waters of
the Hamburg area. Throughout nature he seeks evidence of the
goodness of Providence.[30] The book was an immediate success,
as was the very erudite *Theology of Water, or Essay on the
Goodness, Wisdom, and Power of God* revealed in the creation of
water, published in 1734 by Hans Albert Fabricius, a professor
in Hamburg.

In France, this new sensibility was rooted in the practices of
devout humanism.[31] In the 1650s, remember, contemplation was
in fashion; after the unrest of the Fronde, the taste for solitude in
nature and what Henri Brémond described as the 'sanctifying of
the landscape'[32] were growing. The beauty of the environment
was expected to incline hearts toward lamentation, repentance, and
conversion. Thirty years later, many of the features that expressed
this religious sensitivity were to be found in authors like Bouhours
and Fénelon. This sensitivity reached its culmination in *Spectacle
de la Nature*, published between 1732 and 1750 by the good

Abbé Pluche. One of the most widely read books in the eighteenth century, this work, like that of Nieuwentijdt, was actually a clever, popularized presentation of the latest scientific theories cast in the mould of natural theology. For the author, the issue was not to prove the existence of God – that would be superfluous – but rather to justify all the attributes with which theology endows him. This work gave rise to an extensive heritage: in 1749, Paul-Alexandre Dulard published *La Grandeur de Dieu dans les merveilles de la Nature* [The Greatness of God in the Wonders of Nature]. In both France and England, poets fell into step with theologians. During the same period, Cardinal de Polignac spent more than thirty years mulling over his *Anti-Lucretius*.

This pious interpretation of the spectacle of nature and the harmonious image of the Earth after the Flood must be taken into account in order to understand the new way of appreciating the sea and its shores that was taking shape at the dawn of the eighteenth century. Natural theology was contributing successfully to erasing the original repulsive images.

'The Lord is admirable in the waters,' proclaims Psalm 52.[33] In this respect, however, the beauty of the spectacle gave way for the physico-theologians to admiration of the Creator's sovereign might. A gradual shift occurred from the image of a terrible God who unleashes the torrents of the heavens to the reassuring sovereign who chose to harness the ocean and establish boundaries to it. Authors never failed to quote at least a fragment of the sacred texts that refer to this clear manifestation of divine power: 'Thou coveredst it [the Earth] with the deep as with a garment;[34] the waters stood above the mountains,' proclaims the Psalmist. 'At thy rebuke they fled ... Thou hast set a bound that they may not pass over; that they turn not again to cover the earth.' Or again, '[God] layeth up the depth in storehouses.' The prophet Jeremiah,[35] exalting the Creator's might, makes him say: '[I] have placed the sand for the bound of the sea by a perpetual decree, that it cannot pass it; and though the waves thereof toss themselves, yet can they not prevail.' The book of Job was even more frequently quoted: 'Hitherto shalt thou come, the Creator commanded the sea, but no further.'

St Basil, like St Gregory of Nyssa,[36] St Ambrose of Milan, and St Augustine, gazing on the Mediterranean from the shores of Ostia, evokes the splendour of the sea; and he adds a striking image in his commentary: 'However the sea might rage,' asserts this Father of the Church, it breaks 'upon a grain of sand' on which it sees God's command inscribed and 'withdraws out of respect, *bowing its waves*, as if to worship the Lord who has appointed its limits.'[37]

Among French eighteenth-century poets such as Louis Racine, Le Franc de Pompignan, and Cardinal de Bernis,[38] the theme lends itself to grandiloquence. These verses by Dulard depicting the tempestuous sea are typical: will it overrun the shore?

> Non, ne le craignons point. Un frein impérieux
> Enchaîne, ô fière mer, tes flots séditieux.
> Le doigt du Tout-Puissant a tracé sur le sable
> Un ordre redouté, barrière impénétrable.
> Ton onde audacieuse, à cet auguste aspect,
> Tombe, et pleine d'effroi, recule avec respect.[39]

> [No, let us not fear it. An imperious brake
> Enchains, oh proud sea, thy seditious waves.
> The Almighty's finger has traced on the sand
> A fearful order, an insurmountable barrier.
> Thy daring waters, at this august sight,
> Fall back and, full of terror, respectfully withdraw.]

In England, Richard Blackmore developed the same theme at some length in his *Creation: A Philosophical Poem which Proves the Existence and Providence of a God.*[40]

This soothing view of the shore, which reassures mankind of the impossibility of being submerged again, should be taken very seriously. The biblical text magnifies the paradoxical strength of the sands. It focuses attention on the shoreline, and fills it with meaning. Nowhere else do the power and goodness of the Creator appear so clearly as on the beach that still bears the imprint of his finger. This most astonishing miracle is constantly being fulfilled. For the Christian, the threatening wave is merely a reminder of misfortune and the Fall. The line where it breaks arouses *astonishment*, and fosters admiration and thanksgiving. For the Christian, the curve of the wave in the presence of divine omnipotence, as it becomes calm and ebbs away, evokes a gesture of respect.

God in his infinite goodness arranged the ocean and its shores with man's welfare in mind. The composition of sea water contributes to the Creator's aim: salt prevents it from spoiling, and consequently ensures the survival of fish and the salubrity of its coasts. Moreover, this salt could be used to preserve food-stuffs. It prevents freezing, which would disrupt both fishing

and the life cycles of marine creatures.[41] In addition, God made
the sea salty, asserted Abbé Pluche, because he wanted it to
bring this substance so vital to human beings 'right up to their
dwellings'.[42] Finally, salt weighs down the waves, and 'limits
evaporation', thereby regulating the circulation of water in the
atmosphere. As for tar, this substance forms a coating that
prevents the sea from eating away at the land that lines the
bottom of its bed.[43] Thanks to these two substances, the shore
was perceived as a supremely healthy place, provided it remained
subject to the balanced action of the winds and was thereby
prevented from stagnating.

The relief of the coastline also corresponded to the Creator's
aims. It was God who spread sand on the sea-shores in order to
form a barrier.[44] From this point of view, the beaches and dunes
were not seen as the results of erosion, but as elements of an
architecture erected in the aftermath of the Flood. The pattern
of gulfs and bays[45] was also part of God's design: its purpose
was to provide shelter for ships and to allow foodstuffs to be
conveyed directly within the continents. Rocks and reefs, asserted
Fabricius,[46] had been placed to ensure the defence of naval strong-
holds. Pontoppidan wrote that the offshore islands, kept the
Norwegian coasts safe.[47] Bouhours, for his part, claimed that
Providence created islands 'for the traveller's comfort',[48] and
Thomson sang the praises of the ocean as a divine rampart built
by God to protect Great Britain against the risks of invasion.[49]
As for Nieuwentijdt, he asserted that 'the purpose of gulfs and
bays is to admit the rivers', thereby preventing flooding and
facilitating the mixing of salt and fresh water. God made the
coastal lands low so that they would not block the smooth
flow of the rivers. Nieuwentijdt believed that they operated as
drains.[50] This neo-Hippocratic obsession with drainage left
its mark on the way in which the shape of the shoreline was
perceived.

By the same token, the purpose of the tides was to sweep
the shores and, through their movement, to prevent even the
deepest waters from putrifying. In addition, these flows drove
out the waters of the rivers, carrying ships toward the ports. In
short, tides facilitated navigation.[51]

The sea winds were created by God in order to cleanse the
waters, to drive ships, and to cool lands overheated by the sun.
Even storms had a purpose, as did volcanoes and earthquakes.[52]
The turbulence they engendered, though it might seem pointless,
was intended to adjust the air or purify it by replacing it.

This inexhaustible discourse leads to a virtual hymn in praise of navigation, which brings men closer together, allows sailors to admire the entire Earth, and fosters trade. Above all, it justifies the undertaking of missionary endeavours.[53] The disconcerting expanse of the ocean, insists Fénelon, was willed by Providence to facilitate and accelerate travel.[54] Fournier, following in the footsteps of St John Chrysostom, proclaims that the seascape contributes to the glory of God. Abbé Pluche in his turn sharply criticizes the Horatian conception of the ocean as 'unsociable' (*dissociabilis*).[55]

The fertility of the ocean is as infinite as the Creator's might:

> Immense comme lui, toujours pleine et féconde
> Elle donne toujours sans jamais s'épuiser;
> Et sans jamais se diviser
> Elle répand partout les trésors de son onde.[56]

> [Immense like him, always full and fruitful
> It continually provides without ever being exhausted;
> And without ever dividing
> It spreads everywhere the treasures of its waves.]

Since the sixth day of Creation, when God gave man dominion over the fish of the sea, the waters have provided nourishment for the poor coast-dwelling peoples.[57] The hand of God, Nieuwentijdt emphasizes, is manifest in the depths as well as on the waters' surface. The movement of the tides, claim Pluche and Dulard, is part of this great design: the waves obligingly withdraw, inviting man to collect the creatures that the sea has left behind for him on the shore.

The spectacle of the sea in works inspired by natural theology did not arouse the same enthusiasm as that of rich harvests and pleasant vales. The Edeno-Arcadian code of appreciation remains very significant; once again, it was difficult for an individual of that time to feel and express admiration for the view of the infinite waters. Yet this should not lead to the conclusion that all viewers remained totally insensitive. A few voices are to be heard here and there, expressing fragments of those themes that were to develop at the end of the century, when the age of the aesthetics of the sublime arrived. Literature inspired by natural theology tends to rid its rhetoric of a sea full of nymphs, sea gods, and divine processions. It popularized biblical images[58] which became combined in eighteenth-century descriptive poetry with the stereotypes of the Virgilian storm. Above all, it was discreet about the beauty of the marine world, expressing *wonder* at the *sea's riches*. It intoned a hymn that celebrated the abundance of treasures envisioned by the imagination more than the actual glistening of the water's surface.

The diversity of sea creatures was infinite, proclaimed John Ray.[59] These were not monsters, but the beings created by God on the fifth day of Genesis. The shiny shellfish, the colourful coral, and especially the pearl's purity all symbolize the splendour of Creation. God laid out these treasures to adorn the sea-shores; through them, nature 'makes merry and frolics'.

This shore on which the sea displayed some of its wealth provided an inkling of the abundance to be found in its abysses. There, asserted Fabricius, as on earth, one could find mountains and valleys, hills, fields, and plains.[60] Some of the animals living there were replicas of those that God placed on the earth, while many other species lived only in those salty waters. But all these beings, it is important to note, were considered more perfect than those surrounding man.[61] 'As St Ambrose taught, the beasts that are horrible and cruel on earth are beautiful and gentle in the sea.'[62] Beneath the sea are gardens, orchards, forests, and meadows. The authority of the Fathers of the Church and the accounts of sailors strengthened the belief in an underwater nature full of magnificence, forming a moving replica of earthly paradise. Far from being a dark, merciless abode of cruel monsters, the bottom of the ocean was perceived as the invisible receptacle for the perfection of Creation and a repository for innocence. It was the flip side of the exposed earth, and, paradoxically, it was more luminous and more colourful. As Robert Boyle had just discovered, even the most terrible storms could not upset the serenity of the sea's depths.

Pious scholars and poets also found a way to resolve the problem of evil in this area. As Dulard conceded, sea creatures did devour one another; but they did so to maintain a biological order willed by God. 'These eternal battles they wage among themselves/Are a physical good, regulated by thy wisdom.'[63] In England, Woodward, who assigned a crucial role in his theory of the Earth to the water of the great gulf situated in the centre of the globe, considered, unlike Burnet, that the present seas and oceans were similar in form and size to those that existed before the coming of the Flood. As he was especially sensitive to the variety in the sea's image, he believed in the existence of a watery paradise, without which he considered that the world would have been 'quite forsaken'.[64]

Such admiration for the beauty of the sea appears sporadically throughout this literature. The most significant text in this respect comes from Bouhours. Ariste and Eugène, the characters he uses, reveal their sensitivity to the variety in what they see, the beauty of ships on the waters, and already the sublime character of the abyss occurs as a striking image of the enigmas of Providence. In the same period, Fournier praises the colours of the winter ocean. At the beginning of the following century, Brockes describes the beauty of the North Sea's shores, though this is only a minor theme in his work. Abbé Pluche, too, admires the sea, but through the intermediary of painting, a subject which will be addressed later.

Bouhours, writing in 1671, was not unaware that people came to stroll on the beaches of Flanders to indulge in conversation, and that visitors even ran the risk of meeting troublesome individuals. His characters walk up and down the shore, the backdrop to their meditations. They fall to dreaming as they settle on the rocks: 'Then Eugène and Ariste sat down near the dunes in order to gaze at the sea as it softly went out, leaving on the sand as it went the trace and shape of its waves, with foam, gravel, and shells; they both spent some time dreaming, saying almost nothing to each other.'[65]

In the middle of the eighteenth century, the system of representation grounded in theology began to die out. The learned world gradually turned towards other means of evaluating nature, drifting away from providentialism. But the sweet vision of the earth regulated by God down to its smallest details was revived over and over again by preachers and authors of edifying works; it worked its way deep into the popular consciousness. For some time it remained in keeping with the spontaneous sense of finality in simple souls, and it dictated a way of looking at the environment. [66]

The fixed conception that attributes to the Creator the staging of the natural world's spectacle continued to make its way insidiously. From time to time it sprang up again, adapting to new times. For a long time, it permeated pious literature and religious poetry, and weighed heavily in the poetic vision of the world. The most obvious example of this resurgence can be found at the dawn of the nineteenth century in Bernardin de Saint-Pierre's *Études de la Nature* and *Harmonies de la Nature*. Of course the last works of the author of *Paul et Virginie* contain much more than a naïve providentialism derived from natural theology. By extolling the harmony established between man and natural objects, Bernardin was returning to ancient Pythagorean and Neo-platonic ideals. He enjoins man to grasp and restore this harmony, to return things to their original place, and above all to respect the moral harmonies inscribed in the book of nature. In so doing, he outlines a neoclassical ethic ignored by the physico-theologians of the late seventeenth century. But the theological understanding of the world that forms the basis of this work is actually very close to that of Abbé Pluche and his so-called naïve sense of final cause. [67]

> When nature also wanted to dig basins for the seas, it neither lined out their shores nor made them round; but it laid out deep bays, sheltered against the ocean's general currents, so that, in the midst of storms, rivers could safely flow; legions of fish could find shelter at all times, could lick the alluvia from the land that are deposited there along with fresh waters; could spawn, in their majority, by going up the rivers, where they seek shelter and food for their young. It is in order to maintain these conveniences that Nature has fortified all the shores with long shoals, reefs, huge rocks and islands, that are laid out at suitable distances to protect them against the fury of the ocean. [68]

On the divine boundary of the shore, the slope of the 'basin's excavation' was determined by 'infinitely wise laws'. The cliffs on the coast appeared to Bernardin to be solid structures, and the reefs, fortifications rather than ruins. Volcanoes served as light-houses for sailors, and from the earth's origins, nature maintained and repaired the islands it had designed. Consequently, islands were not the result of the continents' destruction. Each one 'has its fortification, that is proportionate, if I may say so, to the danger it is exposed to from the waves of the ocean'. [69]

It would not be pertinent to dwell at such length on the work of Bernardin de Saint-Pierre were it not for the fact that he is probably the writer who has contributed most to the aesthetic and moral appreciation of the sea-shore. He provided a theory about the superiority of beaches over mountains, and was the first French writer to extol systematically 'the inexpressible harmonies that Nature has spread over the sea's shores'. [70]

THE ADMIRABLE ROAD TO SCHEVENINGEN

In the West, travel to Holland paved the way for the rising admiration of the sight of the ocean and the desire to stroll along its shores. For tourists of the classical age, this country was closely identified with the sea.[71] Two key images govern this national identity: the Dutch had tamed the fury of the oceans, and had successfully subjugated it to their mercantile aims, symbolized by the bulging hulls of their ships. Amsterdam above all, but also Rotterdam and other great ports, henceforth became so many microcosms in which the goods of the planet were brought together, and it was God who made possible this mastery over the waves and this abundant wealth. The miraculous affluence of the shores reinforced the image of Holland as a country blessed by God. The fecundity of herrings rewarded the labour of the poor living along the coast, just as the prosperity of the fleets compensated the audacity of rich shipowners. This two-pronged view of economic activity confers a religious value on the bond that links the Dutch to the immensity of the sea.[72]

These two images were systematically extolled for political purposes by an oligarchy anxious to escape the influence of Burgundian art and create a national culture. The Dutch seascape as a pictorial genre is the outcome of a drive to glorify the energy of a social class. Through their commissions, the provincial State and the trading companies intended to exalt their fleet, just as the municipal authorities vaunted the prosperity of their towns, which seemed to rise out of the sea when viewed from afar.

From the middle of the seventeenth century, a great number of Englishmen going on the grand tour and Frenchmen visiting the northern countries stopped to visit Holland. The itinerary, the stages, and the commanding aesthetics of this progress of the classic journey were established as early as the seventeenth century. 'Holland is a marvel,' wrote the Duke of Rohan when he mentioned the trip he made in 1600. He found the very topography of the land wondrous. The Dutch had managed to set limits to the sea; they had not disturbed the Creator's work, but rather, with his blessing, had successfully completed it.[73] In order to appreciate fully the admiration which this artificial strand aroused in those who saw it, this miracle of waters contained, one must take into account the thrill caused at the time by the boundless might of the waves.

This boldness of mankind in attempting to perfect the outline of the coast that was drawn by the divine hand even made some tourists uneasy. The feeling that harmony between God and man prevailed here was not enough to wipe the permanent threat of the water from travellers' minds. The French in particular, who had little appreciation for the technologies used by the Dutch or the risks involved, were terrified at the thought of staying in that 'land of floods'. 'Yet people do sleep in this country,' noted Diderot with astonishment. Caught up in the aesthetics of the sublime, he wrote in 1773[74] that the sight of the ocean on the same level as the earth 'will make you wonder and shudder'.

Travellers also came to Holland with their heads full of war scenes and images from seascape paintings in order to enjoy the sight of a nation struggling with the sea.[75] Indeed, they analysed the landscape itself as the outcome of a battle.[76] In this respect, a sensibility to the agonistic character of the spectacle proved to be of prime importance. This is radically different from the impression produced by Venice, Amsterdam's rival. The city of St Mark displayed a much more peaceful visage. Naturally sheltered by an offshore bar and a lagoon from what Misson calls 'the real and ancient sea',[77] Venice, symbolized by the sumptuous marriage between the Doge and the Adriatic, did not live by the same conflictual relationship with water. Venetians had in fact neither altered nor completed the outline of the boundaries established by God; at most, they had contributed to the creation of an illusory sea.

Dutch seascape painting, from its origin at the end of the sixteenth century until around 1635, then after 1665, systematically dramatized the relationship between man and the sea. This art is the work of technicians accustomed to the sight of the ocean's cruelty. Seascape painters accompanied naval fleets on their expeditions, painted battles on the spot, and were familiar with the winds and the handling of ships. As initiators of a style emphasizing the pathos of the sea, they celebrated battles, recounting events that gave more importance to warriors than to the liquid element. Their paintings magnify the danger: whether dealing with a storm or a naval battle, they continually suggest to the spectator that the alternative lies between survival and destruction.[78] This kind of painting grew out of the power of the United Provinces, and was linked to a historical imperative: it sought to symbolize the threats hanging over the young republic. The emblematic nature of Flemish Renaissance sea-pieces, depicting the instability of the world, gave way to symbolic representations of the terrible and unending perils that face a nation blessed by God.

Here, of course, no painting of the sea is without a meticulously drawn ship, and man is always closely involved in the scene. Whether steering the ship through a storm, handling it as it comes into port, keeping it afloat, or carrying out some ceremony on board, he is installed, often implicitly, at the centre of the picture,[79] and he never yields to the elements. This seascape painting, which looks to the violence of emotion, chooses the boundless ocean as its stage. Except for a brief golden age between 1635 and 1665, which will be discussed later, seascape-painters did not include the coastline, and certainly no other image of land; they stressed the dramatic element in the struggle between man and the elements by showing the evident unsteadiness of the backdrop to the battle.

Paradoxically, though, Dutch painting and the practice of travelling to the United Provinces actually contributed an apprenticeship in viewing the sea-shore.[80] In Holland, tourists discovered a bright landscape that was in keeping with the code of classical aesthetics. As early as 1636, Ogier compared Holland to the Elysian fields. In 1772, the sights that unfolded between The Hague and Amsterdam suggested an earthly paradise to Voltaire.[81] It is important to remember that Dutch landscape-painters began by praising the beauty of the countryside, of woods, rivers, and canals, before turning to that of the sea coast.

Simultaneously, by means of a complex detour, the experience of travelling to Holland played a propaedeutic role in the collective appreciation of the spectacle offered by the sea.

All travellers emphasized the infinite flatness of the Dutch landscape. Some even complained of its monotony.[82] In 1688, Misson spoke of how tiring the sight of this never ending prairie was to his eyes. Pilati, too, wearied of its uniformity.[83] The impression of monotony was further accentuated by the smooth course of the horse-drawn barge and the slow drifting of the gaze which it induced. Madame du Bocage, for one, considered the ride much too dull.[84] The mastery which the Dutch exercised over nature was in fact visible in a transportation network and a degree of comfort that were far superior to what travellers of the times could find in other countries. As they slipped smoothly along rivers, streams, or canals, these horse-drawn passenger barges offered a pleasant observatory, and the travellers, whose contemplation was not interrupted by the incidents that marked any journey by hackney carriage, took time to look at the scene that slowly went by.

The flatness of the landscape was somewhat relieved by the twisting canals and the outline of the dikes that streaked the surface of the plains. The verticality of the steeples (which took on the role in landscape painting that was later given to masts in seaside scenes), the horizontal cities stretching along the horizon, and the numerous boats that enlivened the scene[85] all made the monotony of the relief more tolerable. Holland seen from the dikes, a sea of prairies or an icy mirror,[86] taught viewers to look at the sky: the tourist need only look back over his shoulder to begin to learn to admire the surface of the sea. By achieving the miracle of being simultaneously gay and infinite, Holland offered a transition between the classical mode of appreciating a landscape and the taste for vastness.

The vision of the countryside and the landscape painting that it engendered[87] led imperceptibly to an admiration of the ocean. In fact, the elements overlapped so continuously that the sea was often part of the composition of the pleasant paintings that classical travellers were accustomed to appreciating. Of all the impressions produced by this country, this was indeed what struck travellers most. Looking at the Dutch microcosm resulted in a continuous review of the constitutive elements of the world. Jean-Nicolas de Parival imagined he saw houses and steeples rising from the sea.[88] 'From the harbour', writes Misson about Rotterdam, 'you are astonished to see a rare confusion of roof-tops, tree branches, and mast pennants. You do not know whether it is a fleet, a city or a forest; or rather, you see something unheard of, the combination of these three things, the sea, the city, and the countryside.'[89] David Hume conveyed the same impression in 1748 when he wrote, after visiting the same port, that the mixture of houses, trees, and ships produced a happy effect, and brought together the city, the countryside, and the sea in a single picture.[90]

Some special views, which became compulsory stopping points on the journey to Holland, lend themselves particularly well to this touching shorthand reconstitution of the world: the Pont-Neuf bridge in Amsterdam, or the mouth of the river Meuse in Rotterdam. It is impossible to understand the fashion for the sea-shore that became widespread at the end of the eighteenth century without taking into account this desire to be present at the point of contact between the elements. In Holland, the interpenetration of the waters, the land, and the sky adds to the giddiness engendered by the sight of the confining walls.

One of the stops insisted on for tourists by the authors of travelogues was The Hague, 'the most beautiful village in Europe' according to Abbé Coyer.[91] From there, foreigners went to look at the sea – especially the French, who were unaccustomed to the view of the open sea and were eager to carry out what in their own country had become an initiatory rite, whether in The

Hague or in Dieppe, for the upper classes. Moreover, at The Hague, hospitality required taking guests to Scheveningen:[92] the purpose in this case was not to contemplate the element tamed by man and see waters mingling with meadows and dwellings, but to view the waves of the North Sea stretching away as far as the eye could see. This outing was all the more attractive to tourists because of another, negative stereotype in accounts of travel to Holland: they feared the fumes and miasmas that rose, especially in autumn, from the stench of the canals.[93] By contrast, with few exceptions, visitors enjoyed the healthy atmosphere of the North Sea coast.[94]

From The Hague, visitors could go to Scheveningen by the most pleasant route in the world. This route, about half a league long, 'cutting across the dunes',[95] consisted of a road for carriages and two paths for pedestrians; it was bordered with 'the thickest, best tended trees'.[96] Benches were placed at regular intervals. According to Samuel Ireland's belated account (1789), 'the ride is delightful,' and the 'foot way' shady and 'commodious'. Above all, 'the sea, breaking upon the eye, just at the extremity of this avenue of flat land, forms a noble and picturesque object';[97] for, as L'Honoré remarked in 1779, 'nothing is more pleasant than a beautiful avenue with the sea at the end of it'.[98] Diderot admitted to having made the excursion to Scheveningen many times, and Ann Radcliffe also expressed her admiration in 1794.[99]

The beauty of the walk led visitors to enjoy that of the shore. The Scheveningen road ended in a fishing village overlooking the beach. There the eye could indulge, once again, in a review of the elements: as early as 1688, Misson marvelled at a view taking in woods to the north, prairies to the south, ploughed fields to the east, and the sea to the west.[100]

Inhabitants of The Hague came to Scheveningen as families or in groups to eat fresh fish.[101] When the weather was fine, women and girls of the working class made the trip in bands to drink bad tea in one of the fishermen's taverns whose smell revolted the tourists. 'On their way back to town, they do not stop singing and gibing at passers-by.'[102] Joseph Marshall remarked as early as 1768 that people bathed in the sea there.[103]

The eighteenth-century tourist, by contrast, comes primarily to plunge into a 'beach scene' like those whose image has been burned into his mind by visits to museums and by engravings.[104] Flemish and later Dutch painters and engravers offer a model for the appreciation of dunes, sand, beaches, and views of the coast. This requires a brief digression: beginning in 1602, and until the end of the 1620s, according to Wolfgang Stechow, exceptional events such as a prince setting sail, the return of a delegation, or a grounded whale served as pretexts for painting the sandy shores bordered with dunes. The nature of the occasions accounts for the social confusion that prevails among the crowds of onlookers. One engraving by Golzius of the beach at Scheveningen offers an early example of this commemorative art.[105] It depicts the corpse of a beached whale being cut up and processed; horsemen move amid the tents, carriages, and carts scattered along the beach.

From 1623, the representation of the shore in the work of Jan Van Goyen began to develop into a painting genre whose stereotypes became quickly defined. This type of painting focused on the sandy beach and the strand, and its aim henceforth was to provide a realistic celebration of labourers on the shore. The seaside here teems with fishermen, tradesmen, and teamsters. The religious significance of this art is clear: the shore of Lake Tiberias underlies these paintings of the beach at Scheveningen or Egmont-am-Zee. The artist is crafting an ode to the fertility of the sea and the daily miracle of the multiplication of herrings.

The work of these painters had nothing to do with mythological subjects or the eroticism of bathing. These artists were at home among those who laboured by the sea; they were familiar with their language and gestures, and enjoyed their fellowship. Jan Van Goyen made repeated trips along the coast in their company. He hired a boat especially for this purpose, and filled several sketch-books during these excursions. Here, the fishermen's role was not merely to enliven the paintings with action.

The backdrop to this genre-painting is not the sea, but rather the beach and the strand. This meant that the artist had to abandon the bird's-eye view so dear to Pieter Brueghel and invite the viewer to come down with him on to the sand, where the hard-working activity was taking place. In this respect, the *Beach of Scheveningen* painted by Jan Van Goyen in 1632 is very significant. Two years later, the artist painted himself sitting on the sand: this was his way of expressing his intention to exalt the pictorial value of the beach. Jan Van Goyen and Solomon van Ruysdael, another painter who specialized in this type of genre-painting, inspired many imitators, who began regularly depicting the shore at Scheveningen.[106]

Towards the middle of the seventeenth century, the meaning and the content of the beach scene began to change. Simon de Vlieger, Jacob Van Ruysdaël, and a few years later Adriaen van de Velde[107] all sought to express with greater subtlety the gentle breeze and translucent moistness of the shore, and in so doing, they began gaining altitude and adopting another point of view. The genre scene was transformed into a coastal landscape that could encompass the span of the shore and the path of the excursion to its fullest extent. At the same time, the social significance of these pictures began to change. Admittedly, the beach remained the site of the fishermen's labour and an extension of the public space of the village; but it also showed the culmination of the ritual which was the urban outing. The shore was scattered with elegant bourgeois walkers engaged in flirtatious conversation or dashing horsemen, while other visitors came to gaze at the open sea. [108]

Between 1635 and 1665, Dutch seascape painting enjoyed a golden age. The desire to dramatize the struggle between man and the elements gave way, for a few years, to the description of the peaceful shore. Cuyp's and Jan Van de Capelle's serene, luminous paintings of the seaside also extol the labour of seamen and bargemen, and they help to glorify the beauty of shores whose lively atmosphere other painters[109] endeavoured to re-create.

At the end of the seventeenth century and throughout the eighteenth century, while tourists flocked to Scheveningen, painters of the shore returned to anecdotal images, depicting the bustling crowd once again. They mingled scenes from popular life with images of commerce.[110] In England during the same period, artists influenced by the Van de Veldes (who had settled there in 1670) began applying the stereotypes of Dutch painting to British shores.[111]

This makes it easier to understand why travellers rushed to Scheveningen in order to watch the activity on the beach and mingle with the fishermen whose praises had been sung so often. Those who, like Samuel Ireland, went down on to the beach, would stroll as if they were characters in a Simon de Vlieger painting [see plate 3], and might even do some drawing themselves.[112] They would ask the sailors to take them on fishing trips along the coast.[113] Diderot's admiration encompassed both the beauty of the Scheveningen shore and the simplicity of the fishermen.[114] By the end of the eighteenth century, the village was equipped to profit more fully from this beneficial influx: André Thouin remarked in 1795 that four or five tradesmen 'display in front of their shops shells, stuffed fish, marine plants, artificial flowers, and above all, little models of ships, launches, and other objects related to seafaring'.[115]

This brief overview hardly exhausts the rich subject of the journey to Holland. It is important to remember that the Republic of the United Provinces also constituted a political wonder. In the Age of Enlightenment, this land of tolerance and freedom was perceived as a model of good government. Diderot looked on it as the climax of the 'philosophical journey'; but this takes us away from our subject. The journey to Holland was an event around which a classical pattern evolved for appreciating the beach and the strand, that uncertain, ephemeral, luminous boundary. Human beings and their activity remained the focus of attention. Yet, by placing the spectator on a level with the water, almost at the point where the waves come crashing in, Dutch seaside-painters offered a visual grasp of infinite horizontal space.

The seaside scene generated a social model for the use of the sea-shore. It contributed to popularizing a ritual which will be discussed again later, as part of the era of seaside holidays. At the very end of the 'sublime road to Scheveningen', and along the Neapolitan marina (*Chiaia*) as well, practices were being instituted that surreptitiously foretold the rise of a collective desire.

THE PLEASURE OF INTERPRETATION: THE PILGRIMAGE TO THE SHORES OF CAMPANIA

> César foulait ce même sable.
> ['Caesar trod this very sand']
> – Bérenger

For Romans, the sea-shores of Campania are a symbol of beauty.[116] The plain and the hills of volcanic earth clustered against Vesuvius outline an enclosed landscape to the east that is a pleasure to the eye. The mild climate is in harmony with the splendour of the harvest and the neat vineyards. The Romans enjoy letting their gaze wander over the shoreline that gave its name to the region. As he retreated from Capri, Tiberius 'gazed at the beautiful curved outline of the coast'.[117] In the heart of the pleasantly shaped bay, the lazy waves lose their fury.[118] This peaceful sea becomes the echo of the Campanian countryside, tamed and mastered by man. The construction along the shore and the villas above confirm that the waves have been enchained. The shape of the coastline and the view of the sea lend themselves easily to an image from the golden age suggested by abundant crops and flourishing herds. This is where, according to legend,

Virgil was moved by the gentleness of the Tyrrhenian Sea and composed the seascape pieces that counterbalance the rustic scenes in the *Georgics*. [119]

Despite the disintegration of the monuments and the upheavals that had altered the site, travellers who stopped in Naples between 1690 and 1760 felt themselves to be in harmony with Latin writers. The visit to Campania, one leg of the Englishman's grand tour and the continental inhabitants' journey to Italy, provided one of the few possible occasions to appreciate the Tyrrhenian sea-shore. Tourists actually found the Italian coast repulsive, [120] and the itinerary they chose led them at the very most to stop in Genoa and admire its location before rapidly crossing through Livorno. [121] But the Bay of Naples was more splendid than any of the rare coastal resting places; moreover, it lay at the very end of the journey. Before the last decade of the eighteenth century, it was not customary to continue beyond Campania, where Europe seemed to end. Apulia, Calabria, and Sicily were then considered to be torrid African lands. [122]

Thus the journey to Italy would reach its conclusion in the heart of the most beautiful scenery in the world. 'The various prospects one discovers from this height', exclaimed Misson in 1688 from the Chartreuse St Martin, 'make the mind pause in admiration.' Before the viewers' eyes, all orders of beauty seemed to come together.

> One can distinctly consider the greatness and the outline of Naples, with its castles, its harbour, its mole, its lighthouse. It is a pleasure to look at the gardens that surround it ... If you cast a glance in another direction and look along the sea-shore, the undulations that intermingle with the small capes washed by this peaceful sea and the pretty little villages with which this coast is scattered, comprise a most pleasing subject. A little further down, the air becomes thick with the horrible smoke of Vesuvius, and one has a full view of that hideous mountain. [123]

Richard Lassel declared in 1670 that 'this is as fine a prospect as Europe can afford, excepting that of Greenwich'. [124] President de Brosses (1739) felt 'delightfully surprised' by the 'wonderful view' [125] he discovered from atop Mount Vesuvius, which was undoubtedly the most spectacular view to be enjoyed in Europe. He especially admired the country houses bordering the coast. Elsewhere, he remarked that in Naples, 'the bay is so compact that one can take it all in at a single glance'. When the sea could be enclosed in such a tableau and stripped of its monotonous immensity, it became possible to find a certain beauty in it. In 1755, Abbé Barthélemy admitted as he sat on the terrace over-looking a friend's casino that he had 'never seen a finer sight'.[126] Twenty years later, Abbé Coyer declared that he marvelled at the prospects he discovered from Capri and at 'the whole *circuit* of the Bay of Naples'.[127]

Because navigation techniques were so poor, travellers who arrived by sea had plenty of time to gaze at the Bay. Those who came from Rome after driving through Gaeta and Cumae emerged directly into the heart of the panorama. But for these tourists, the shore at Naples offered many other attractions besides those of the bird's-eye view made famous by Pieter Brueghel as early as 1556.

Reading authors from antiquity was the basis of the educational system in England, and educational programmes were inspired by the concept of imitating the Greeks and, even more important, the Latins.[128] A knowledge of ancient languages was considered indispensable in developing good taste: this was the only way to participate in the aesthetic quarrels that governed literary criticism at the time. Reading Horace and Virgil was one of the pleasures of adult life. The ancients thus became intimately involved in the life of the ruling classes; they offered models for good conduct, and, when the time came, they helped the dying to achieve a peaceful end.[129]

And what can one say about the hold of antiquity on French culture? Following Daniel Mornet's example,[130] innumerable his-torians have studied the period when the ancients were the sole masters of European schools. These authors have analysed the *Ratio studiorum* used in Jesuit schools, the slow penetration of French into the classroom at the expense of Latin between 1720 and 1770, the intense debates aroused in the eighteenth century by the educational issue of imposing a '*plan d'études*' or curric-ulum as a literary genre, and the emergence of the idea of a national education programme, followed by its belated imple-mentation. Daniel Milo has recently undertaken a very detailed study of the development of the repertory of authors and the models of writing and eloquence that shaped classical canons, whose impact extended well beyond the field of education.

All the specialists, particularly Jean de Viguerie, the analyst of humanistic teaching, show how the gaze of members of the cultivated elite was riveted on the last years of the Republic and the Augustan period. These classicists sought to ignore the richness in the evolution of antiquity. All that they borrowed from this civilization was in accordance with this process of reduction and simplification. In this respect, the people of that period were in a way maintaining an attitude that was already solidly rooted in late antiquity, influenced as it was by a cultural world dominated by Virgil and Cicero.[131]

A natural addition to all this was the passion for classical archaeology. This interest was already well developed before the great excavations at Herculaneum undertaken in 1737. Until then, as the writings of Montfaucon and of Caylus show, those known as 'antiquaries' considered their work to be primarily enumerative and descriptive.[132] This was in keeping with a journey to Italy whose main aim was to compile visual memories and collect monumental images. In 1699 Addison considered that the list of remains that tourists simply had to see was already so well known that he refused to present it yet again.[133] During the second half of the eighteenth century, spreading neoclassical aesthetics, developed under the influence of Winckelmann, altered the way in which these emotions were expressed, and the enjoyment of amateurs came to be based on a new understanding of ancient art.[134] Made richer by the visit to Herculaneum, then to Pompeii and the museum at Portici, the experience of the journey to Naples acquired a new significance. A revealing example of this change of direction is the wonder with which Abbé Barthélemy described certain new discoveries in 1755.

This intellectual trend, which constituted one aspect of the Enlightenment, concerns a European elite whose core was composed of gentlemen travellers gathered together since 1733 in the society of *dilettanti*.[135] Supported by a plethora of ambassadors and patrons of the arts who both discovered and organized the looting of classical treasures, the amateurs of the neoclassical period helped broaden the geographical range and the epistemological horizons of archaeological investigation.

At the same time, as early as the end of the seventeenth century among the English aristocracy and gentry and somewhat later in France among members of the academies,[136] a public of connoisseurs of landscape painting was developing. A new complicity was established between these enlightened readers and the authors of guidebooks and travelogues who placed increasing importance on the spectacle of nature.[137] Studying Italian and Italian-trained painters of the seventeenth century, especially Salvator Rosa, Claude Lorrain, and Poussin, shaped the tastes of a constantly widening public. The literary culture of these connoisseurs allowed them to debate the relationship between the various arts and

to participate in the quarrel arising over the commentary on
Horace's *ut pictura poesis*. By 1740, a knowledge of painting was
a must for every fashionable Englishman. The man of good taste
was defined by the pleasure he took in interpreting works of art.
The desire to collect, the proliferation of copies, and the pop-
ularization of painting techniques all fed the new fashion which
instituted the grand tour as an indispensable complement to a
young gentleman's education.

Shortly after the Treaty of Ryswick, the tradition developed of
sending such young gentlemen, often accompanied by a tutor, to
visit the countries of the Continent, and especially the treasures
of the Peninsula.[138] Between 1698 and 1740, the number of British
tourists increased by leaps and bounds, while in France the fashion
of the journey to Italy took hold.[139] Visiting the Peninsula made
it possible to associate literary reminiscences with the contemplation of works of art and to become visually familiar with the
landscapes that had inspired ancient writers and seventeenth-
century artists.[140]

Thus the already very old tradition of the journey to Italy
underwent a first revival, and was constantly enriched by an
accumulation of attractions. The prestige of taking the waters,[141]
religious pilgrimages to the sources of Christianity, time spent
in universities, visits to museums, and the English need for re-
creation were all gradually reinforced or replaced by a vague
longing to enjoy the world of nature. This longing would turn to
delight when the age of neoclassical aesthetics and the picturesque
voyage finally arrived.[142]

Supreme refinement of the culture, the journey to Italy was
something to prepare for. Before sailing, Addison reread the
appropriate authors, noting that he made a compilation of those
that he thought he might need. Half a century later (1763),
Gibbon preferred to plough through the two volumes of Cluvier's
Italia antiqua and Nardini's *Treatise on Ancient Rome*. 'A man
of letters', he wrote, 'likes to know these famous countries
down to their smallest hideaways, countries in which the tiniest
village is well known in history or poetry. It is in order to prepare
myself both for my journey to Italy and my future studies that
I undertake this reading.'[143] Collecting paintings and engravings
and engaging in academic or desultory conversation helped can-
didates for the journey to outline the objects they were to contemplate and to trace the range of pleasures they could expect.

The journey to Italy required a specific strategy, which was
developed in detail in a profusion of standard-setting literature.[144]

Social encounters were of prime importance: at each city in which he stopped, a traveller had to find his way into various 'assemblies' or 'conversations' that, as luck would have it, were largely open to foreigners.[145] If possible, he was to be introduced at the princes' courts. In any case, he must visit the celebrities. This necessitated obtaining letters of reference. Enjoying the vestiges of the past, especially when visiting Rome, also required obtaining the services of a good antiquarian and, even more important, getting oneself fitted out. Already at the end of the seventeenth century, a watch, a compass, an astrolabe, and above all field-glasses were part of the necessary gear for a conscientious tourist. Ideally, the visitor should be accompanied by a good sketcher, even if he was an artist himself.[146] Otherwise, it was advisable to collect engravings in order to keep alive one's memories.

The itinerary, too, was strictly codified. On the return trip, the similarity of the roads taken fostered complicity among tourists, encouraging them to compare their emotions. Travellers focused their attention on the cities, the views they offered, the institutions by which they were governed, and the monuments they contained. This primacy was particularly evident in Rome, whose 'semantic absolutism'[147] has been so justly emphasized by Michel Butor. This concentrated attention was heightened by the great craze for *vedute*.[148] Between the various urban stops, tourists relaxed their attention, displaying a sort of blindness as they crossed the countryside. In the carriage, they would reread the appropriate authors in order better to savour the richness of the next stop, or they might indulge in conversation.

This blindness was not continuous, however. Travellers did occasionally look at nature, but only in accordance with an elaborate code which shaped the classical model for evaluating the sea-shore in particular. The traveller's vision was placed in the service of ancient texts engrained in memory, texts which the tourist had admired since adolescence. This mediation established the group of amateurs, made the social recognition of its members possible, and contributed to the aim of distanciation revealed by this social pattern.

David Hume, who found himself in the Plain of Lombardy in 1748, was moved at the thought that Virgil had trodden that soil; at Mantua, he kissed the ground on which the poet had stood.[149] On the outbound journey, Addison made the trip to Campania in Horace's company. 'The greatest pleasure I had as I went from Rome to Naples', he wrote, 'was to see so many

fields, cities and rivers that have been described by so many
classical writers, and that witnessed so many great deeds, for the
whole of this road is extremely sterile in curiosities. It is worth
casting a glance upon Horace's journey to Brindisi when one
comes through this place.'[150] On the way back, he decided to
follow in the footsteps of Aeneas because, in his opinion, Virgil's
description was very accurate. In Campania, Président de Brosses
also let himself be guided by the sixth book of the *Aeneid*,[151] a
text so well known that he did not dare to quote from it. As
for Abbé Coyer, he wrote: 'I had my Virgil in my hand, and I
followed Aeneas as he approached Italy.'[152] At the end of the century, when neoclassical sensibilities became triumphant, President
Dupaty nevertheless remained devoted to this form of travel.
When he visited Virgil's grave, he wrote, 'I recited Gallus's
eclogue; I read the beginning of the fourth book of the *Aeneid*, I
uttered the names of Dido and Lycoris; I cut a branch of laurel
...'.[153] He imagined the faithless Cynthia, who made Propertius
weep, and Seneca's wrath against the debauchery of Baiae.
And when Swinburne travelled through the two Sicilies at approximately the same time, he examined the customs of the people
living on the southern coasts of Italy in the light of ancient
authors.[154]

By the same token, classical aesthetics determined the selection
of itineraries and the catalogue of objects to be admired.[155] Travellers were to retain that which dealt with humanity. Among
the landscapes, their gaze would pause on pleasant, fertile sites,
hillsides, orange groves, and garlands of vines. Such places suited
the harmony that defined the classical beauty of nature, and,
above all, they lent themselves to inspiring quotations. In travellers' writings, the praises are most heavily concentrated on the rich
plains of northern Italy, the shores of the Clitumne, and
ancient Campania. Addison marvelled at the sweetness of Capri,
but the rocks of Sorrento were unsettling to him: they were not
yet a customary place to visit, wrote Abbé Richard in 1766.
Tourists also turned their eyes to the scenes of historical events,
captivated by the crossing of the Rubicon and the banks of Lake
Trasimeno. In general, the image of man was always present
behind this admiration for landscapes.[156]

Even more fascinating were the places where a Latin writer had
been, especially if the author in question was Virgil, Horace,
Cicero, Seneca, or Pliny, who were the most frequently quoted in
travel literature. Roland de la Platière made a point of exactly
modelling his visit to the shores of Syracuse on Cicero's journey;

in so doing, he was compelled to privilege topographical and monumental descriptions. [157]

A commonplace spot, no matter how tiny, could attract great attention if it had been ennobled by the gaze of one of the ancients. This explains the intense semantic charge of the Campanian shore. Addison paused before three pointed rocks next to one of the grottoes in Capri because he thought they were the *Sirenum Scopuli* mentioned by Virgil. [158]

On the other hand, all visual perception was lost in the face of sites not mentioned in prestigious texts. Before 1750, the touristic eye was not yet trained to analyse colour or other substantial qualities.

In Naples, the figure of Virgil composing the *Aeneid* cast a long shadow, and the first and sixth books of the poem determined the itinerary and how the scenery was viewed. Travellers were haunted by the image of the Tyrian port, said to have been inspired by Pozzuoli; the Sibyl's grotto and the Elysian fields visited by the Trojan hero; Cape Miseno, ennobled by the death of the pilot Palinurus; and the coastal route followed by Aeneas on his way to Latium; and they could not resist making a pilgrimage to the poet's grave as they approached Posilippo. [159] The ancient splendour of Baiae, which Misson described as 'the most magnificent place in the world', [160] also exerted its own fascination, though it is difficult to see how this came about. Four images determined ways of seeing with regard to holiday resorts: the activity of the Pozzuoli sea-shore and its contemporary decline; Tiberius's scandalous stay at Capri, which provided opportunities to unleash erotic fantasies; [161] the delightful house belonging to Pollius Felix, Statius's friend; and the tunnel built by Domitian and criticized by Seneca. Occasional visitors recalled that Sanazzaro resided on the coast of Posilippo in the sixteenth century. [162] The pleasant *Chiaia* excursion and, later, volcanology, as made fashionable by Hamilton, also contributed to define the network of interest which viewers found in the Campanian countryside.

The subordination of vision to Latin texts helps to explain the absence of a genuine descriptive style. [163] Why try to depict what has already been so well described? This is what Addison asked himself as he left to Horace, Virgil, Statius, and Lucan the task of depicting the landscape. Here, writing consisted in arranging a collection of borrowed elements and evoking reminiscences and emotions that arose from the reading of ancient texts. It was a filtering achieved through the ample corpus of possible quotations, a means of selective reappropriation that

defined the author's specific sensibility. In fact, this borrowing
of descriptions was subjected to original selection procedures
which involved not only the texts chosen, but also the objects
admired and the reasons for this admiration. Consider the example
of Campania. Travel accounts borrow from Virgil above all else
for the storms, the incidents during the coastal journey, and the
impressive adventure that begins for the hero of the *Aeneid*
inside the Sibyl's grotto. On the other hand, the same authors
close their eyes to the seascapes in the *Georgics*;[164] yet the theme
of precarious harmony was to bear heavily on European literature
after 1750, especially in England, as filtered by Thomson.[165]

The authors of travel literature pay no attention to Cicero's
contemplation of nature and the comfort he offers to citizens weary
of politics. Above all, Lucretius is almost totally absent from the
corpus. In short, the coast is not portrayed here as a place for
meditating on enigmas; the beach is not considered a privileged
spot for philosophical retreat. Before 1750, the joys found in
ancient holiday sites were poorly understood,[166] and were more
often criticized than praised, as seen through the writings of
Propertius and Seneca. Addison is completely lost when he admits
that he fails to understand the seasonal rhythm of the ancient
otium.[167]

On the whole, the narrowness of the network of references
is as striking as the poverty of the range of borrowings. A
small number of stereotypes, very limited when one considers
the richness of ancient literature, weave a monotonous discourse,
and generate a fairly barren emotional strategy.[168] This parsimony
serves a purpose. For the period, it facilitates social complicity
and recognition, and limits the danger of misunderstanding.
References to texts familiar to everyone hinder the emergence of
a descriptive style, but they also provide a literary shorthand.

The process of distinction induced by the classical voyage
gradually became more complicated. The philological knowledge
expected of an end-of-the century amateur and the aesthetic sensibility he was required to display at a time when the pictorial
code was sharpening visual analytical skills reveal a cultural
horizon of quite another breadth. As a result, the composition
of the public created around this travel literature was modified
at the same time.

Equally striking is the absence of criticism of the ancients' way
of seeing. This renders all the more interesting for our purposes
an exceptional text by Moisant de Brieux, written in 1668. This
hermit of Bernières, accustomed to the sight of the English

Channel and knowing nothing of the Mediterranean, confesses that he is baffled by one line in the fourth book of the Aeneid: 'The wild seas had found their rest.' He charges Virgil with inaccurate observation: 'This description of the night is beautiful,' acknowledges Moisant; 'the whole of Nature seems asleep ... the sea is the only aspect in which I cannot understand this rest, as both at night and during the day, it is always the scene of restlessness and inconsistency. It has its ebb and flow ...'.[169]

The emotional tactics and the pleasures induced by the classical mode of reading landscapes trace a coherent system which is difficult for modern-day readers to comprehend. This pleasure comes first of all from identifying places; this is the emotional origin of individual discovery. The game is to summon forth the most unexpected hypotheses, in order to have the enjoyment of interpreting a text through the vision of nature.[170]

Failing an emotion aroused from inventing an unexpected correspondence, there is always the satisfaction of verifying a confirmed concordance. In this case, the pleasure is caused simply by the feeling of congruence between the traveller's emotion and that of the ancient writer before a landscape that has been considered remarkable for so many centuries. And of course there is the additional hope that sometimes creeps in of discovering something else that might add to the beauty of a site so often exalted. Addison savoured this harmony that became established, beyond the passage of time, between a traveller and a dead genius. 'I must confess', he wrote, 'that studying these various descriptions on the very spot and confronting the places and their situations with the descriptions of them left us by the poets was not the least of the discussions and pleasures I enjoyed during my trip.'[171]

Adults also took pleasure in seeing images come to life that had been born in their adolescent minds. The journey to Italy created an opportunity to grasp the unity of the self while at the same time revealing a harmony that developed among viewers across the centuries.

Travellers knew that the Campanian countryside which they had before their eyes was not that with which Virgil was familiar. The earthquake of 62 and the eruption of Vesuvius in 79 had altered it considerably; they knew that since the death of that poet, sulphur springs had bubbled up, and that the construction works undertaken by Agrippa and Domitian had also wreaked havoc with the profile and outline of the sea-shores. At the same time, some travellers endeavoured to re-create the vanished scene in their imaginations;[172] their minds gradually gave shape to the notion of a landscape in ruins, although they were not overcome by the nostalgia that was to plague Romantic travellers. It was without nostalgia that President de Brosses dreamt of the 'golden gondolas', the sea 'covered with roses' and marked by the wakes of little boats 'full of pretty women in elegant scanty gowns', or the concerts on the water in ancient Baiae. In so doing, he was sketching an early picture in the purest neoclassical style. By contrast, at the end of the century, Swinburne felt overwhelmed by the contrast between the present decay of Taranto and Sybaris and the past attraction of the enchanting abodes described by Horace.[173]

During the last third of the century, under Winckelmann's influence, the idea evolved that the aim of the journey was not only to evoke, confront, and interpret, but that, thanks to a better understanding of the ancients, it was also possible to *bring the texts back to life*. This is a familiar notion to modern-day travellers, but it emerged rather late in the day – no doubt because this new desire called for a more refined aesthetic communion between the traveller and the artists of antiquity. On 17 May 1787, Goethe wrote to Herder that the *Odyssey* had not become 'a living expression'[174] in his mind until he had come to know the Mediterranean and its shores. Swinburne's strategy was more subtle: through a complex back-and-forth movement, he expected both that his journey would bring about a revelation of the ancient texts and that these works would enlighten his understanding of what appeared before his very eyes.

The journey to Italy satisfies the intellect: it teems with aesthetic emotions. With the exception of developments geared to gallant conversation, there is virtually no mention of the body or of coenaesthetic impressions. Travellers were merely touched by the pleasure of strolling up and down the memorable shores, of climbing the slopes of Vesuvius to gaze into the horrible crater into which Pliny the Elder threw himself, or of delving into the grottoes made famous by the Sibyl's gestures and Tiberius's debauchery. The Campanian sea-shore, one of the gates of Hell, incited travellers to make a journey into the depths, rendered even more fascinating by the proximity of the volcano. The fact remains that the greatest pleasure here was in fact to model one's trip on that of Aeneas, to experience the perils and brave the storm that he had had to overcome. For Addison, the horror of the coast that separated Rome from Cape Miseno revealed the temptation of the sublime. For him, the itinerary of Aeneas became a series of emotions. Travellers took pleasure in hearing – or conjuring up? – the howling of wolves and the roaring of lions born of the nocturnal sounds of the sea on the rocks lining the steep coast.[175]

Though it is less apparent in the actual texts of travel accounts than in the abundant literature generated by aesthetic arguments travellers sometimes multiplied the range of interpretative pleasures: they no longer commented on a landscape solely in light of ancient texts, but also by referring to modern works modelled after the older ones. Before the appearance of Abbé Barthélemy's *Les Voyages du jeune Anacharsis*, the French were inspired by Fénelon's *Télémaque*, though admittedly less often than by Virgil's poem. This mixing of references justifies the tactics adopted by Jean Houel who, as he neared Malta and approached what was said to be Calypso's grotto, watched, book in hand, not for the Homeric scenery but for that of *Télémaque*.[176] In this particular case, the approach through references within references could be justified by the luminous description of the ancient seas found in this work. Calypso's grotto overlooking the shore combines the harmony of an Arcadian painting with the charm of an ancient seascape. Better than any cool copy of the Virgilian storms, the image of the waves crashing around the foot of the Egyptian tower where Telemachus is kept prisoner foreshadows the sublime appreciation of the violence of the sea.[177] In 1699, Fénelon began to reinterpret the ancients, and composed the first sea-pieces in French literature. Consider the impression of motion depicted through the ship's course on leaving Alexandria: 'The shores of Egypt receded far away from us; the hills and the mountains grew gradually flatter. We began to see nothing but the sky and the water, while the sun, as it rose, seemed to bring out of the sea its glistening lights; its rays gilded the mountain tops that we could still faintly distinguish on the horizon.'[178]

A knowledge of seventeenth-century Italian art, especially Claude Lorrain's painting, contributed to organizing the traveller's gaze as he beheld the sea-shores.[179] Lorrain was able to interpret the classical tradition in the light of Christianity. From Virgil, and even more from Ovid, he borrowed a world of harmony whose beauty and mystery are rooted in the divine source that governs the universe, though they conceal a dark side. Yet at the same time, his painting integrates the hymn of the Fathers of the Church with the beauty of the world, and proposes to reduce the elements observed in nature to Christian symbols.

Claude Lorrain's coastal scenes, painted under the influence of Paul Bril and Agostino Tassi, helped to make him famous.

By the end of the seventeenth century, they were being massive-
ly exported to England. Lorrain designed them as companion
pieces to his inland landscapes and, through this dual vision,
attempted to re-create the unity of the image of the world. The
artist incorporated the sea's horizon, its light, and the sun's rays
upon the waves into the orderly and harmonious earth depicted
in his painting. Yet to the viewers' eyes, the coastal scenes also
presented the Christian symbols of embarkation, the perils of the
open sea, the coming into harbour, and all the episodes that mark
human destiny. Art-loving travellers from the beginning of the
eighteenth century liked to indulge in this hermeneutical play
of mirrors. The mediation of Lorrain's painting simply increased
their pleasure.

This multi-faceted reading was part of the complex range of
tactics that ensured a richness in the classical interpretation of
landscapes, despite the apparent poverty of descriptive style. The
tourist who contemplated the beautiful Campanian landscape at
the furthest point of his voyage was completing a pilgrimage at
the same time as he performed a ritual.[180] He relied on the joys
of evocation and interpretation. His delight in the text grew
with this confrontation, and confrontation in turn helped to
deepen a culture that encouraged social complicity through similar
itineraries and ways of travelling, and by codifying the emotions
aroused by the journey. In addition to the beauty of masterpieces from antiquity and the perenniality of human kind, travellers
contemplating the famous Bay were also implicitly celebrating
this socio-cultural cohesion of elites. This cohesion was to be
further reinforced by neoclassical aesthetics, despite the use made
of them by French revolutionaries.

The way of looking at the sea from the end of the seventeenth
century until the middle of the eighteenth is part of a classical
epistēmē that combines Judaeo-Christian tradition, Hellenistic
philosophy, and Latin literature, and does so confusedly, though
with apparent order. On the whole, what dominates is the fear
of the sea and a repugnance towards visiting its shores. The
image of the terrible ocean, that chaotic vestige of the disasters
buried in mankind's past, and the unpredictable wrath of its
moving, lugubrious immensity are in keeping with the perils and
the pestilence of the enigmatic beach, a fluctuating line subject
to all kinds of incursions, along which are deposited the excreta
from the abyss.

Admittedly, people occasionally dreamt on the sand as they
gazed upon the manifest signs of the Creator's might or the

beauty of his work. They went to Scheveningen to admire the labour of the fishermen who benefited from the miraculous fruitfulness of the northern seas. On the canals of Amsterdam, they came to savour the harmony that reigned among the elements in the heart of this microcosm blessed by the God of the Bible. Admittedly, travellers journeyed to Naples for the pleasure of strolling up and down the most beautiful of classical shores, which shone with universal harmony and a complicity that spans generations. But the centre of these shore scenes, full of symbols and confused signs waiting to be interpreted, continued to be occupied by man. The representations of the waters and the sand continued to be haunted by recapitulation of the world's constituent elements and by evocations of episodes in human destiny. Nowhere, except in a few rare individuals with premonitory sensibilities, was any admiration expressed for the infinite expanse of the waves or the delight that arises from a visual analysis of actual substance; nowhere did anyone express a desire to feel the power of the waves or the touching coolness of the sands against his body. Coenaesthetic impressions are absent from the range of both experience and speech.

The irresistible awakening of a collective desire for the shore arises in the period from 1750 to 1840, the time covered by this book. This was when the coasts of the ocean began to appear as a recourse against the misdeeds of civilization, as the place where it was easiest to grasp the new sense of time proposed by scientists, and experience the dissociation of mankind's history from that of the earth. This was where the sublime beauty of the ocean and the pathos of its storms unfolded. Here, more than anywhere else, the individual found the means of encountering the elements and enjoying the brightness or transparency of the water.

Part II

The Pattern of a New Pleasure

3

A New Harmony between the Body and the Sea

THE SHIFT OF ANXIETIES AND DESIRES

Cure-takers began rushing toward the sea-shore around 1750 in order to relieve an old anxiety: this was one of the tactical weapons used to combat melancholy and spleen. It also served, however, to assuage the new anxieties that were looming up one after another throughout the eighteenth century among the ruling classes. This, in fact, is what determines the richness of medical language describing the virtues of cold sea water and especially the indications for bathing in the waves and taking holidays at the seaside. Physicians and hygienists speak of alarm and desire at the same time as they discuss scientific knowledge. Their discourse generated, handled, or codified practices that would gradually escape from their control.

Before examining the background of these behaviour patterns, it will be useful to review the forms of melancholy, no doubt the most complete chapter in the modern history of mind-sets.[1] Medieval *acedia*, a diabolical gangrene of the soul that drove its victims to despair of attaining salvation, had provided fuel for the diatribes of ecclesiastical authorities long before physicians began to strengthen the theologians' position as they revived the ancient theory of humours.[2] During the sixteenth century and into the beginning of the seventeenth, melancholy was fashionable. Later, in the France of Louis XIV, this morbid delight began to wane. The gentleness of St Francis de Sales's spirituality, the vigour with which the Jesuits fought against *acedia*, the fascination with the Ciceronian art of living, and the 'therapeutic sociability' displayed at Court combined to offer effective means to combat the sundering of body and soul that characterized this insidious disease. French classicism devalued the melancholy temperament, and Molière's audiences laughed at the misanthropist.

Across the Channel, however, the ruling classes remained under the sway of what all of Europe would soon describe as 'spleen'. In 1621, Robert Burton published his *Anatomy of Melancholy*,[3] which was to exert an enormous influence on the behaviour of British aristocrats. To combat the disease, the author lays out a complex strategy combining concern with the environment, hygienic precepts, and body care with a subtle therapy for the soul, or rather the mind. Burton begins by inviting the melancholic person to choose judiciously the place where he is to stay. He is pleased to note in this respect that the gentry are very preoccupied with the quality of the environment in which they decide to dwell.[4] In keeping with Hippocratic tradition, the melancholic person knows that 'such is the air, such are the inhabitants'.[5] He should opt for a dry soil and an area that is 'uneven and scattered with hills'[6] and free from sources of putrefaction. He should choose to settle on 'rising ground overlooking an extensive horizon'.[7] To support his hypothesis, Burton points to the excellent health and surprising longevity of the inhabitants of the Orkneys, 'due to the sharp, purifying air that blows from the sea'.[8] That said, the author, who was himself of a melancholy temperament, did not advise Englishmen to follow the Neapolitan example and build their abodes on the strands of their damp shores. It would therefore be an exaggeration to regard him as the 'inventor of the sea'.

Like Celsus in times past, Burton stressed the advantages of diversity; he recommended travelling, alternating between stays in town and country, and striking a balance between bodily and intellectual exercise. 'A good prospect alone', he wrote, 'will relieve melancholy.'[9] A varied landscape was therefore part of the remedy he prescribes. The inhabitants of Genoa, Naples, or Barcelona, for instance, like those of Aegina and Salamis in olden days, had before them a prospect that delighted their souls, as the Mediterranean introduced variety into the landscape. On the slopes of the natural amphitheatres that overlooked a bay, dwellings were built from which the inhabitants could enjoy a beautiful sea view. There the sight of the islands and the motion of ships would add to the pleasure of glimpsing the bustle in the streets or contemplating restful green gardens.

Burton's book, a huge catalogue of topoi, rehashes a number of ancient precepts. Like all things in the world, man needs movement. This is why exercise is beneficial to him. The author advocates rural sports, recommending horseback riding, fishing, swimming, football, bowling, and a range of other activities that were part of the leisure pursuits in which the common people indulged. Learning to emulate their gaiety and vigour was part of a therapy that relied on the circulation of practices among different social levels. On the subject at hand, it is important to point out that until that time, river or sea bathing was considered to be an immoral pastime better left to the ill-mannered lower classes. In Burton's time, it became an acceptable activity. In 1622, Henry Peacham even considered that a true gentleman should know how to swim. The *Anatomy of Melancholy* is packed with references to ancient hydrotherapy, and thus it fostered the rise of the fashion for therapeutic bathing and the prosperity of the spas that spread throughout the inland regions.

Even if a hunter kills no game and a fisherman catches no fish, they both none the less enjoy the pure air, the sweet scent of flowers, and the song of birds, all of which cure melancholic individuals. Burton actually returns to a tactic developed earlier by St Bernard, encouraging walks between the woods and the river in 'pleasant places'. The list of pleasant places that he draws up complies with the classical code of landscape appreciation, and it is widely recognized that the so-called English garden was to become part of the struggle against spleen. Burton focuses on rural settings, and neglects the sea-shore. He acknowledges that the seaside offers healthy air and a pleasing view, but he does not yet think of choosing it as a place for excursions or holidays.[10]

Following the book's publication, new anxieties stimulated the longing for refuge. Remember that, between 1645 and 1660, the English aristocracy, experiencing a radical threat to its political and social power, rushed to the countryside in order to enjoy the compensatory pleasures of retreat. At the same time there was a revival of the old diatribe against city life that had formerly ensured the success of Isaac Walton's *Compleat Angler*.[11] In fact, since the thirteenth century voices had been heard complaining about London's bad air. In 1578, so it was said, the environment had become so unhealthy that Queen Elizabeth had been induced to leave the capital for this sole reason. In the seventeenth century, collective complaints against sulphur and carbon fumes became more numerous, in Sheffield and Newcastle,

as well as on the banks of the Thames. John Evelyn led the
war waged by public opinion against this scourge. In short, the
theme of urban blight took hold long before its counterpart
which spread across France at the end of the *ancien régime*.[12]
Significant in this respect is the work of Tobias Smollett,[13] which
was contemporaneous with the creation of the fashion for sea
bathing. The author of *Roderick Random* brings together all the
elements of this criticism, loudly proclaiming his indignation at
the pollution of the capital's air and of the water of the Thames
by dust, smoke, and refuse. He was revolted by the spectacle of
stinking crowds crammed together in the ballrooms and bathing
establishments of Bath. The theme of excrement, omnipresent in
his work, supports his denunciation of the degradation of city-
dwellers' tastes. His writing displays a classical drift from hygiene
to morality: in the city, criteria for appreciation are reversed;
a perverse taste for viscous and adulterated foods triumphs, and
a deliberate stench is allowed to spread, because it is too easily
tolerated. Social mobility and ostentatious luxury were merely so
many signs of the form of collective suicide that was devastating
the city.

Smollett was fascinated by the purity of water. He was an
untiring apostle of hydrotherapy, devoting his first book to it as
early as 1742.[14] An adept of cold bathes, he would swim in the sea,
and made long stays on the French and Italian coasts. In short,
his behaviour demonstrates the connection that was emerging
between disgust with sticky city life and the longing for the shore.

The fact remains that the emergence of this diatribe, like the
taste for retreats and, later, the longing for relaxation and rest,
came long before the Brighton vogue. It is interesting to examine
the reasons for this delay. For a long time, the attraction aroused
by nature was purely literary. This appeal was focused on the
countryside and on gardens, 'the synthesis of the tame and the
wild'.[15] By contrast, it was impossible to subdue the ocean, which
could not be tamed in any way. Man could find no genuine
shelter there, nor build for himself a secondary residence. Because
it was irremediably wild, the liquid element represented the primitive state of the world. The seaside itself was beyond man's
control; it did not easily lend itself to the carefully constructed
apparent disorder of the English garden. Above all, a site that
cannot be altered by force preserves no trace of human history:
sand and water erase any sign, just as they frustrate any design.
With the exception of ports and dikes, which were themselves
often precarious, and the vision of moving sails, the seaside

offered no image that could demonstrate mankind's mastery of nature. Before the sea-shore could enter the range of attractive places, the desire had to arise for visions of the sublime, and the therapeutic necessity had to make itself felt.

In fact, in eighteenth-century France and England, anxiety was taking on new forms. These were in keeping with the rise of the sensitive soul. Physicians tended, *contra* the Hippocratic tradition, to consider the diaphragm as the regulator of sensitivity and the 'centre of internal turmoil';[16] with the brain, it exercised an unstable diarchy. As the seat of anxiety and the trip-wire that triggers spasms, this organ expresses 'the essential deterioration of vital equilibrium'.[17] This explains the importance attributed to it at a time of growing interest in the nerves and the various disorders of the soul, which were gradually to develop into a range of disturbing forms. Vapours, hysteria, nymphomania, and menstrual problems all fuelled a proliferating literature which stressed the specificity of both female nature and woman's social status.[18] Doctors dwelled complacently on the risks of puberty, maternity, and menopause; they emphasized clear-cut stages of life that were replicated across the social spectrum. This prolix literature also secretly informs us about the troubles afflicting man's psyche: as a fascinated spectator of the disorders caused by mysterious telluric forces that overwhelmed the nymphomaniac,[19] the individual himself became increasingly threatened by hypochondria, a modern extension of classical melancholy.

Spleen drives its sufferers to travel; it stimulated the mobility of individuals that accompanied the growing circulation of goods in the eighteenth century. Anxiety fosters curiosity and a sense of adventure. It haunts travellers on the grand tour, such as young Breton aristocrats eager to perform great deeds, whose mobility is stressed by Jean Meyer.[20] In accordance with the theory of climatic influence, these new anxieties exerted a stronger influence in the north, and this was precisely where a new type of recourse took shape, fanned by the temptation to return sincerely to nature.

In fact, people were frightened by excessive frailness and pallor. The ruling classes, believing that they did not have the vigour which the working classes enjoyed thanks to their labour, felt that they were being consumed from within. The elite of society feared their artificial desires, their listlessness, and their neuroses. They felt threatened by social death from their own particular types of fevers and passions, because they were unable to participate in the rhythms of nature.[21] This is the perspective within

which the sea-shore began to develop its appeal in the middle
of the eighteenth century. Remember that, even more than the
countryside, the ocean represented indisputable nature which was
more than just scenery, and which remained unaffected by false-
hood. This explains how the paradox developed on which the
fashion for the beach is based: the sea became a refuge and a
source of hope because it inspired fear. The new strategy for
seaside holidays was to enjoy the sea and experience the terror
it inspired, while overcoming one's personal perils. Henceforth,
the sea was expected to soothe the elite's anxieties, re-establish
harmony between body and soul, and stem the loss of *vital energy*
of a social class that felt particularly vulnerable through its sons,
its daughters, its wives, and its thinkers.[22] The sea was expected
to cure the evils of urban civilization and correct the ill effects of
easy living, while respecting the demands of privacy.

Physicians began to favour fortifying prescriptions. They set
about restoring energy and tonicity in human fibre; they over-
whelmingly adopted the vigorous therapeutics advocated by
Sydenham at the end of the previous century. The untameable and
infinitely fruitful sea (especially in northern realms) could sustain
vital energy, provided that mankind knew how to handle the
terror it inspired. On its shores, people found renewed appetites,
slept, and forgot their worries. The cold, the salt, and the shock
to the diaphragm from plunging abruptly into the water, along
with the view of a healthy, vigorous people, including many
centenarians, and a varied landscape were to cure chronic invalids.
In addition, patients taking a cure could enjoy themselves amid
the brilliant society that quickly began to frequent the fashionable
resorts. In 1804, after seventeen years' practice, Dr Buchan published a book which took the form of a retrospective[23] in which
he theorized about the connection that was developing between
a horror of the city and the pleasures of seaside holidays. He
heavily praises the Thanet peninsula, which cures the listlessness
that is due to excessive 'originally feeble and delicate stamina';
it ensures the serenity of young girls during puberty, holds sexual
passions in check, and prevents men of little virility from becoming effeminate. It is a cure for an excessive use of the mind;
it is also a means of avoiding sunlight, whose action 'is aug-
mented by the infinitely multiplied reflections from brick walls,
and burning pavements'[24] in the city. The beach toughens those
who are slaves of comfort and who can only walk on carpets.
Above all, the taxonomy of the seaside's virtues in Buchan's book
becomes the reflection of a hymn to the longevity of hard-working

fishermen[25] and the productive sailors and Nordic peoples. Englishmen enjoyed bathing in the frothing waves because they still vaguely shared the Greenlanders' instincts.

For a scientist of that time reading *De liquidorum usu* from the Hippocratic collection, bathing could not be considered a casual experience. As Maret clearly explained in 1751 in the *Encyclopédie*, physicians expected a good deal from immersion in water, whether cold, cool, or warm; the liquid was believed to have a mechanical effect on the tissues. It either tightened or loosened fibres, according to its temperature. It modified the consistency of the humours and the rhythm of blood circulation, and had an influence on the working of the nerves.

This element of the therapeutic arsenal took on a different significance in the eighteenth century with the rise of cold bathing. In this respect, the vogue for sea bathing was merely a final step; bathing in salty water at between 12 and 14 degrees often appeared to be an attenuated means of resorting to cold which was reserved for faint hearts,[26] children, women, and old people. Of course, this new fashion was deeply rooted in ancient beliefs: 'Bathing the body in cold water fosters long life,'[27] asserted Francis Bacon in 1638; while Hermond Van der Heyden, a leading medical authority of Ghent, published a well-received book on the subject. It was, however, Floyer's *History of Cold Bathing*,[28] written in 1701 and 1702, that was to establish the new fashion. Before then, Englishmen did not appreciate cold water. The author, who compiled an impressive panoply of ancient references, reminded his readers that Antonius Musa prescribed cold bathes to the Emperor Augustus, and that Pliny and Seneca recorded the spread of this practice, which had been strongly recommended by Hippocrates, Celsus, Coelius Aurelianus, and Galen, to mention only the best-known figures.

The benefits of this rediscovered treatment were firmly established in a book called *Contraria Medicina*. 'Heat succeeds cold and cold succeeds heat in a natural way,' stated Floyer.[29] Cold water closes the pores and refreshes, compresses, and condenses the air contained in the humours. Thus confined, this air increases in quantity, and becomes more elastic. This creates a sensation of heat, and even of burning. Moreover, cold bathing is among the toughening processes that are particular to northern peoples, and it ensures their exceptional longevity. 'A cold regimen suits cold countries':[30] individuals must build up their bodies so that they function in harmony with the surrounding air. This therapeutic injunction took on a moralizing tone: cold bathing,

declares Floyer, corrects depraved living habits. It abates the
vehemence of the passions, and must therefore be part of young
people's education. Above all, the British physician relies on the
terror and shock caused by sudden immersion in cold water to
cure disorders of the soul. And he takes on a virtually religious
tone when he praises the virtues of this substance:

> Cold bathes cause a sense of chilness, and that, as well as the
> Terror and Surprize, very much contracts the Nervous membrane
> and tubes, in which the aerial spirits are contained, and they
> being kept tense and compressed, do most easily communicate, all
> external expressions to the Sensitive soul. Not only the external
> senses are more lively in cold water, but all our animal actions and
> reasonings are then more vigourous by the external compressure of
> cold air.[31]

By purifying man's animal spirit and compressing its excessively irregular fluctuations, baptism by immersion once prepared the
soul to receive the divine mark. Floyer deeply regretted that
the Church had abandoned this beneficial practice, though it had
long been maintained by northern peoples, especially the Picts,
the Scots, and the Welsh. The book concludes with a chorus of
praise for the energy and morality of northern people. Floyer, like
Burton before him, contributed to the establishing of a myth
concerning the longevity of the inhabitants of the Orkneys.
Hardened to the cold, surrounded by a sea whose abundance
was inexhaustible and from which savages sometimes emerged,
engendered by the icy waves,[32] the Nordic aboriginal was thought
to show better resistance to death than townsfolk weakened by
comfortable living. This was why Englishmen were advised to
avoid tobacco, coffee, tea, wine, spices, and all such southern
inventions which could only jeopardize their ability to adapt to
the climate in which they were meant to live.

Floyer then sets forth a series of precautions, and draws up
a catalogue of indications that outline, as early as 1702, the
features of what was to become a medical attitude toward sea
bathing. He advocates, for example, that immersion in water
at a temperature of less than 10 degrees be accompanied by
exercise in cold air. He recommends walking and horseback riding.
Eager to test the virtues of each type of immersion in cold water
for himself, he tried sea bathing, and recommended it for para-
lytics. It was therefore something of an exaggeration for Michelet
to claim that Dr Richard Russell was 'the inventor of the sea'.[33]
Sea bathing was part of a logical evolution in experience. The

fashion for cold bathes actually became widespread after 1732.[34] On the Continent, Maret explained twenty years later how this practice increased animal warmth, and in 1763, Dr Pomme believed that cold bathes would soothe vapours and lessen nervous irritation. The practice of bathing in the cold water of rivers and streams then began to develop in France. But this is enough about a subject that digresses from the theme at hand.

In fact, the therapeutic values of sea water were being rediscovered at the time, and physicians were recalling that Cribasius had already advocated it in cases of scrofula. One of John Speed's aphorisms clearly explained the approach of medical practitioners: 'To bathe in the sea is to have not only a cold bathe, but a medicinal cold bathe.'[35] The therapeutic use of the sea also had its own prehistory. In a form of medicine based on contradiction, it was traditional to plunge sufferers from hydrophobia into the waves, and the shores periodically resounded with the cries of raging victims supported by the vigorous arms of volunteer bathers.[36] In 1667, Dr Robert Wittie also advised gout-sufferers to bathe at Scarborough.[37] This resort was able to make a gradual transition from traditional hydrotherapy to a recourse to the virtues of sea bathing. It so happened that from the side of a cliff overlooking the beach, there flowed a mineral spring that had attracted patients to take the waters since at least 1627. Because of the infiltration of sea water, the source at this spa had a salty taste. It was considered to have purgative virtues that were emphasized by Celia Fiennes in 1697.[38] Already at that time, those taking the cure at Scarborough combined drinking this salty water with excursions along the beach and the strand at low tide. In this way they pioneered what was to become one of the major experiences of seaside holidays.

In 1748, more than forty years after Floyer's isolated experiments, Dr Richard Frewin sent one of his young patients to drink sea water and bathe at Southampton. On 17 November the patient began taking daily dips in the sea. As of the fourth day, his condition started to improve. On 30 November his appetite returned, and on 12 December the patient had recovered his vigour and liveliness. He continued to bathe every day until 11 January, however, by which time his health was completely restored. He then began spacing out his bathes, and finally went home on 8 February. This first report of a sea cure[39] reads like an account of a miracle, a tone that would soon become commonplace in this type of literature. The experiment took place between late autumn and the middle of winter, and so was in keeping with the fashion for cold bathes. This therapy combined drinking sea water with immersion.

Frewin strengthened Dr Richard Russell's convictions. Russell had already noticed that coastal fishermen frequently used sea water as a remedy and that sailors took it as a cost-free purgative. All that remained was to translate these virtues into theory. In 1750 Russell revealed the results of his reflection and experiments conducted over almost twenty years in a book in Latin that was translated in Oxford three years later.[40] The author was a student of Boerhaave, and it was from the teaching of the Leyden master, and especially from informal conversations with him, that Russell drew most of his convictions. He was visibly influenced by physico-theology, and proclaimed his allegiance to a natural medicine that is worth attempting to define here. At a time when increasing amounts of research were being done on antiseptics, this contemporary of Pringle, Lind, and MacBride believed that the Creator had provided natural means of defence against decay and putrefaction. The sea, for instance, accomplished this great design of Providence in many ways. First, it contained salt, which prevented its own putrefaction[41] and checked the decay of the body. In addition, the Creator had endowed sea water with the power of dispersing vitiated humours in the glands. On this point, Russell based his ideas on the concept of 'the natural alliance' of secretions:

> For as the author of the human Nature foresaw that some of these would be obstructed by various Accidents, and less fit for the laws of circulation, he formed the Body with plenty of Secretions for the Purpose of mutual Assistance; that when one of these are impeded, the Body might be relieved by the help of the others. And indeed Nature cures many diseases by her own Power after this Manner, and for this Reason she is properly stiled the *Healer of Diseases*.[42]

By thus ensuring spontaneous regulation, Nature played the role of healer. All the physician had to do was to copy the methods he observed nature using when left to its own devices.[43] Russell, whose approach was in keeping with the interest displayed by several eighteenth-century doctors[44] in glandular secretions, thought that sea water could check an excessively rapid process of putrefaction inside the body, dissolve 'indurated humours', and 'cleanse and protect the glandular system as a whole against impure viscosities',[45] and perhaps do all these things simultaneously. In short, it allowed physicians to control secretions. Sea water also possessed the virtues attributed to cold water: it 'invigorates all parts and gives strength and vigour to the body as a whole'.[46]

The therapeutic strategy was a logical extension of these convictions. Patients were to bathe once a day and drink half a pint of sea water in the morning and a glass on coming out of the water; if necessary, they were to be massaged with freshly collected seaweed from the rocks and to shower (fomentation) with cold sea water that had been heated.

A whole series of indications could be deduced from the catalogue of virtues. 'The sea washes away and cleanses every human stain':[47] these words from Euripides' *Iphigenia in Tauris* summed up Russell's certainties. They appear as an epigraph in his book, and he had them engraved on his tombstone. The healthy gums of fishermen demonstrated the antiseptic qualities of salty sea water, which was the best of toothpastes. It was to be prescribed in all cases related to ancient 'putrid fevers'. It was also advised for all glandular illnesses. And lastly, it proved to be beneficial to sufferers from asthenia, who could be trusted to choose a beach adapted to their temperament and observe strict hygiene in their way of life.

Russell's message met a widespread demand. Within a few years, Boerhaave's student had developed an enormous practice. Brighton, where he lived in a sumptuous residence, soon became a fashionable resort. Fifty years later, some high-ranking noble families were still faithfully carrying out the prescriptions given by the prophet of the sea. Admittedly, his book gave rise to endless controversy. But his critics were more concerned with advising caution, modulating the prescriptions of their colleague, and debating some of his instructions than with radically contradicting him. His opponents endeavoured to develop a normative approach rather than to deny the benefits of the sea.[48]

In 1766, the Académie of Bordeaux organized a competition on the subject. Dr Maret received the first prize for his *Mémoire sur la manière d'agir des bains d'eau douce et d'eau de mer et sur leur usage* [Memoir on the Workings of Fresh-water and Sea-water Bathing and on their Use].[49] According to the author, sea bathing had all the virtues of cold baths (12 to 14 degrees). By reducing the humoral mass, it increased the secretions already stimulated by the salt. Maret consequently attributed a great diuretic virtue to sea water; it effectively removed obstructions from glands and viscera. But the sea had other virtues as well. More than Russell, Maret stressed the beneficial effects of the shock caused to the nervous system by immersion. To do so, he adopted a tone that harmonized with the rise of the aesthetics of the sublime.

> The depth of the sea can only be imagined together with the horror engendered by the fear of being engulfed in it. However, the sight of the sea has little effect on men when they do not think they have cause to fear that they will be thrown into it; but if they are *hurled down into it*, before they can prepare or prevent their immersion, a prodigious upheaval occurs throughout the whole body; the *soul, surprised* by such an unexpected event, *startled by the fear of disunion from the body that it thinks is close at hand*, lets the reins of government of the body over which it presides drop from its hands, so to speak. This results in irregular irradiations of the nervous fluid and a new modification in the organs of thought and the whole nervous system. The disorder is all the greater, and the modification all the more varied, the more cowardly the man is who has been thrown into the sea, or the more convinced that his life is being threatened.[50]

It is easy to understand why Maret advocates sea bathing for those who suffer from 'phrenzy', nymphomania, and hypochondria. In addition, he refines Russell's injunctions, and outlines the model of therapeutic sea bathing as it was to be used until the middle of the nineteenth century. Patients were to rest before facing the waves. They should choose to bathe shortly before sunset,[51] if possible in a shady spot. They should throw themselves energetically into the sea. They should come out of the water as soon as they felt shiverish for the second time, and at most, after half an hour. A bather should have ensured that as soon as he emerged, there would be 'people ready to assist him as required'.[52] A comfortable bed was to have been prepared so that he could rest after his exertions. Faithful to the old Hippocratic and Galenic injunctions, Maret prohibited bathing when in a sweat or on a full stomach; women were to avoid swimming in the sea during their periods, and Maret advised against bathing during epidemics. He envisaged series of thirty to forty bathing sessions for patients, and autumn seemed to him to be the best season for going to the beach.

One year after the Bordeaux competition, Dr Awsiter[53] called for caution, and strongly stressed the dangers of the open sea. This English doctor, a rival of Russell's, advocated installing hot baths at seaside resorts. This would enable patients of fragile health to receive treatment all year long. Moreover, Awsiter stated that it would be a pity to deprive patients of the beneficial effects of hot baths. These caused the pores to open, thereby allowing an active cleansing of the organism. This thesis heralded the decline of the fashion for cold treatments that had set in at the end of the eighteenth century. On the other hand, Awsiter, like Montesquieu[54] before him, praised the sea's surprising life-giving powers. The mythical body of Venus, born out of the spermatic foam of the waves, began to haunt the language of thalassotherapy. Sterility became yet another indication for sea bathing, and soon doctors were advising impotent men to swim in the waves and make a regular habit of eating fresh fish, which made sailors so 'essentially procreative'.[55]

Consequently, shortly after the middle of the eighteenth century, the sea bathing fashion developed out of a therapeutic objective. Doctors prescribed veritable cures modelled on those offered by spas, which were then very much in vogue. Sea bathing quickly emerged as a strictly regulated experience, and in every resort, a municipal establishment was soon constructed so that water temperature could be adjusted and all the services made available that were required for carrying out the medical prescriptions.[56]

Doctors at the beginning of the nineteenth century did not radically alter the therapeutic sea bathing model. This new treatment found several justifications: the focus on diseases related to lymphatic temperament and disorders of the neuro-vegetative system, the importance then attached to the concept of scrofulous diathesis which was invading diagnostic science, and the somewhat phantasmic fear of congestion of the glandular system. The beaches became covered with scrofula-sufferers.[57] Where the vitalist trend exerted its influence, it encouraged this recourse to the sea's reinvigorating faculties, while belief in the water's pharmaco-dynamic powers also progressed.[58] At the same time, there was a rising sense of the dangers of etiolation. Urbanization was increasing, and this together with the importance that the ruling classes attached to the valve of their genes, rendered pre-existing anxieties more acute. More than ever before, the sea was expected to make rickety children straight and vigorous, to put colour back into chlorotic girls, and to restore hope to barren women. Sea bathing, which cured so-called white flowers, was also thought to make the menstrual cycle more regular. Above all, it was considered one of the only effective treatments for cases of neurosis.

The understanding of how the body heats itself was also growing. In 1797, Dr James Currie carried out some remarkably precise experiments:[59] he plunged men and women into the sea, measured their reactions, and concluded that after an initial drop in temperature, the organism warms up again, before finally cooling once more. Physicians were therefore right to set a limit to the duration of bathes. The pharmacological qualities of sea water were in turn confirmed by Dr Balard, who analysed their substance and detected the presence of iodine and bromine.

Very soon, the discovery of the virtues of sea water led to the invention of the beach. Doctors were inspired by neo-Hippocratic climatology for the same reasons as they were by balneotherapy, and the image of the salubrious beach came to comprise a voluminous chapter in the medical landscape that was developing in the West. Russell himself outlined its initial features. In a letter to his friend Dr Frewin,[60] he described the model seaside resort. First of all, it should be 'neat and tidy'. Moreover, it was to be far away from any river mouth, so as to ensure that the waves would be high and the water sufficiently saline. A good beach was 'sandy and flat', making it easier to cross in a Bath chair. The shore should be bordered by cliffs and dunes suitable for walking and horseback riding. For Russell, it was important to associate a marine cure with exercise of the type advocated long ago by the physicians of antiquity and more recently by Sydenham. Oddly enough, however, he failed to mention swimming.

His fellow doctors continued to refine the definition of the salubrious beach. More than Russell, they stressed the qualities that were necessary in the soil and especially the air. Significant in this respect is the 1761 book by Dr Anthony Relhan[61] devoted to the advantages of Brightelmstone. Thanks to its chalky soil and the absence of swamps or forests, emanations there were reduced to a minimum. The cliffs sheltered the resort from northerly winds, while preserving for it the advantages of beneficial sea-breezes that drove away mists and fog. Demography seemed to confirm the salubrity of the spot: according to Dr Relhan, who personally calculated the birth and death rates, people who lived in the vapour of the sea and continually paddled in the water had exceptional life spans. An increasing number of medical studies were devoted to coastal areas. An inexhaustible literature drew comparisons among the merits of each fragment of the shore, and offered an analysis of microclimates that claimed to be thoroughly scientific. Having the corresponding medical publications seemed to be indispensable to creating or promoting any resort established along the British coasts. This procedure aroused Jane Austen's irony and the derisive laughter of the American Wilbur Fisk at the beginning of the nineteenth century.[62]

It would be most interesting to undertake a specific analysis of the extreme precision of physicians' vocabulary: the sharp coenaesthetic attention that it reveals corresponds to the increasing precision of clinical language devoted to bathers' reactions. Over the years, pretentious divisions became established. The superiority of the southern coasts of England, especially those of Sussex, gradually became generally accepted. They were sheltered from northerly winds, but exposed to sea-breezes; they were more luminous and enjoyed a better exposure, which helped the morning mists to dissipate quickly. The eudiometer also determined this classification of resorts.[63] Some sites, such as Brighton, Eastbourne, Newquay, Scarborough, and Yarmouth, were said to enjoy particularly bracing air. The weather there encouraged exercise. Most bathers preferred these beaches. Other resorts offered soothing air and were conducive to rest; Bournemouth, Falmouth, and Torquay were recommended to convalescents and more generally to all those suffering from symptoms of debility.

Within a given resort, scientists delighted in practising their art of classification and division. A delectable literature compares the merits of the different parts of the Brighton area. Similar, and perhaps even more lively, debates took place in Germany after 1793, during the founding of the first resorts there, as specialists weighed the merits of the Baltic and the North Sea, of Doberan or Norderney.[64] Twenty years later, in Restoration France, physicians who preferred Dieppe were still arguing, though less energetically, with the partisans of Boulogne.[65]

Over the years, growing attention was paid to air quality, while the merits of the waters came to be less highly valued. The triumph of Lavoisier's theories created a privileged position for oxygen after 1783, and anxiety over pulmonary consumption increased. Suddenly, what counted most was to breathe well. All these factors favoured the beach. Long before Dumas and Boussingault, Ingenhousz's learned calculations demonstrated that sea air was the purest and most heavily oxygenated, and henceforth physicians were unanimous in advocating boat rides and stays on the islands. The therapeutic virtues of an island sojourn were exalted in terms that coincided with one of the most pressing longings of the Romantic generation. Indeed, the landmark book on the Thanet peninsula published by Buchan in 1804 is permeated with this atmosphere. Patients taking the waters came first of all to enjoy the well-being associated with breathing good air. There, in contact with the elements, they expected to be cured primarily by the qualities of the air, and secondarily, by those of the sea. As for the sun, its only function was to purify the atmosphere and foster evaporation; after ten o'clock in the morning, patients feared it would cause congestion and fled its beams. Shadiness also made its way into the catalogue of a resort's advantages.[66] The land, however, was rife with dangers; none of its exhalations were beneficial. In fact, chalk seemed to be the most salubrious parent rock precisely because it gave off very few emanations.

Gradually, the portrait of the ideal beach became more detailed and its coloration changed. With the rise of picturesque values, doctors were driven to praise the virtues of 'splendid seascapes'. Sensibility to marine aesthetics increased. In this respect, the ambitions of the founders of the Bournemouth resort are significant. Eighteen years later, when James Atherton started New Brighton, he was very careful to build houses offering a view of the sea.[67]

At the same time, the moralizing aims also become more intense. A patient who takes the waters far away from the miasmata of the city and breathes the purest air, enjoys the sight of the immense ocean daily, experiences the shock of repeated immersion, and takes tonic exercise will easily give up illicit pleasures and forswear his disorderly life-style without regret. The beach, noted Dr Le Coeur, was the site of a 'return to innocent inclinations'.[68] Grand ladies and even serious gentlemen are not ashamed to gather little pebbles, blades of marine grass, and seaweed. At the seaside, the conscience awakens, and scepticism collapses. Then, on the wet or dry sand, the naïve wonder of the theologians converges with the longing for regression detected by Ferenczi.

When in the early 1840s railroads throughout Europe began to reach the coast, and a new profit-making system started to alter the character of resorts, Brighton was already the world's leading *sanitorium*.[69] At that time, the image of the beach started to blur, the myths became confused, and the stereotypes piled up in disorderly rivalry. The attribution of merits came to be determined by a shifting assessment of the qualities of the respective elements, the characteristics of the topography, the efficiency of hospital installations, the scope of the social network, and the wealth of cultural life. In order to help the would-be seaside holiday-maker, Dr James Currie published a directory in 1829 which was also a guide to all the western shores and the various tropical climates.[70]

It is difficult for modern readers to understand this abundant and confusing literature. Since then, the ways of appreciating the sea, the earth, the air, and the sun have changed, as have the thresholds for enjoying heat or cold, and even the sense of well-being. Above all, a revolution has occurred in the very means of coenaesthetic perception.

BATHING, THE BEACH, AND SELF-EXPERIENCE, OR THE AMBIVALENT PLEASURE OF SUFFOCATION

Bathers and physicians both agreed that the sea should offer three major qualities: it should be cold, or at least cool, salty, and turbulent. Pleasure came from the whipping of the waves.[71] The bather delighted in feeling the powerful forces of the immense ocean. Bathing among the waves was part of the aesthetics of the sublime: it involved facing the violent water, but without risk, enjoying the pretence that one could be swept under, and being struck by the full force of the waves, but without losing one's footing. This explains the preoccupation with safety. Everything – the detailed medical prescription, the services of expert bathers, the attendants,[72] and a hard sandy soil whose slope had been carefully assessed – contributed to removing any peril, leaving only the emotion.

On the other hand, the patient had little freedom of choice: the physician prescribed the season, the time, the duration, and where exercises were to be carried out; he also set the number of times the patient must bathe during the season.

That said, there were two opposing models that reinforced the sexual dichotomy which medical discourse was striving to cultivate. For women, young girls, children, chronic invalids, convalescents, or those of faint heart, the emotion of sea bathing arose from sudden immersion, which was repeated until the patient shivered for the second time. The 'bathers' would plunge female patients into the water just as the wave broke, taking care to hold their heads down so as to increase the impression of suffocation. This custom aroused the anger of several practitioners,[73] though it prevailed at Dieppe, Doberan, and on the English beaches. The bathing attendants were very keen on it because it justified their presence. In fact, their role was to carry out a prescription that could even detail the number of immersions the patient was to undergo. With time, the medical prescriptions became more refined, and by the same token, the duties of the professional helpers became more complex. At Boulogne, Dr Bertrand indicated the manner in which 'the wave must be received' on various parts of the body. In the 870 pages of the manual published by Dr Le Coeur in 1846, readers entered into all the finer points of marine therapeutics. Yet it should be noted that, at that time, many patients were shaking off medical authority, and practices were becoming more dispersed, even as bathing itself was becoming more strictly codified.

Brutal immersion head first in water at a temperature of 12 to 14 degrees brings about an intense shock. This practice was part of the technique for toughening the patient, as Michel Foucault once described it. It is reminiscent of tempering steel. The process acts on the diaphragm, considered the seat of sensibility. The patient, with her breath taken away, suffocates while the bathing attendant tries to revive her with vigorous massages, until her reaction makes the feeling of cold disappear. This was the same goal pursued by dumping cold water on the head of a patient as he prepared to enter the sea. Servants or the expert 'bather' would fill about ten buckets and proceed to drench the patient on the strand.[74] When the bathing guide felt that his patient was sufficiently hardened, he would let her jump, splash about, and rub herself down in the water.

Such exercise usually lasted from five to fifteen minutes, occasionally twenty, and never more than half an hour. Bathing provided an opportunity to experience 'a multitude of sensations'.[75] The description of women and, even more so, that of young girls bathing became surreptitiously transformed into a depiction of voluptuous episodes. The female bathers held in the arms of powerful men and awaiting penetration into the liquid element, the feeling of suffocation, and the little cries that accompanied it all so obviously suggested copulation that Dr Le Coeur was afraid that the similarity would render sea bathing indecent.[76] At any rate, this was a means of toughening young girls who suffered from dangerously pale complexions; it got them accustomed to being exposed to the elements, and prepared them for the emotions and pains of puberty, as well as the sufferings of childbirth. At Dieppe during the Empire, even before the municipal facilities were built, several ladies acquired the habit of 'taking their young daughters to bathe in the sea'.[77] Some physicians also hoped that the rhythm of the tides would restore that of the menstrual flow. Well before Michelet, Dr Viel relied on the power of the waters to regulate periods.[78]

For both children and young women, the thought of being abruptly plunged 'under the crest of the wave' caused anxiety and sometimes even terror. The first bathe in the waves took on an initiatory character. Sometimes physicians would forbid sea bathing to faint-hearted children, for fear of provoking convulsions.[79] Another option was to resort to ruse. For example, at Boulogne in 1826, Dr Bertrand had to attend to a young girl who trembled at the mere thought of the waves: 'we took her on the river, it was the harbour; we led her imperceptibly to the sea view.'[80] Fictional literature enriched the catalogue of medical cases. One of the young heroines in Jane Austen's *Sanditon* is terrified at the thought of having to bathe in the sea; she loudly calls for the moral support of her friends. Léopoldine Hugo, for her part, provides a minutely detailed description in her correspondence of the intense shock experienced by her younger sister, who bathed with her for the first time at Le Havre in September 1839. 'Dédé was terribly moved, she cried, shouted, trembled, scratched, and asked to go back so strongly that she was immediately taken back to her cabin, where she got dressed again.'[81] A letter from Pierre Foucher indicates that the child finally managed in 1843 to overcome what her uncle considered to be faint-heartedness.

It must be admitted that this hardening of young girls proved effective. Frances ('Fanny') Burney did not merely immerse herself in Brighton's brilliant society in 1782; early in the morning, she would go to the seaside with her hosts. Thus, on Wednesday, 20 November: 'Mrs and the three Miss Thrales and myself all arose at six o'clock in the morning and "by the pale blink of the moon" we went to the sea-side, where we had bespoke the bathing-women to be ready for us, and into the ocean we plunged. It was cold, but pleasant. I have bathed so often as to lose my dread of the operation, which now gives me nothing but animation and vigour.'[82]

Hardened men, on the other hand, had a different way of bathing. Those who made use of an expert bather did so as a mere precaution, yielding to snobbery.[83] In any case, the relationship that was established between the two men remained distant; the patient behaved independently, and the 'guide's' main function was to give advice and ensure safety. When necessary, he could fulfil his mission with authority. The 'bather' assigned to the Prince of Wales, fearing that King George III might reproach him with failing to perform his duty, decided to haul the future Regent back to shore by the ears because he found the boy's way of swimming too reckless.[84] In fact, male patients remained free to pit themselves against the waves and to test their energy against that of the ocean. The thrill of being whipped by the rollers[85] and of pretending to be crushed as one dived under a wave transformed bathing into a make-believe drowning and a victory over the elements. Some young Romantics enjoyed facing the sea in complete solitude: Friedrich von Stolberg on his Danish island and Byron on the shores of Aberdeen both appreciated this joust that left the bather exhausted on the shore.

This means of pleasure coincided with the way people swam. In fact, for the theorists of that time, swimming did not consist in merely being able to loll, play about, and slip between two waves in a feeling of dynamic communion with the liquid element. Swimming then was a total effort, a fight against being engulfed, and a display of energy. This 'violent, active practice'[86] which Gaston Bachelard perceives as a cosmic challenge to the liquid element was primarily justified by the conviction that the swimmer must above all remain on the water's surface. This activity corresponds to the state of general motion that physicians found so useful in bathing. The way people swam was dictated by an incessant fear of drowning heightened by the firm belief that man could neither float nor swim naturally. The theorists were obsessed above all by avoiding submerging the breathing organs and by maintaining the regular rhythm of the lungs. In their view, swimmers could support the weight of their heads only through violent, learned motions, modelled on the repeated jumping of a frog. The kind of swimming they advocated demanded that the bather strain and exert himself; it outlines a virile image of aquatic copulation that finds its counterpart in the feminine emotion at being abruptly submerged.

This masculine model of sea bathing, known as 'bathe and swim', coincided with the nascent practice of exercises whose aim was to make the individual feel dynamic; it viewed the body as a network of forces,[87] and it also reflected the impact that pneumatic chemistry and the physiology of breathing then exerted on the minds of scientists. It procured a particular delight that Bachelard calls 'the coenaesthetic joy of violence'.[88]

In this prehistory of sea bathing, two distinct roles are played out on the sea-shore under the eyes of spectators equipped with lorgnettes. Women were tremblingly confined to the shallow water, ready to retreat as if they feared they would be attacked, and only pretending to face the breaker; they stood close to their

Bath chairs, or sometimes under the canvas of their tents in the company of their 'bathers'. The fact remained that tent flaps, or in some cases palanquins, along with the danger of prying eyes and the fear of indiscreet lorgnettes all raised the stakes in this game. The strict code of decency that was beginning to prevail[89] foreshadowed the intensity of a sensation that could be either painful or titillating. It was something incredible for a middle-class woman to leave the realm of privacy, even if this were only in a Bath chair, and to find herself in a public place with her hair down, barefoot, showing the shape of her hips – in short, in the attire generally reserved for the man with whom she shared her intimate life. In order fully to appreciate this, it is necessary to bear in mind the intense erotic power of women's ankles and hair. The mere contact of a bare foot on the sand was already a sensual invitation and a barely conscious substitute for masturbation. Even more than aristocratic ladies accustomed to the social circuit, bourgeois women trapped in their homes found in the medical prescription an unexpected freedom that offered undreamt-of pleasures.

Men, on the other hand, acted out a scene of bravery: they hoped to emerge as heroes for having faced the staggering blows from the sea, felt the scourging of the salty water, and overcome it victoriously. The virile exaltation that a man experienced just before jumping into the water was like that of an erection, and it was quickened by the proximity of women, a potential audience for his boldness and one that he could see, exceptionally, in a semi-nude state. That the women were gathered together and confined to one area of the beach in a new version of the harem, in order to prevent mixed bathing, did not make any difference. Or rather, this abundance of young women bathing was made even more provocative by the presence of young virgins all in a flutter. All this justified the game of leering through lorgnettes, which everyone agreed was the favourite male pastime on the beaches of the day. At Brighton, according to a 1796 guide, indiscreet men were always on the look-out for women bathers, 'not only as they confusedly ascend from the sea, but as they kick and sprawl and flounder about its muddy margins, like so many mad Naiads in flannel smocks'.[90]

The fact that some virile bathers managed in this way to escape partially from the therapeutic code raises a question: was it not possible that among the ruling classes an autonomous experience of sea bathing founded in a quest for pleasure preceded the fashion created by medical prescription? Are historians

victims of their sources when they decree, in accordance with
Michelet, that the 'invention of the sea' dates from 1750? A
certain amount of evidence suggests that this is the case, while
at the same time it delineates the limits of this premonitory
behaviour.

In England, swimming in rivers belonged to the range of
rural sports practised by the seventeenth-century gentry, and was
soon accompanied by a similar activity on the sea-shores. People
were already using bathing-machines on the Scarborough beach
in 1735. The following year, in a fundamental document, the
Reverend William Clarke discussed the pleasures of his summer
in Brighton: 'We are now sunning ourselves', he wrote to a
correspondent on 22 July 'upon the beach at Brighthelmstone ...
My morning business is bathing in the sea, and then buying fish;
the evening is horse riding, viewing the remains of old Saxon
camps, and counting the ships in the road, and the boats that
are trawling.'[91] This clergyman, a solitary pioneer who could
be assumed to be a follower of the physico-theologians, brought
rural sports to the sandy beach, and in so doing, invented the
seaside holiday. In 1795, Cambry enjoyed 'delightful bathing'
on the golden sand with his hosts at the end of a journey in the
Finistère region.[92] But clearly these spontaneous, sporadic actions
would soon be overshadowed by a global system, and, by the
same token, become codified. The fashion for therapeutic bathing
was intended to prevent the mingling of the sexes, and considerations of social status overpowered this anarchic individual
behaviour. The common people were left with customs that were
already strictly limited.[93]

The primacy of therapeutic bathing determined the range of
ways in which the beach could be used. People did not come to
the beach in order to expose themselves to the sun's rays, which
were known to cause congestion, dry out the bodily fibre, give
the skin a working-class coloration; and in any case, these
rays caused distress. People did not lie down on the sand; they
walked across it, or sat upon it. The beach was a place for strolling
and for conversation; it was an extension of the excursion circuit
that began on the dunes or along the paths above the cliffs. On
the sand, the primordial family circle could be formed once
again. For men fond of horseback riding, the strand offered a
smooth surface on which they could let their mounts run free.
Fashionable beaches were criss-crossed with galloping riders,
and for many, the daily ride was the greatest attraction of the
stay.[94]

This explains the great attention paid to the quality of the sand, which patients taking the waters could discover with their bare feet as they headed toward the water, if they were not carried down in a Bath chair.[95] Describing a beach necessarily implied evaluating the consistency of the sand. It should be firm, allowing riders to gallop without danger and giving an impression of safety to bathers as they prepared to jump into the waves. The ideal was the 'fine hard sand' described by Thomas Pennant at Scarborough.[96] The sand should be comfortable, which is to say free from 'the muddy element', that unpleasant silt on which bathers' feet could slip and get dirty.[97] In this respect, more importance was given to the nature of the strand and the evenness of its slope than to the quality of the beach itself.

All the same, tourists did not like sand that was too loose; for this reason, Ireland hated the beach at Scheveningen because it tired the walker. At this stage, bathers were not bothered by pebbles. Sometimes they even seemed to appreciate bits of gravel mingled with sea shells, as this was an easy surface on which to walk. The ideal beach, however, remained one of the proper consistency, whose compact sand did not conceal any sharp stones.

If a bather happened to be far from one of the beaches whose area had been marked off by the authorities, on which customs were clearly codified, he was advised to seek out his own bathing spot – that is to say, one that corresponded best to his temperament. Smollett's behaviour when he came to the coasts of Boulogne in 1763 to take care of his health is a good illustration of the importance that Britons attached to selecting an individualized site.[98] Physicians, obsessed with the notion of idiosyncrasy, encouraged this quest for affinity. Dr Le Coeur, for instance, advised bathers to visit the area they had chosen at low tide before taking a dip when the tide came in.[99]

On the beach, the behaviour of an individual who had decided to go for a swim revolved around anxiety about facing the waves. There was a distinct difference between the walker and the bather. At the appointed hour, once the bather had donned his attire, he did not linger, but marched resolutely towards the sea. This was a moment for concentration, and the bather took all sorts of precautions. He feared the cold and the sun's rays; he wanted to be comfortable. The model of therapeutic bathing was designed for members of the leisured class who were accustomed to being attended by servants. These people were often frail, and they came to find vigour in nature; but their desire for contact with the sea had to be in keeping with the luxurious living to which they were accustomed. Physicians codified a series of practices for their benefit by carefully taking into account the sociology of sensibilities. Scientists at the time were convinced that social status, along with sex, age, and temperament, determined the reaction of the person plunged into the water, and they endeavoured to adapt their prescriptions accordingly.[100]

Respect for privacy and the imperative of maintaining social distance were preserved on the sea-shore, as is demonstrated by the way people dressed and by the use of bathing-machines. The bathing-machine appeared as early as 1735 in an engraving of the Scarborough beach.[101] The model designed by the Quaker Benjamin Beale for the Margate beach soon carried the day in England. The spread of the bathing-machine was rapid; in 1768, it was introduced at Lowestoft. At Weymouth, Margate, and Scarborough, and later on at Ostend, the machines were covered so as to protect the privacy of lady bathers; this was not the case at Brighton, where, as was seen above, the telescope was an important piece of beach equipment.

On very crowded beaches, bathers had to wait for their bathing-machines in an uncomfortable hut.[102] This heightened the anxiety of timid souls who were afraid of going into the sea. The comfort of bathing-machines varied with the models. All of them featured a bench, often covered with velvet. Once the bather had hoisted himself inside with the help of a small ladder at the back of the vehicle, he would find towels, a dry bathing costume, and sometimes a cloak or burnous that he would slip on as he left. A scrub brush for the skin, a bootjack, and a mirror completed the equipment.[103] The vehicle went into the water to reach a depth of about twenty centimetres. The 'bathers' would sometimes spread a drill tent in order to protect their lady customers from the excessive heat of the sun or from indiscreet lorgnettes.[104] They would then help her down the few rungs of the ladder before seizing her. On the way back, there was much fussing about in the jolting vehicle, what with drying off, rubbing down, and quickly tidying up.

Decency and the fear of being dishonoured by prying eyes determined appropriate bathing dress.[105] The first women to swim in the sea slipped on thick woollen dresses that preserved body heat. In France during the Restoration, this is what women wore to bathe. At Royan[106] in the early 1820s, men and women wore frieze gowns over a pair of wide trousers. For a long time, a range of individual apparels could be found. Then the bathing-costume gradually became standardized and uniform on the basis of three imperatives: morality, therapy, and exercise. The history of the bathing-costume follows the reinforcement of standards of decency which, according to Norbert Elias, is part of the process of civilization.

In the early 1840s, wearing trousers became a necessity for women bathers; as waves could lift up their garments, the dress alone had become indecent. The most common bathing-costume then consisted of a shirt and a pair of trousers meeting at the waist and kept together by a single belt, so as to comprise a single garment that could be opened in front by a series of buttons. The sleeves were quite short, usually stopping at the elbow. In order to facilitate the swimmers' movements, an opening was sometimes made under the armpit. To this ensemble, young ladies added a little petticoat that was fitted to the belt; its purpose was to conceal the fulness of their hips. It was considered inappropriate to place too much emphasis on this promising element of a young woman's charm. The outline of feminine curves could only be revealed once they had been made decent by the full bloom of motherhood. There was another combination for women that was also fairly widespread: the bather would wear a separate pair of trousers and, instead of a shirt, would slip on a blouse or tunic that was cut full around the chest. At the beginning of the 1840s,[107] singlet bathing trunks appeared, made of a knitted woollen fabric which was generally brown, since white was too transparent. 'These', Le Coeur pointed out, 'are real swimsuits or jerkins in one piece. They are very light and comfortable for swimmers ...'. Unfortunately, they also emphasized the figure too much, and the good doctor had his doubts as to whether 'they will ever be adopted as bathing costumes for women'.

Many English women bathed bare-headed; they plaited their hair, rolled it up around their heads, or wrapped it up in a scarf with combs and pins. As she came out of the water, a woman would sometimes make a gesture of great freedom and let her hair down to dry its wavy locks in the sun. More prudish bathers, or those who were more afraid of the sun's rays, would protect themselves with a waterproof oil-cloth bonnet or cap, always white. On the Continent, this habit spread fairly quickly. Women who were particularly afraid of the sun would deck themselves out with broad-brimmed hats made of rough straw, inspired by the 'cottage hats' or 'cottage bonnets' that became widespread in France during the reign of Louis-Philippe. Le Coeur once observed a woman entering the water with a silk mask covering her face.[108]

Caution required bathers to put a piece of cotton wool in each ear and to prepare beforehand a fur-lined coat in order to protect themselves from biting air or excessively bright sunlight. If necessary, bathers also wore wooden shoes or ankle boots, especially if they had to go on foot across a thin fringe of gravel and shells or a slippery band of seaweed.

Many working-class women could not care less about such insistent modesty, and the profuse literature devoted to sea bathing may actually camouflage the existence of deeply rooted vulgar behaviour. In fact, along the coasts of the Baltic, the North Sea, the English Channel, and the Atlantic, traces of popular bathing are to be found, sometimes in confrontation with the rigorous therapeutic practices that aroused so much emotion in the ruling classes. It also becomes quickly apparent that between the two systems, there was an insidious sharing of practices. The model of popular bathing, which could be characterized as northern, corresponds to altogether different aims from those of strictly codified 'sea bathing'. It was an extension of the free splashing of childhood and adolescence; sometimes it was part of the range of exercises or contests that formerly served as models for the gentry's rural sports. It occurred within the framework of collective activities, festive or playful, and always noisy, of which coast-dwelling people are fond.[109] As its participants were less worried about prying eyes and less imbued with a sense of privacy, popular bathing allowed the sexes to mingle. Unfortunately, it is difficult to follow the trace of a practice that was so quickly adulterated by the dominant model, when it was not prohibited by authorities hostile to anything that might impede the fulfilment of patterns set by the leisured class. Luckily, this social stratum occasionally approved of collective customs that enable us better to understand popular sensibilities.

Inhabitants of the English shore bathed well before 1750, so it was said, at Deal, Eastbourne, Portsmouth, Exmouth, and Brighton.[110] Dr Le François confirmed in 1812 that the children of coastal people got into the habit of swimming in the Channel ports when they were as young as 6 or 7.[111] Every year on the last Sunday in September, whole villages of Basques living in the mountains would come down to paddle in the sea at Biarritz.[112] Bathing had been a common custom on that coast for a long time. Early in the seventeenth century, Pierre de Lancre, president of the Bordeaux Parlement, described the surprise expressed by travellers at the sight of 'grown girls and young fishermen' who 'mingle in the water', then go and 'dry off in the Chamber of Love that Venus has put there on purpose, on the sea-shore'.[113] The way these beaches were used was in no way inspired by the British model. Descriptions often give a hedonistic, playful image of bathing, with less sharp divisions. In the confused mingling of the sexes, bathers yielded more willingly to the tossing of the waves. Travellers confessed that the exercise was marked by latent eroticism. A good deal of familiarity was possible in the *cacolettes*[114] that carried bathers down from Bayonne. The legend of lovers tragically drowned in the 'chamber of love' contributed to the erotic atmosphere. For town-dwellers, collective bathing was nothing more than the culmination of a country outing.

Sometimes different models coexist together: on the shores of Boulogne at the end of the eighteenth century, the locals would bathe in the summer in order to cool off; a few Englishmen, accustomed to taking the waves, would join them. Dr Bertrand, who related the occurrence,[115] clearly detected dissimilarities in behaviour. There, an occasional practice, stimulated by the weather and driven by a spontaneous desire, found itself joined by circumstance to the strictly codified prescription bathing followed by patients used to taking a dip in the sea in all seasons. By the same token, the pleasures of bathing enjoyed by the inhabitants of Le Havre in the early years of the Restoration, amidst a mingling of the sexes and the social classes on the beaches of the area,[116] stand in contrast to the more ritualized habits that were beginning to take shape at Dieppe. Southwestern physicians and surgeons, accustomed to recommending Barèges and Cauterets, soon began to prescribe sea bathing at Biarritz. In August 1765, the sub-delegate encouraged the city authorities to facilitate access to the resort for foreign patients.[117]

This social juxtaposition of customs sometimes made it necessary for the authorities to intervene. Male nudity, for instance, raised a problem. In the eighteenth century, it continued to be accepted for some time, until it was gradually displaced by the requirements of therapeutic bathing. As late as 1778, tourists from The Hague came and bathed nude at Scheveningen,[118] while the young women of the area held their clothes, gave them a rub down as they came out of the water, and helped them to get dressed. According to Pilati, the nauseating smell of these young people made social differences so keenly obvious that nobody ever tried to seduce these fishermen's daughters. A similar custom aroused criticism from visitors at Biarritz and misgivings on the part of the authorities, who were anxious to preserve morality.[119] By the end of the reign of Louis-Philippe in France, a clear division had been established: male nudity was forbidden on beaches frequented by bathers, though it was allowed elsewhere within the limits set by the municipal authorities.[120] The same came to be true at Ostend in 1859.[121]

It occasionally happened that without going so far as to permit nudity, the authorities allowed the sexes to mingle inside or outside the official bathing area. Paquet-Syphorien, who was hardly accustomed to such liberties, rushed into the water without a moment's hesitation in 1811 in order to strike up conversation as quickly as possible with the young ladies of Ostend.[122]

The common folk also bathed on the often repulsive shores of the Mediterranean; but this was an entirely different exercise that found few adepts for the time being among members of the ruling classes. Mediterranean bathers, who were often divers as well, were not trying to recover lost vigour; nor were they seeking to provoke or fight with the sea. They simply splashed about in the clear, cooling waters. This practice, seldom a solitary one, always took on a playful air. Groups of bathers would spend long moments paddling about in the sea like bands of dolphins. This is how Joseph Hager described the summer hours devoted by the Palermitans to dozens of aquatic games.[123] In 1783, Bérenger described children bathing on the beaches of Marseilles to his correspondent: 'Groups of naked children dive into the sea, swim on their backs, or busy themselves on the rocks pulling off shellfish. The groups become more numerous, and the hollows of these caves serve as shelters for these merry bands standing half-covered by the water as they give themselves over to a thousand frisky games.'[124] These activities sometimes became a part of festivities. On gala days, Abbé Coyer remarked in 1764, the people of Marseilles headed off in different directions: some 'go and seek purer air on the high sea, and more lively entertainment',[125] and others bathed while revelling took place on the shore.

As a rule, bathing along the Mediterranean was reserved for males. At Saint-Tropez, only little boys played about in the sea.[126] Observers failed to mention the sensuous groups of naked bathing women to whom Vernet was then devoting so many successful paintings. Here, the pictorial stereotypes do not seem to be based on reality. A sense of modesty prevented such antique or picturesque scenes from being staged, as Bernardin de

Saint-Pierre was well aware: the young Paul dug a hole in which the pubescent Virginia could swim, so that her nudity would no longer be exposed to his gaze; in so doing, he put an end to the fraternal bathing that had been a delight in their unforgettable childhood, and somewhat unconsciously took the first steps toward their inevitable separation.

During the last third of the eighteenth century, travellers from the North marvelled as they discovered this model of bathing that was so attuned to neoclassical aesthetics. Tourists regarded the behaviour of these bathers and divers as a form of art. They praised the beauty of the naked ephebes, who were for them like so many sea gods or dolphins in ancient seascapes. The memory of Tiberius and an aesthetic exaltation tinged with latent homosexuality even encouraged some observers to view them as obliging 'mercenaries' whose pleasing appearance fitted well into the Virgilian landscape. In July 1787, Goethe stayed at Posilippo in the country home of Hamilton, the scholarly knight. 'After dinner', he wrote, 'a dozen young boys swam in the sea, and it was a handsome sight. How many different groups they formed, and all the poses they struck in their games! He [Hamilton] pays them for this in order to have this pleasure every afternoon.'[127]

Some tourists who were adepts at bathing in the waves were tempted to imitate them. Jean Houel, fascinated by the transparency of the Tyrrhenian Sea, was fond of sleeping on the Sicilian shore and walking across the seaweed and the glittering sand. His voyage around the Mediterranean was punctuated with solitary bathing sessions. Others bathed out at sea, jumping from their boats, and mingled with the frolicking of new sea gods. 'We find everyday proof', wrote Brydone in May 1770 during a stay in Naples,

> that bathing in the sea is the best remedy against the effects of the sirocco wind, and we enjoy this advantage with all the pleasure it is possible to wish. Lord Fortrose has provided himself with a very comfortable large boat for this purpose. We gather every morning at eight o'clock, and after having sailed for about half a mile, we undress in order to bathe ... Milord has hired ten sailors who are really some kind of amphibious animals, since they live in the sea for half the summer.[128]

These sailors looked after the bathers, and would dive forty or fifty feet under water to gather shellfish. 'In order to accustom ourselves to swimming on all occasions', Brydone added, 'Milord has bought a garment that we each wear in turn ... We have learnt to undress in the water,' something that could prove useful if the ship sank. When he reached the Syracuse area, the traveller confided: 'We have discovered a very comfortable spot for bathing; it is always one of the first things we seek because this exercise constitutes one of the chief pleasures of our expedition.'[129] Nor was the therapeutic excuse totally absent, since the author remarked that but for bathing, 'We would all be in as sorry a state as the French marquis.'

The fact remains that it is difficult to assess just how widespread these practices were among such elite travellers. Certainly they were indicative of a wish not only to commune with the clear water, but also to recreate scenes from antiquity and to imitate commoners' behaviour. The spread of such forms of behaviour was in any case restricted by the threat of pirates and robbers and the torpor that prevailed on the unhealthy Mediterranean shores. While the northern sea-shores were already teeming with energetic bathers, mass visits of patients taking the waters were still unknown on the Mediterranean beaches. In France, the Sète coast played a pioneering role in this respect. It is important to remember that the vitalist physicians of Montpellier and the physicians of Bayonne were the first in that country to recognize the virtues of the sea. Already during the First Empire, Professor Delpech relied on it to cure consumption. Not without some disappointment, he sent his patients to swim among the waves.[130] Despite the fact that the amenities on this beach remained rudimentary for some time, scrofula-sufferers flocked to Sète in large numbers during the following decades. This was when[131] Dr Viel began to sing the praises of the beneficial effects of solar radiation, which he claimed strengthened the organs and invigorated the whole body. The nascent vogue for the first great seaside resort on the French Riviera accompanied a growing fashion in medical literature for lukewarm and hot baths. At the same time, on northern beaches the qualities of the air were beginning to replace the advantages of cold water in the minds of theorists.

CARING FOR ONESELF AND SEASIDE HOLIDAY-MAKING, OR A LIFE-STYLE RICH IN OBSCURE JOYS

A whole life-style was taking shape on the sea-shores, and the aspects related to sociability will be considered later. While on one level collective customs were being codified and strategies for establishing distance and distinction were organizing the social scene, at a deeper level individual ways of caring for oneself were developing; these revealed new frameworks for assessment and engendered unprecedented models of behaviour.

The rise of the figure of the invalid, someone who suffers or thinks he suffers from chronic disease, fits into the process of specification of the ages of life that had been taking place since the middle of the eighteenth century; it bears modest witness to the impact of narcissism, whose various manifestations within the social body I have attempted to identify elsewhere. This trend goes hand in hand with the growing confidence of the medical profession. Improvements in clinical observation, the increased concern for hygiene, and the normative ambitions that this increase implies all combined to shape the invalid's existence, like that of children, young girls, or women. With one voice, physicians, poets, and philosophers all urged greater preoccupation with and attention to self. The figure of the invalid that emerges from this convergence reveals the intensity with which co-enaesthetic preoccupations were to obsess the leisured class in the nineteenth century. At the seaside holiday, even more than in the humidity of the spas, detailed behavioural patterns were codified as part of the quest for well-being. Invalids learn to savour the pleasant sensations provided by the smooth functioning of their bodies: good circulation, refreshing sleep, and renewed appetite. Coenaesthetic attention to oneself stimulated individuals to write about themselves, and this practice took over from examination of one's conscience or spiritual diary keeping. The history of diarists would be incomprehensible without reference to these countless accounts of therapeutic sojourns[132] and to travels undertaken to combat spleen, and without taking into account the prolific literature of medical case-studies that reiterate this material. This twofold record was in keeping with both medical topography, then paradoxically in its heyday, and the proliferation of voluminous hygiene manuals obsessed by concern for normative behaviour. Proust and his saga were to emerge as the epitome of this point of view.

A handful of major beliefs determined this means of constructing – or destroying – the self. In keeping with the thoughts of the ideologues, especially Cabanis,[133] a bather who was already an invalid or who was guided by a preoccupation with not becoming one, would endeavour to take 'organic symphathies' into account: he would devote careful attention to the slightest sign from his organs. The importance attached by the physicians of the day to the idiosyncrasy or irritability[134] peculiar to each individual fostered an evident hyperaesthesia,[135] especially among the ruling classes. This phenomenon, particularly in women, was encouraged by practitioners who were often at a loss therapeutically and so devoted themselves to assembling a forbearing medicine for the soul. This explains the importance of the figure of the young girl, presented as the archetype of a sensitive, frail individual.

By the same token, paying attention to oneself and elaborating individual (and always somewhat neurotic) rituals were suddenly promoted to the level of medical indications. Because of his idiosyncrasy, each individual, asserted Dr Viel (who is but one example), has his own way of responding to the demands of sea bathing and to breathing sea air.[136] Each individual, decreed Dr Le Coeur, must discover his own way of bathing.[137] Each individual must develop his own style and network of walks.

As a result of this careful, continuous attention, which corresponds to the sharpness of the clinician's gaze, a collaboration developed between the invalid and the practitioner. The frequent practice of questioning the patient was one of the high points in this indispensable alliance: this oral examination determined the first prescriptions, and launched a continuous back-and-forth movement between confession, description, and the progressive refinement of the treatment. The precision with which Dr Bertrand prescribed walks[138] and Dr Le Coeur's invention of 'fractional bathing'[139] with strictly timed rhythms constitute striking examples of the growing complexity of individual practices.

This strategy results in an obsessional self-centred arithmetic not unlike that aroused by close attention to sensual pleasure. Just as others count their orgasms, the invalid notes the number of bathing sessions and even the successive immersions in order to ascertain that he is effectively carrying out the prescription. Female bathers vie with one another and compare their scores, just as Flaubert[140] and his friends do their exploits in brothels. The seaside holiday was evolving into a contest, a form of competition.

Stays at the seaside have engendered an inexhaustible literature of medical case-studies, a genre inaugurated by Frewin and Russell and pursued in Germany by the indefatigable Vogel. In Roman Catholic France, this genre was modelled on the recounting of a miracle; there, bathing sessions were counted with the same exactitude as rosaries and novenas. In July 1826, for example,

Dr Bertrand observed that the condition of a 13-year-old child suffering from 'mesenteritis' improved upon his forty-eighth sea bathe. The following year, he was cured after the hundred-and-eighth session.[141] 'In general', wrote Dr Viel, 'bathers are proud and happy when they can present a great number of bathing sessions to their account, and when they can say, my season numbered thirty, forty sessions.' 'People forget about differences in temperament and states of health and have but one aim, that of being the first to reach such-and-such a number of bathing sessions.' This could prove dangerous, especially in young girls, whom it was necessary to 'tone up' without making them irritable.[142]

It would take too long to follow the series of invalids presented in the literature through their cures, as they are so numerous. For seventeen years at Margate, Dr Buchan divided his attention between his own internal organs and the clinical observation of his patients. Richard Smollett, another physician, devoted the account of his travels[143] to describing the vicissitudes of his organism and to detailing the tactics he used to cure his ailing lungs. For Boulogne, where he bathed every day in 1763, he described his favourite bathing spots. When he settled in Nice, he drew up a climatic table for the place, and analysed the concomitant variations in his condition. At the end of his stay, he made a conscientious assessment, endeavouring to calculate the improvement in his health, as the authors of spiritual diaries used to do regarding the progress of their souls. Accounts of these health-oriented holidays weighed the effects of time, and accompanied the rising and waning of hope. They were the forerunners of the broader approach that would be used by Amiel and the great diarists of the following century, obsessed by the need to keep a statistical account of their withering away through anxious coenaesthetic attention. This kind of book seems to have found avid readers. Smollett, for instance, contributed largely to popularizing the seaside holiday at Nice among his countrymen. The English invalid was a close cousin of the traveller haunted by spleen, and he undertook an indefatigable, though somewhat disillusioned, quest for the miraculously salutary cure. Before putting various climates to the test himself, he would listen to the sententious advice of his friends, anxiously question travellers, and go over medical topographies again and again. The invalid knew he must analyse himself endlessly if he hoped to achieve his own recovery.

The frail invalid's cautious life-style stood in sharp contrast to the prodigal life of the gentleman hunter, passionately fond of horseback riding and rural sports: his greatest pleasure was to plunge unrestrainedly into the virile, often brutal and hard-drinking sociability of the English aristocracy of that time. A tension sometimes developed between the two models. In this respect the discomfort of John Byng, Earl of Torrington, is revealing. In August 1782 he settled in with his family at the royal resort of Weymouth. The dashing baronet did try to adapt to the rhythm of the cure-taker's life; for a few days, he managed to establish a stable timetable, which he presented sketchily: 'My life now begins to be arranged in a regular way: I rise at six o'clock, buy fish, read newspapers, walk the beach, visit my horse; at nine o'clock return to breakfast; ride at ten, dine at four; in the evening walk beach again till the rooms begin, cards till ten o'clock; light supper, bed.'[144] But the main thing for him remained bodily exercise. He was very forthright about liking not the sun on the sand, but rather the sea air that whets the appetite, the salty contact of the sea-breeze on his skin, the feeling of tiredness early in the morning before breakfast, the rough horse rides on the strand, and the fresh fish he bought from fishermen on the beach after testing their viscous consistency. Torrington hated the social life of the resort, and avidly devoured all these sensations. Using a complex emotional strategy, he tried to adapt the landed aristocrat's joy of living and the pleasures of the antiquary and the aesthete to the beach. He enjoyed taking a solitary walk, riding to visit monuments in ruins, or contemplating the seascape; he drank thermal spring waters and compared their tastes. After a few days, aggravated by the feeble virility and snobbery of the resort, he left his family in order to broaden the range of his rides in the countryside. The kind of behaviour he adopted in 1782 differs both from the classical voyage and from philosophical travel. It was to become widespread when war with France once again disrupted the grand tour and compelled would-be travellers to discover thoroughly the wild beauty of their islands.

The most fascinating text I had occasion to read in the course of my research remains, however, the *Journal Kept on the Isle of Man Relating the Weather, the Winds and the Daily Happenings Occurring for over Eleven Months*,[145] the account of a visit made by Richard Townley, Esquire, in the first year of the French Revolution. The author, an invalid belonging to the leisured class, had a weak chest, and a spasmodic cough was spoiling his existence. Luckily, a long cure at Boulogne had virtually restored his health when he settled in Douglas on the

Isle of Man on 9 May 1789. In previous years, driven by the desire to improve his condition, he had stayed in the Netherlands, in French-speaking Flanders, and in Picardy. At Boulogne, he had kept a diary that he could compare with the one he undertook to write at Douglas. He stayed there with his 'good spouse', but his schedule, to which he was strictly faithful, compelled him to behave as though he were alone. Townley was of a religious disposition; he frequently attended the Anglican church, whose comfort and congregation he enjoyed, mingling there every week with 'decent, smart, well-dressed people'.[146] His only regret was the lack of fresh air during the services.

In the spirit of physico-theology, Townley referred to the psalms in order to sing the Creator's praises through the spectacle of nature. He was well acquainted with the English poets of the beginning of the century, especially Pope, Gray, and Thomson. He worshipped Virgil's work, which haunted him. These authors, and of course Shakespeare and Milton, shape the network of his references, and guide the view he takes of nature on the Isle of Man: he was moved alternately by its sublime beauty and its gay charm. Townley's way of life was that of a man who defined himself as 'a lover of exercise and rural sports'. His diary records both the variations in the weather and the vicissitudes of his condition, and he scrupulously relates the exercises that resulted from this twofold series of observations.

The weather is actually the main character in the book and the one that determines events. The author pays the foremost attention to the quality of the air and the wind, endeavouring every day to define them with the greatest possible accuracy through their effects on his senses and his soul. Townley, for whom the wind could be 'pleasant', 'soft', 'balmy', but also 'harsh' or 'unpleasant', constantly sought a cooling breeze. He was most fond of the delicate sea-breeze that came in with the tide. Mists and fog are described from a similar psychological perspective. By contrast, the author does not dwell on the precipitation or its quality; he merely reports whether it is raining or not. An 'awful' storm, which he renders in artistic terms showing Thomson's influence, does, however, manage to merit a few lines. Townley did not like the heat, and he feared the bright glare of the summer sun on the sand. He was apparently little bothered by it during the eleven months of his stay. On the other hand, he praised the vivifying, invigorating warmth of the winter and spring sun, which he called a fountain of beauty.

Townley also observed the state of the sea, whose calm was what he appreciated most. The rising tide 'with its soft murmur', often accompanied by a cool breeze from the high sea, was for him the most delightful moment of the day. He also loved the sea when it was covered with boats, a scene he enjoyed on the return of the herring fishing fleets. When writing about storms, he spoke essentially of their pathos and the imminence of ship-wrecks which would darken his sleepless nights.

The invalid does not describe equally the different parts of the day. With the exception of a single hymn to the serenity of the evening, he speaks of little else besides the morning. This is when the observation of the weather leads him to define the day's forecast. A few scattered remarks about the noises that fill his sleepless nights may serve to start the outline of the day. Townley is particularly sensitive then to the distant echo of the sea and the hammering of rain on the roof.

Townley constructed a strictly regulated life-style for himself. Though he never bathed, he was glad to observe that the fashion for bathing was on the rise on the Isle of Man that year. On the other hand, he juggled lovingly with an extensive range of excursions, each of which had a distinct identity. His walks generally lasted two hours. 'I took a breath of morning air' or 'I went for a morning walk' are the words that often begin the account of his day. His aim on these outings was 'to meet the freshness of the morning breeze, brought to the sandy shore by the approaching tide'.[147] A satisfying walk was for him a feast for the senses. On the morning of 4 August 1789, for instance, after climbing up a hill, he noted, 'I very much enjoyed the freshness of the morning breeze, and the shining prospect of a peaceful sea. Afterwards varied the scene, by taking a walk of two or three miles, up into the country.'[148]

Sometimes, he would content himself with a short walk[149] on the gravel beach, very rarely in the streets of Douglas, or more often to the end of the pier, from which he could gaze at the sea. On occasion, when he was feeling sprightly, he would scramble up the rocks. In fine weather, he would set out on one of the long outings that made him completely happy. Three times during the year, he organized a party with some friends, but this did not seem to please him as much.

The invalid's personality can be defined by the network of his favourite walks, always carefully indicated in a few words. Their destinations, however, are diverse, as primary importance is attached to variety. In most cases, Townley would go to one of the beaches whose appearance he described; he would specify whether it was sandy or gravelled, with a rocky or steep coast. He alternated among the various kinds of shores, but seems to have had a special penchant for walks 'across the sands' or 'pleasant strolls around the beach'. He noted one day that some of the grottoes on the shore offered comfortable hideaways to lovers on the island; the only note of eroticism he finds is suggested by the figures of Dido and Aeneas. In his writing, the Virgilian image of the grotto emerges as the archetype of privacy and seclusion.[150]

Townley also liked walking around the harbour and through the fish market. He enjoyed watching the mail-boat, the Liverpool packet-boat, the colliers, and the few sloops belonging to gentlemen. During the herring season, he marvelled at the return of the fishing fleet, despite the olfactory discomfort he suffered in the vicinity of the drying nets. When climbing up the hills, Townley would admire the broad seascape, and especially the ample panorama of the bay of Douglas. The infinite expanse of the waves was apparently still repugnant to him, but he liked to gaze at the 'horrible' or 'romantic rocks', and in this respect, he reflects the contemporary impact of picturesque values. Nevertheless, whenever he could, he returned home through the Arcadian scenery of a pleasant inland valley. He was attracted by 'Druidic' and Danish ruins, like any good antiquary in 1789. On 11 June he undertook a journey by land and by sea, to one of the little islands off the Isle of Man coast. Moreover, he refused to leave the main island before having made a complete visit of it. In short, his behaviour fits the classical pattern of a Robinson Crusoe-like adventure: the wish to become master of the land by meticulously exploring its entire circumference.

All that remains to be examined are the means by which he evaluates his country walks. In keeping with sensualist philosophy, what he seeks above all are things 'very pleasing to [his] senses' and those which stimulate his appetite. To this end, he focuses his attention on 'the impression granted by the atmosphere' and on the effects of the wind on his breathing. Early in the morning, his habit is to open his windows; he sniffs the scent of flowers, which he believes attests to the quality of the air. As was seen above, what he appreciates most in all things is freshness and the sensations by which it is implied. He likes to listen to the murmur of the incoming tide. In winter, he listens from his bed to 'the tide receding from the rough, pebbly shore, near the pier-head'.[151] He recalls 'the murmuring dying away gradually, till the percussions became so very languid, that sound was scarcely distinguishable from silence'. Then he feels nostalgic for the pleasant sensations that had accompanied this soft murmuring in the previous summer.

Like most cultured tourists,[152] Townley took a great interest in marine plants and sea-birds. As winter approached, he would rise very early in the morning in order to go to the end of the pier to listen to the 'favourite' bird whose song he had identified. He even observed insects and reptiles occasionally. Only the cormorant was revolting to him, because Milton had made it a winged creature of Hell.

Once the cycle of the seasons is complete, the book concludes with an overall evaluation of the island's salubrity. This chapter is supported by a demographic study, considered indispensable in this kind of work. There, as at Boulogne, Townley insisted on carrying out his own investigation in the archives and in the small romantic cemeteries where he deciphered the epitaphs. This traveller concludes in the usual manner:

> No place can afford better natural conveniences, for sea bathing, than this island; having so many charming bays, and where the water is so pure, so pellucid, and so free from every kind of sullage, and the shores so safe, so comfortable, and devoid of every entanglement; affording so many snug little recesses or creeks, for dipping in; where the water will be found in a quiescent state; even when there happens to be a pretty strong breeze at sea.[153]

It would be absurd to claim that the sea, unlike the mountains, was not appreciated in the West during the 1780s, were it not for the fact that such assertions were made only recently by specialists who were obsessed by a single type of text and whose focus was primarily visual. In actual fact, a prolific literature including medical case-studies, travel accounts, and descriptions of cure stays, abundant correspondence, and constant word-of-mouth publicity from the period all testify to the intense appeal of the seaside. Through the accumulation of such accounts, an emotional strategy is constructed, and an unprecedented manner of enjoying the sea and the beach emerges and begins to spread. Already at that time, this new interest was making its way discreetly at Boulogne, Ostend, Scheveningen, and Doberan along the Continental coasts.

This means of appreciating the sea-shore admittedly owes a great deal to Thomson, to a rereading of Virgil that his works inspire, and to Macpherson. It assumes a sensual commitment of the diarist's entire being and close attention to emanations, to imperceptible sounds, and to the slightest whiff of breeze. This is something completely different from the wish to 'see the sea' that stirred members of the French aristocracy, assiduous viewers of salon seascapes.

Seventy years before Michelet's stay at Saint-Georges-de-Didonne, Townley, from his retreat at the water's edge, was attuned to the soft murmuring or the roaring of nocturnal waves; in the course of his innumerable walks on the sand, he, too, intently observed the colour of the seaweed, the sea-birds' songs, and the fish. Like Michelet later, he was anguished by the thought of the perils of the sea, and was moved by the occurrence of shipwrecks. Yet he did not have a Romantic appreciation of the beach. He remained faithful to providentialism, and sang the praises of the Creator, and he viewed nature with the eyes of a classical poet. His account is almost totally lacking in the scientific considerations that mark the maritime works of Michelet and Victor Hugo. Above all, his book respects the neo-Hippo-cratic correspondence that was developing between weather and health. From this point of view, Townley's real achievement consists in transforming medical injunctions into a life-style full of obscure joys.

On the sea-shore a model of self-care was developing that succeeded or coincided with that of the rustic retreat. The in-valid, who could be compared with the Germanic walker or the traveller tormented by spleen, enjoys this life-style. The great rush towards the sea-shores proclaimed by Cowper[154] cannot be explained solely by the fascination of Brighton's dazzling pleas-ures. At the edge of the ocean, modern man comes to discover himself and to experience his own limits in the face of the ocean's emptiness and the availability of the shores. In this sublime place, the ego thrills as it confronts the waves or the salty whiff of the sea, or contemplates the solitary vision of the storm. Here, adults fond of horseback riding on the strand could also indulge in regressing, and might find themselves digging in the sand in search of shells, hiding in grottoes, or exploring the boundaries of an insular territory.

At the seaside, sheltered by the therapeutic alibi, a new world of sensations was growing out of the mixed pain and pleasure of sudden immersion. Here, for the benefit of the leisured classes, a new way of experiencing one's body was developing, based on rooting out the desires that disturb it. On the shores of the ocean, visitors try to allay the anxieties aroused by their loss of vigour and the etiolation, pollution, and immorality typical of urban life. Yet this groping quest for harmony between nature and the body paradoxically leaves no place for hedonism. The sea makes it easier to renounce sensual pleasures; the life-style that emerges along the sea-shore is also part of a process of emphasis on health that accompanies the refining of the way individuals pay attention to themselves.

4

Penetrating the World's Enigmas

By the middle of the eighteenth century, the shore no longer appeared solely as something utilitarian, a remedy whereby users hoped to allay their anxiety. It also became reconnected with an ancient role, and once more became a focal point for the world's enigmas. People came there to wonder about the earth's past and the origins of life. There, more than anywhere else, it was possible to discover the multitude of temporal rhythms, to sense the duration of geological time, and to observe the indecisiveness of biological borderlines, the vagueness of different kingdoms, and the surprising transitions that linked them to one another. The proximity of zoophytes and fossils thousands of years old demonstrated the intense meaning that emanated from a long-abandoned place.

A series of ambulatory quests were encouraged by the *libido sciendi*. In their own way, they demonstrated the growing appeal of the shore. These experiences generally corresponded to a great number of ambitions, combining aesthetic enjoyment, the pleasure of scientific observation, and the satisfaction provided by physical effort. Along the beaches of Western Europe, various ways of appreciating the shore, forms of contemplation, and habits sprung up in this way, and formed a system. The shore took shape as the laboratory for a cluster of practices whose coherence has long since been forgotten.

THE ARCHIVES OF THE EARTH

Even more than to mountain tops or hillsides, people came to the coasts to browse in the archives of the Earth. Several factors determined why the point of contact between the elements should be the site of investigation. First of all, there was the considerable role then attributed to the sea in fashioning the earthly globe. To most people of that age, the forms of its relief were the outcome of the Great Flood. As seen earlier, the image of this disaster willed by God shaped the representation of the sea-shore in the classical period, and it should be remembered that many eighteenth-century scientists were quite satisfied with the simple explanation of its morphology offered by Genesis.[1]

This perception of the coastal landscape began to crumble, however, over the course of the century. Gradually, both here and in other areas, the separation between science and religion deepened. A great number of Enlightenment scientists became hostile to cosmologies whose adoption of the idea of divine intervention rendered the universe incomprehensible to man, convinced as they were that the world was governed by an intelligible order. They began to break away from the Genesis account.[2] Some sought manifestations of a nature with vast creative faculties, still at work, in the surrounding environment.

On the subject of the Flood, however, caution must be exercised. It is true that criticisms of the literal interpretation of the biblical account abounded at the time. The account was disqualified by too many improbabilities to satisfy anyone eager for scientific explanations. Critics wondered whether enough water to submerge all inhabited lands could have fallen in just a few days, and furthermore, then suddenly receded. There was debate over whether the disaster had been global, and no one was really sure whether the Americas had also been affected. Did the cataclysm remain confined to areas inhabited by man? If so, how could the topography of the rest of the planet be explained? There were arguments over the intensity of the Flood's geological action: could it have been responsible for the cracks in the earth's crust, or was this crust dissolved by the chemical action of the waters? Had the earth been through a long antediluvian history? Could fish have survived the disaster, or was it necessary to accommodate them inside the Ark? Was the ship big enough to shelter all potential refugees? Where did the dove find the olive branch it held in its beak on the fortieth day?[3] All these questions gave rise to endless debate, especially during the first half of the century.

Opinions varied on all these points. Even excluding those who clung faithfully to the Genesis account, the majority of the enlightened minds of the time accepted the Flood as historical fact,[4] although some refuted the veracity of the details of the events related by Moses, or minimized the scope of the disaster and, by the same token, its geological consequences. In fact, the Flood runs as a common theme through the oral traditions of all ancient peoples; indeed, it is the main example of such a common theme. In his search for 'the misfortunes of the earth',[5] Nicolas Boulanger examined various cultures, and succeeded in drawing up a long catalogue of monuments from the disaster. Scientists of the century were fond of comparing ancient texts with archaeological ruins; quite naturally, they adopted this approach when they attempted to establish a link between the letter of the Scriptures and the traces inscribed in the memory of the ground.[6]

A number of scientists, however, though only a minority, considered the Flood to be a mere fable. In his *Telliamed*, Benoît de Maillet claimed to have proved its unlikelihood.[7] Rainfall lasting forty days seemed insignificant given the amount of water that would be necessary to submerge mountains. A disaster so limited in time could not have laced thick layers of earth with shells. At most, it could be assumed that there was very local flooding. Twice in his *Dictionnaire philosophique*,[8] Voltaire barely concealed his scepticism, writing ironically about the greatest of miracles. D'Holbach repeatedly asserted that the shapes of the Earth's relief could not be explained by the disaster described in Genesis.

Still others, conscious of both the difficulties involved in interpreting the biblical text and the risks incurred by overly sharp criticism, opted for silent respect. 'Miracles should hold us in awe and in silence,' advised Buffon in his 1749 *Histoire et théorie de la Terre*.[9]

In the opinion of scientists convinced that present causes had been acting on the earth without interruption since its creation, the disaster simply provided an unnecessary explanation. For them, the Genesis text was completely discredited. This theory, put forward clearly by Hutton and even more radically by G. H. Toulmin,[10] had found a number of adherents at the end of the century, before the French Revolution reached its climax. Hutton, reconnecting in this respect with physico-theology, considered that God had created a self-sufficient system operating smoothly through the action of natural laws. The planet was the scene of continuously functioning processes.[11] In place of the image of a heap of ruins, Hutton proposed that of an active earth, capable of endlessly producing new forms and continually making up for its own losses. Within this dynamic perspective, the Flood seemed impossible, unless it was assumed that there was a flaw at the very heart of the order instituted by God.

Increasing criticism and the difficulty of reconciling the observation of phenomena with the Mosaic text suggested other possible explanations, and led to new theories of the earth. While distancing themselves from traditional diluvialism, these theories nevertheless attributed an important role to the sea, its currents and tides, and the variations in its level. Biblical imagery persisted in the figure of the primordial or 'primitive'[12] ocean, which cast its shadow over the emerging science of geology. In the footsteps of Benoît de Maillet, 'Neptunian' scientists remained persuaded that the sea 'is entirely responsible for all the physiographical, lithological, and structural characteristics of the earth's crust'.[13] Buffon, too, in his *Histoire et théorie de la Terre*, borrows extensively from Benoît de Maillet; he popularizes actualism, and glorifies the action of sea currents.[14] The ocean, in his view, explained and represented the earth's past; in this respect, Buffon was taking a position that preceded scientific observation.[15] According to him, the mountains were the outcome of activity in the depths of the abyss.

After 1770, the Neptunians asserted that the study of fossils was indispensable, and later, the succession of geological strata.[16] The attention that many scientists began paying to sedimentation reflected a desire to temporalize the natural order. To this end, they developed the hypothesis that what was now dry land had spent a long period submerged in the depths of the sea. The huge popularity of the German Werner and the success of his gigantic, primordial ocean endowed with powerful chemical properties helped to implant the Neptunian theory firmly in the minds of scholars. From that time onwards, the ocean came to represent the primitive force responsible for the shapes of a topography that was no longer perceived as the outcome of brutal destruction, but rather as the product of long sculpting under water.

Werner's success explains the power of the resistance put up by what was then known as 'catastrophism', by contrast with 'actualism'. Compared with the initial form of diluvialism, however, the character of the theory had changed. A distinction emerged between geological flooding and the biblical Flood.[17] In fact, observation made it necessary to assume that there had been numerous cataclysms; this, so it was thought, should make it possible to reconcile science with the Scriptures. If it was simply assumed that there had been a long period of underwater sedimentation between the Creation and the Flood or, better yet, that mankind had been created on the eve of the last great upheaval, then diluvialism could take on a scientific character.

This is what occupied a good many scientists after 1800. Actualists and catastrophists were then engaged in veritable warfare,[18] and a series of factors were operating in favour of the latter. In England, the French threat revived hostility toward the rationalism of Enlightenment thought.[19] Scientists of renown, such as Richard Kirwan and J. A. de Luc, endeavoured to give scientific form to the providentialism of Genesis. A group of clerical geologists attempted to restore what had been a waning faith in the alliance between God and geology.[20] The Reverend Buckland suggested that a careful distinction should be made between 'diluvium', or remnants of the biblical disaster, and 'alluvium', the outcome of subsequent geological processes.[21] The geological moulding which would later be interpreted as the result of glacial erosion was then used in favour of catastrophism. After all, how could one explain, before the publication of Agassiz's writings, the existence of random boulders, the evident disproportion between the size of glacial valleys and the rivers coursing through them, or the terraces that bordered waterways, without resorting to the intervention of some kind of catastrophe? This was certainly what Buckland claimed to show, with the help of observations made along the coasts of Devon and Dorset.

In France, gripped by the concept of revolution, the firm belief that the earth, too, had undergone terrible convulsions, worked against actualism. During the Restoration, advances made in palaeontology, and especially Cuvier's discoveries, provided decisive support for catastrophism. The sudden disappearance of species revealed by palaeontology seemed to underscore the weakness of actualism. Cuvier[22] belonged to the tradition of those who attributed great importance to the ocean, and the personal diluvian theory he proposed seemed capable of being reconciled with the biblical text.[23]

This brief review of the changing directions of scientific thought that preceded the victory of Lyell's actualism and the emergence of glaciation theory makes it easier to understand the ways in which people then viewed nature along the coastline. At the same time as 'a belief in the efficacy of offering explanations of the nature of things or phenomena by means of their history[24] was growing, a new conception of time was emerging. Thinkers were gradually compelled to give up the short chronology that had been accepted until recently, when the history of the earth was still considered to be one with that of mankind. Bossuet's calculations, according to which 6,000 years had elapsed since Creation, or the assertion of the English archbishop Usher in 1650 that the world was created at nine o'clock in the morning on 23 October in the year 4004 BC, eventually lost all credibility in the eyes of philosophers. In addition to short time, which followed the rhythm of alternating days and seasons, and historical time, marked by the succession of centuries, there was now the immensity of geological time, a belated corollary to the discovery by seventeenth-century scientists of infinite space.

Then attempts were made to establish epochs, periods, or phases, revealing a desire to express the extent of the duration. In 1779, Buffon determined that the origins of the earth went back 75,000 years, rather than 6,000.[25] At the end of the century, Giraud-Soulavie introduced a million years as the unit of time in geology.[26] Hutton, for his part, put forward his cyclical vision of phenomena, suggesting that the earth's past was infinite. Geological cycles operated in a series, and it would be vain to attempt to identify their beginning, since they erased all traces of the creative act as they succeeded one another.

By the same token, the history of man and that of the planet became disconnected. From then on, science proposed a series of images of continents and the ocean before the appearance of the human race. The idea of a very ancient earth, indifferent to the beings dwelling on it, attained a new sublimity. The pure, uninterrupted, and endlessly repeating sound of the waves came to be viewed as a proclamation of the eternity of the world.

In the opinion of actualist scientists the present no longer represented the aim of the world's evolution. To use the image of a roller breaking on the shore, it was merely a moment within a continuous series of changes. The contemporary world was sculpted from the ruins of another world that preceded it, and from its own ruins, a new earth would arise. Geology, like Lavoisier's chemistry, which was also in the process of development, struck the imagination with the endless renewal of its combinations.

Obviously such a profound upheaval in the system of representations did not take place in a linear fashion. It is understandable that some pious scholars should begin to attempt in the 1790s to rehabilitate a less dizzying chronology.[27]

Given the decisive role attributed to the ocean's waters in the history of the earth, the coast became the site and the subject of very attentive studies within the framework of this revolution in images of time. [28] The discovery of oscillations in the shoreline, then their measurement, transformed what had been considered an inalterable exposed slope into an uncertain, residual space in which the indecisive struggle of the elements took place. This was the context in which the concept of active tectonics was to emerge.

The ancients were quite familiar with the mobility of the shoreline. In this context, eighteenth-century scientists were fond of quoting Aristotle and Ovid. Contemporary observations quickly reached the firm conclusion that the ocean was gradually drying up. This confirmed one of the fantasies that shaped the imagery of the sea. The ebbing of the waters strengthened the belief in the disastrous decline of the ocean's fertility, about which Western fishermen were complaining.

Two series of observations, in the North and in the Mediterranean, helped found this conviction. The first scientific study of the phenomenon took place on the shores of the Baltic. In 1694, the Swede Urban Hiärne stressed that Scandinavian fishermen had been observing the withdrawal of the sea for a long time and that Viking law already took this into account. In order to obtain a more precise picture of the phenomenon, he decided to circulate a lengthy questionnaire, the answers to which he published between 1702 and 1706. In 1724, Celsius, [29] abandoning all ideas about the Flood, began to undertake a methodical investigation. For twelve years he travelled along the coasts of the Baltic. He questioned old fishermen and analysed toponymy; he enquired about the emergence of certain reefs and the abandonment by fishermen of channels that were no longer navigable. He noted the rocks on which seals had successively dwelled. Finally, in 1731, Celsius decided to draw reference lines on rocks at water-level in order to be able subsequently to assess the extent of the water's retreat. In 1743, he undertook the first scientific measurement of the phenomenon before publishing a work with Linnaeus in which he defended the hypothesis that the oceans were shrinking.

From his observation of the shores, Celsius derived both the perception that telluric operations were still continuing and the conviction that their scope could be assessed. The relationship he established between the duration and the amplitude of the sea's movements introduced a new means of measuring the stages marking the earth's history. This method owed nothing to Scripture; it diverged from the brilliant, but hypothetical, cosmogonies that were fashionable at the turn of the century. Moreover, the works of the Scandinavian scientist reinforced the idea that northern peoples were relative newcomers. The Celsius scale actually suggested that Sweden had emerged only recently.[30] This gave rise to indignation among nationalists and fuelled sharp polemics.

Shortly thereafter,[31] E. O. Runeberg[32] also set about upsetting traditional views. According to him, the ocean, though apparently the most mobile element, was not responsible for the oscillations of the shoreline. A subterranean force was slowly lifting up the rocky base. The notion of telluric uplift was a subject of debate until Leopold von Buch succeeded between 1806 and 1808 in giving the new thesis decisive support. In 1834, at the end of his journey to Sweden, Charles Lyell in turn accepted the hypothesis of a gradual uplift, a notion that Norwegian specialists later worked out in greater detail.

At the same time, a group of Mediterranean scientists, struck by the evident withdrawal of the sea along the Mediterranean coasts, began to engage in similar observations, which harmonized with one another.[33] After Jean Astruc, Benoît de Maillet made the abatement of the seas the backbone of his theory of the earth. The narrator's ancestor, the hero of *Telliamed*, surveyed the shores for over 200 leagues in order to observe the work of the sea. Convinced of the marine origin underlying the relief's shape and of the validity of the theory of actual causes, he hoped from his observations to be able to infer the mechanisms that had presided over the history of the earth.

> With this in mind, he slowly went along the sea-shores, sometimes on foot, sometimes on a light vessel that ran alongside them and often very close to them ... He would stop for hours on end on a shore and observe the work of the waves on a beach as they collapsed at his feet, or the sands and pebbles that the flood-tide brought there, according to whether the sea was calm or rough. Sometimes he would sit on the top of steep rocks that were washed by the sea; and from there, in as much as the depths of the sea made it possible for him to do so, he would observe any remarkable occurrence that took place.[34]

Yet other observations were upsetting the image of the sea's boundaries. Count Marsigli, the father of modern oceanography, placed the coastline within the overall context of the land and the subaquatic relief. According to him, there was a symmetry determining the distribution of mountains and of the ocean's depths. Moreover, he observed that a continental plateau stretched out before the beaches. Consequently, the shore could no longer be viewed as a line, but as an area,[35] which included the beach and the continental platform; it stretched between the mountains and the ocean valleys. This was the axis of the symmetry he assumed.

Fifty years later, the French geographer Philippe Buache[36] met with great success in expounding his theory of basins both on land and under water. According to him, the mountain chains wind their way both under the sea and across the land. In both regions, they outline basins. Buache suggested that the two types of relief should be represented by the same type of orographical mapping. Using fathom curves and contour lines, he tried to provide a single vision of both submerged lands and those that emerged out of the sea. On this point, it is important to remember that the requirements of navigation had led to measuring the depths before the need was felt to measure the heights. Like truth, and like Venus, the means of calculating altitudes rose from the watery depths.[37]

To come to my point, all this scientific labour led to a new perception of the shape of the coast. The view of a sandy or rocky coast and the way in which this landscape was interpreted varied according to the beliefs or, more simply, the scientific background of the beholder. Ariste and Eugène, seeing with the eyes of natural theology, as readers will recall, admired both the traces of miraculous events on the shore and those of the ordinary and even more wonderful order of the world. Their eyes gazed with delight at those unchanging rocks, seen as so many indestructible battlements established by God in order to protect the earth from the roaring ocean, which respected this barrier that prevented it from pouring out its waters and causing another flood. This was the fixed vision of observers blind to the endless alterations of forms in which they could see nothing but the result of God's will as expressed on the morrow of the disaster. It was a cluster of beliefs that made them blind to what today would be recognized as erosion, accumulation, or adjustment.

At the dawn of the nineteenth century, as viewers were finding it increasingly difficult to interpret the coastal landscape in this way, J. A. de Luc, a pious scholar, suggested a variation of this conception of coastal morphology.[38] In his opinion, the shoreline with its capes and its rocky headlands revealed obvious traces of the Flood.[39] All the waves of the sea breaking on rocks could not wear them away. Nevertheless, since de Luc was aware of the adjustment that occurred along the coasts, he admitted that the alluvial deposits he studied had accumulated after the cataclysm. But he saw in them nothing more than minor 'alluvial' changes. At most, the sea was capable of rounding its shores, smoothing their slopes, and filling in the curves of its contours. Beyond that, it could do no more. This represents an intermediate vision which diverges slightly from the fixed concept of the diluvian shore, while remaining generally faithful to it.

The revolution in ways of seeing took place when people began to perceive the coast and the reefs, rocks and cliffs that border it as so many telluric ruins – that is, no longer as chaotic relics of a cataclysm, but as the products of age-old wear.[40] The outline of the coast's visible landscape bore witness to the immensity of time; by the same token, it created the possibility of predicting future transformations. The images of the past, the present, and the future of the coastal landscape had all changed simultaneously.

The three interpretations sketched here in their broad outlines were made simultaneously by men belonging to the same age, and it would be difficult to even attempt to ascertain the extent to which their views were shared. Besides, the shift from the fixed vision of the landscape to an awareness of endless change did not proceed in a linear fashion. The return to catastrophism at the beginning of the nineteenth century caused a temporary retreat of the concept of the coast's wearing away since time immemorial, in favour of a sensitivity to the chaotic traces left by successive cataclysms.

As a result of these interpretations structured by the various ways of conceiving temporal rhythms and geological processes, the vision of the coast varied in its mechanisms and its processes. In a general way, however, there was a growing perception of volume, and a three-dimensional interpretation became common: this interpretation was made possible by a vision more capable of taking layers into account, and it attempted to organize them in order to find more easily the rhythms of the new temporality in them. Two special settings facilitated this comprehension of the volumes that gave meaning to temporal distances. The first was the depth of mines, which provided an immediate picture of stratification. It would be hard to exaggerate the importance of the role of studying and representing strata in an education of vision that sought to describe orderly strata. The spectacular development of German mineralogy, to which Werner's European popularity bears further witness, and the fashion that led travellers of the times to climb down into the depths of wells were contemporaneous with the refinement of the stratigraphic scale in the visual language of geology.[41] The fascination exerted by rocky layers and the desire to explore mines competed successfully with the interpretation of external features of present landscapes proposed by the picturesque code. In the heart of the mine, that archive of the earth,[42] the succession of strata functions like so many texts offering the traveller in search of the sublime an obscure autobiography of the earth's crust.[43]

By abandoning its monopoly of taxonomic objectives, natural history ceased to be a mere inventory process. In the field of geology, this was expressed by the search for a visual language capable of describing the depths.[44] The geological map, the stratigraphic scale, and the divisions they could create made it possible to dispense with verbal description. Above all, they offered an interpretation of the surface. Combined with a knowledge of the terrain, they enabled geologists to visualize the arrangement of strata; they taught geologists how to look at their environment.

What proved decisive in the spread of this form of vision was the role of cliff views, which became increasingly common in Great Britain after 1814. These views borrowed heavily from the views from the sea used for coastal navigation, and they were presented as so many open-air geological sections.[45] In cliffs, stratigraphy was laid bare, revealed, and exposed to the eye. This dispensed with the need to plunge into the bowels of the earth, and facilitated the visual grasp of the layers. By induction, it enabled an interpretation of the structure of the substratum. In short, cliffs offered a three-dimensional spectacle.

Draughtsman topographers very quickly conformed to the imperatives of this new geology, and played a decisive role in the education of vision. From this point of view, it would be interesting to analyse the whole of William Daniell's work and the particularly revealing example comprised of the plates produced by Thomas Webster on the Isle of Wight in 1815 as illustrations for Henry Englefield's picturesque text.[46] In them, scientific theory simultaneously subverts and ordains topographical representation; it surpasses the initial confusion in the natural spectacle by imposing an order that engenders an understanding of structure. Webster teaches the reader to cut through layers with his imagination, to read topography as he would a stratigraphic section, and to browse repeatedly through the order of layers of sediment, which is primarily the order of successive periods.

This model subsequently became widespread. The arrival of geological time soon left its mark on the documentary drawings intended to illustrate travel accounts.[47] Once the work of formalization was complete, representations of coasts gradually took on the appearance of geological sections, with their conventional colours. Artists' perceptions and renderings simplified their subject, ordering it and accomplishing an implicit scientific task, allowing viewers to understand the regularity of the strata. Often, observes Martin J. S. Rudwick,[48] the stratified rocky block seems to exist only to tell the story of the underground depths, which are more impressive than the steep cliffs in which the strata end. Images of human beings or of birds are placed in the service of the scientific project, as they would be on an archaeological plate. A human figure provides a scale for the size of a cliff, while a bird gives a better feeling for the depth of a gaping precipice.

This visual pedagogy focused on interpreting the coast lends support to the praises that William Gilpin sang to the chalky cliffs as early as 1774, when he travelled along the Channel coasts. This same approach leads viewers to accentuate the main features of coastal structure, exaggerating its slope and verticality. This generates a dizzying effect that is in keeping with both the sense of the sublime and the mainsprings of the horror story.[49] As coastal landscapes become more explicitly geological, the pedagogical aim becomes merged with the prevailing aesthetic codes.

Finally, the evolution of scientific thought was changing the vision of the beach itself. Spectators interpreted this line in a new way as they came to realize that it was only temporary, as they reconsidered the surface area that was, for some analysts, the culmination of an erosion cycle. Along northern coasts, this new uncertainty about the shore's outline came to correspond to the ephemeral character of a territory that was liberated every twelve hours by the receding tide. The boundary between the beach and the strand was blurring.

WHERE THE WORLD BECOMES DRY

At the same time, the *libido sciendi* encouraged other interpretations of the sea-shore. This line of contact between the lands that emerged out of the waters and the fecundity of the sea seemed to many observers to represent the site where life began. The indecisive character of the strand's outline with its changing contours corresponded to the ambiguity of the king-doms. Here, more than anywhere else, the links in the great chain of being were connected, and the biological interest in trans-formations could then develop. This stimulated other questing on foot.

Several theories posited that the sea was where life originated.[50] Benoît de Maillet noted, though somewhat erroneously, that Thales, Anaximenes, Anaxagoras, and Homer all considered water to represent the principle of all things. According to Leibniz, life first developed in the sea, and land animals were but the descendants of aquatic species. At the beginning of the eighteenth century, as Marsigli was founding marine biology,[51] *Telliamed* made of a sea animated by God with the earth's vital spirit the source of all life. Remember that, for Benoît de Maillet, as for Bishop Pontoppidan and many other scientists, the ocean deeps accommodated animal species with equivalents among species on land: transitions could be made between the two environments. According to *Telliamed*, the 'terrestrialization'[52] of marine animals occurred where air and water were thrown together. The major problem that arose on such occasions was that of 'acclimatization', or adapting to dry land. In this respect, Benoît de Maillet turns the traditional horror images of the Flood upside down: he reverses the order of the disasters. The catastrophe was no longer due to the Flood, but to the drying out, a repetition of the trauma of birth. Two centuries before Ferenczi, for whom the anxiety caused by the Flood arose first of all from the horror of the receding waters,[53] the author of *Telliamed* explicitly linked the underwater past of animal species with intra-uterine life bathed in amniotic fluid. Seen from this point of view, the 'terrestrializa-tion' of water-dwelling men lost its improbability, through the analogy established between this process and the drying out that characterizes birth and is caused by it.[54]

Water was thus deemed a natural environment for man. Ac-cording to Benoît de Maillet, this explains the beneficial effects of sea bathing,[55] which takes on both a hedonistic function and a therapeutic value. The water-dwelling girl and woman[56] once taken in by the King of Portugal liked to dive into the sea, to the astonishment of the Court. Since the sea was the original womb, since it had an effect on seeds, it must have the power to restore fertility. Quite logically, then, philosophers proclaimed the generative virtues of the ocean's waters.[57]

According to Benoît de Maillet, 'terrestrialization' could prob-ably occur only along northern coasts, deep in watery caves filled with thick fog in the heart of forests whose damp air facilitated the transition. On the coasts of Greenland, Spitzbergen, or Hudson Bay, water-dwelling men could adapt to breathing air, as do the amphibious species who haunt those shifting shores in herds.[58] Then these vigorous barbarians would head south, as they continued to adapt.

Buffon belatedly borrowed here once again from Benoît de Maillet when writing *Époques de la Nature*. He, too, considered the sea to be the great reservoir of Nature's primitive force, and to be especially powerful in the polar regions of the northern hemisphere.[59] This was where, according to him, organized species first appeared. Not only were northern lands the first to have become fertile, 'but it was also in these very regions that living Nature took on its greatest dimensions'. This sheds light on the significance of the images of the herring and the whale in the literature of the time: these were the two major symbols of fecundity and power. In Buffon's view, however, the superiority of the North was effective only at the source; as the barbarians descended toward southern regions, where the forces of nature became dissipated, these men became civilized.

Such theories, then widely accepted, had several implications. In the short term, *Telliamed* extolled the beneficial effects of the shrinking of the seas. This would enrich coastal agriculture – here the image re-emerged of Holland, rising out of the water and blessed by God. Since the sea was also the source of plant life, Benoît de Maillet entertained the hope that algae could be cultivated to remedy food scarcity. But in the longer term, the fantasy of the sea's decline, the collapse of the waters' fecundity, and their inevitable drying up fostered collective anxiety. The abundant literature on the fishing crisis clearly testifies to this.[60]

The image of the ocean as the primordial womb corresponds to what was said earlier about the shore, which was perceived as a recourse against the weakening of bodily vitality and fertility. This image reinforces the assumption that the underwater world swarms with life. It ordains the northern realms, damp cold, and fog as so many catalysts that foster the adaptation of life forces, and in this respect it coincides with the vogue of sea bathing and the fashion for Ossianism and Highlands travel. In the literature dealing with the life sciences, sun, burning, and dryness are associated simultaneously with a lessening or a decline in vital forces and with refinement. Dryness, lost vigour, sterility, and civilization all depict a chain of images that helps to explain both the repulsion aroused by the southern shores and the fascination exerted by their prestigious relics. This contributes to an understanding of how a society could be torn between the appeal of the North, with its promise of restored vigour, and that of the Mediterranean, where ancient cultures blossomed.

THE SCHOLAR'S FEATS OF VALOUR

These numerous, apparently contradictory attitudes actually all express the same shift of point of view towards a longer, profounder sense of time. As a result, a range of quests emerge that both express and sharpen the appeal of the seaside.

Historians know today in which milieux these new practices took root. In Great Britain, as the eighteenth century began, the rise of the figure of the scholarly travelling gentleman[61] accompanied the emergence of the 'man of taste'. The former, fond of investigating on the ground and lacking a real research programme, focused on making inventories if not on actual collecting. Between 1700 and 1710, this pastime was essentially the preserve of the cultured aristocracy, blessed with the leisure provided by their landed status. In this milieu, women, too, were very keen on collecting natural objects. At the end of the century, King George III himself illustrated this elitist model of the scholar.

Across the decades,[62] however, the clergy and the middle class became enchanted with this passion for science. Timidly, a provincial, bourgeois culture took shape, governed by utilitarian aims. The emergence of leisure in these classes and its increasing commercialization, the spread of knowledge, the proliferation of popular works written in the vernacular and illustrated by specialized artists, and the development of numerous forms of cultural sociability all encouraged this expanding curiosity. Between 1775 and 1780, the middle classes devoted themselves even more resolutely than in the past to quests of this type, which were further stimulated by a taste for rural life and a rising feeling of pride in rural nature. The labour of these amateur scholars gave rise to a proliferation of recording procedures. The number of travel diaries began to increase, and topographical literature and landscape sketching thrived. At the same time, a real scientific approach gradually took the place of mere collecting. It is revealing in this respect that geology emerged in practice as an autonomous discipline even before its specialists constituted an organized community.

Daniel Roche has provided a detailed description of the social history of curiosities in eighteenth-century France. He, too, stresses the growth in the number of physics, natural history, and chemistry laboratories, especially in the south of the country.[63] This rise was stimulated by the pedagogical venture undertaken after 1750 by the most eminent learned societies. Gradually, the figure of the 'academic man' began to spread. On the Continent as well, the inactivity of a provincial aristocracy played an important role in altering a culture: study became the necessary condition for noble status. This, should not, however, overshadow the growth in the number of bourgeois involved in this new social and academic practice.

In these milieux, the thirst for knowledge, combined with taxonomic aims, stimulated the pursuit of these quests on foot. People believed instinctively, as Daniel Roche observes after Gaston Bachelard, that a natural history collection could be developed like a library, as opportunities arose.[64] Any trip, even the most trivial journey to a place that seemed insignificant, held open the hope of making a scientific or aesthetic discovery. It was all a matter of looking and being curious. A walk along the shore might prove to be rich in delightful observations, in unexpected discoveries of fossils, seaweed, or algae, in curious sights, and soon, in daydreams. There was still nothing ordinary about taking such a path.

In relation to our subject, Celsius's research or the allegedly imaginary efforts of the narrator's ancestor in *Telliamed* are premonitory. The latter sets forth the model of what was later to become the lively, joyful journey through coastal sites;[65] he inaugurates the practice of asking questions of rocks and sand in the hope of hearing the forgotten message from the dawn of time. Such an approach became widespread in the course of the second half of the century, as geologists turned even more resolutely into hands-on scholars.[66] Revealing in this respect were the expeditions triggered by the debate over the nature of basalt, a debate that developed between 1790 and 1810, before the division between the respective fields of geology and mineralogy became clear.

For many travellers who were also amateur scientists, walks along the shore were guided by the desire to take a vertical journey. Novice geologists sought to experience the thicknesses of the strata with their bodies as well as their eyes. They could be seen alternately climbing, advancing along the cliff side, following a watershed, going up and down the shore beneath a rocky wall, or plunging into caves. They were known to be eager to discover the order of the constituent elements in what seemed at first glance to be chaos, and to reach spots that still spoke of the primeval earth. They exuded a special fascination with underground passages, in which the elements came together in the depths of the earth.[67] There was growing curiosity about a substratum that was no longer merely the primitive womb and grave of mankind, but also the possible revelation of an indifferent earth's secrets. The desire to experience this new sense of depth by actual contact with telluric layers partly discredited the beach, which was a mere surface reference of lesser significance.

Richard Pococke in *The Giant's Causeway*, Dolomieu on the island of Elba, Fabricius on the sea-shores of Norway, and Pictet along the coasts of Great Britain all illustrate this new quest. Two other, later examples will provide a better grasp of the pleasure arising from the journey through the coastal layers. When Faujas de Saint-Fond decided to undertake geological research in Scotland, the Caledonian journey was fashionable, and this scholar had read Dr Johnson, Pennant, and Knox. The lion's share of all these accounts was devoted to the Highlands; none of the authors had undertaken to explore the Caledonian shores, as Faujas yearned to do. Hear how he relates his geological walks along the beaches at Oban:

> I would leave at dawn, a haversack on my back, accompanied by a servant, my faithful companion, who, for his part, was carrying a bottle of wine and some cold meats, that we would touch only after several hours' work.
>
> It was then that we had our frugal, but excellent meal; sometimes on the top of a steep rock, sometimes in some sheltered cave, on the seaside, where the waves, breaking at our feet, gave us the spectacle of an angry sea, on which we were glad not to find ourselves.
>
> In the evening, laden with stones and instructive notes, I went back to my peaceful abode: I laid out all my treasures on a table; I sorted them, I even admired them ... I shared them out in advance among my correspondents and friends, and I was happy.[68]

This happiness implies complex tactics of memorization and recording. Faujas passes them in review during his visit to the isle of Staffa: 'After having written down all the details concerning Fingal's cave, after having made drawings of the objects that interested us most, after having taken the measurements that I was very glad to get, I started to inspect the other parts of the island, and I made a collection of various lavas, zeolites, and other stones.' On his return home, he added: 'I devoted myself for several days to writing out my observations about the isle of Staffa.'[69]

Faujas was driven by a longing to reach the cavern's deepest depths. More daring than other tourists, he wanted to go to the very bottom of Fingal's cave. Barefooted on the damp rock, he managed, 'not without trouble and risks', to slip through the deepest crevices, constantly haunted by 'the thought of coming out again'.[70]

In 1806, 1807, and 1808, the Genevan geologist L. A. Necker de Saussure, following in the footsteps of his ancestor, undertook his own journey to Scotland. By this time, the trip had become commonplace, and on his way the scholar met quite a few mineralogy enthusiasts. Saussure decided to concentrate on the shore. His book gives an account of a scrupulously traced itinerary along all the Caledonian coasts. He outlines the archetype of the journey into the depths described above. It should be noted, however, that Saussure, like most tourists of that time, had many other ambitions: he endeavoured to learn the local dialects, to study popular customs, and of course, to observe the flora and fauna as well as the coastal morphology. The most interesting fact remains that minute trips seem to warrant that he should successively record them, relate them, and publish his account. Like Townley concerning the monotony of his days on the Isle of Man, Necker de Saussure enjoys talking about events that would not have held the attention, under the Third Republic, even of a schoolboy desperately searching for a subject for his essay. Obviously, this traveller experienced pleasure, as did Pictet, in exploring the smallest twists and turns of the coast, jotting down the diversity of his impressions as they came, and making on-the-spot sketches of the details that gave a 'curious' character to what he saw. In the evening, all this material became useful for writing out the travel diary. For months on end, Saussure moved meticulously from beach to beach. He followed the tree-shaded paths, climbed up black rocks, dunes, and cliffs; he took shelter from the rain under overhanging rocks, and visited even the smallest caves. He made an exhaustive journey.

The text of over 1,500 closely written pages corresponds to the author's wish to be perceived as a hero. The real subject of the book is the scholar who risks his life for the advancement of science, and to this we owe the accounts of derisory exploits, which are, however, in keeping with the emotional strategy of the sublime. In the vicinity of Ely, after having covered long monotonous beaches and climbed with his guide up a 'sand hill', Saussure cannot resist the pleasure of going to the foot of the steep slope in order to gather fragments of basalt columns. 'Without thinking of the danger', he writes, 'I wormed my way as best I could by clinging to tufts of dry grass that were growing there in abundance, and by way of steps, I used the tops of columns that sprung from the earth.'[71] In this way he manages to reach his goal. The traveller then describes the return trip: 'With considerable difficulty I reached the summit, where I found my guide, who seemed quite surprised to see me back safe and sound.'

On the island of Rhum, the desire to take samples creates another occasion for feats of valour, and indeed heroism. Saussure climbs up 'the steep embankment that separates the rock from the sea', running the risk at any moment of being hit by one of the stones that indifferent sheep knock loose from the summit. On his way back, he writes: 'I came down laden with the precious spoils I had just picked up ... I was often compelled to sit on the grass and let myself slide down at very great speed.'[72] On the Isle of Skye, he faces an even graver peril, and the hero emerges all the more glorious. A clumsy goat slips and falls down stone-dead 'on the very spot where, one second earlier, I was busy taking some samples of zeolites; [it] would inevitably have made me share the same fate if one of my companions had not warned me in time of the risk I was running; and I had just time to look up and throw myself by reflex under the shelter of a big block whence I saw the dying goat fall down next to me, as well as the two chunks of rock that followed it.'[73] The motion in this account enables the reader to experience the depth of the relief.

Compared with the behaviour of invalids along the beaches, the physical activity arising from the *libido sciendi* is the most characteristic feature of this scholar's epic of the Caledonian shores. It is evident that the diffusion of such travel diaries helped to propagate a model of conquering behaviour.

The fascination that these travellers feel – and describe – for sea-birds goes naturally with the measuring of coastal layers. A bird's wing serves simultaneously to reveal, stress, and deride the yawning void, while its cry, thrown back by the cliff, accentuates the vertiginous depth. With Thomas Pennant, the ornithological quest goes hand in hand with geological observation. This alliance determines the way he listens to the sea-gull, which makes him dream. 'I often rested', he writes,

under such rocks, paying attention to all the varied sounds that
mingled over my head; and this mass of mixed sounds, accompanied
by the deep, muffled howling of the slowly swelling waves coming
out of the lower caves, had a surprising effect on my ear, that had
its own charm. The gulls' piercing voices, the guillemots' twitter-
ing cries, the penguins' loud notes, the herons' sharp cries, together
with the cormorants' periodic cawing that acted as the bass among
the others, have often provided me with a concert of an unprecedented kind.[74]

This is a kind of sensibility that soon became ordinary in England.
At the end of the century, bird-watchers armed with their optical instruments were legion on the Isle of Wight.[75] This hobby,
combined with the pleasure of describing vertiginous slopes, is
also reflected in the self-satisfaction with which travel accounts
linger over the exploits of nest-collectors clinging to the sides
of vertical cliffs.

At the same time, scholarly travellers began to peruse the
stretches of sand uncovered by the ebbing tide. This territory of
the void had been surrounded by a further void in terms of any
discussion until then. It was henceforth to exert a growing fascination, and to become in turn the scene of quests on foot,
built on the model suggested by walking fishermen. Of course
occasional oceanographers had been going there for ages to study
the tides;[76] but to my knowledge, no source until now had mentioned the journeys motivated by the mirage of the strand alone.
On this uncertain surface, however, individuals strongly experienced the interlocking solar and lunar rhythms. The alternation of
days and nights became coupled with and complicated by that
of the ebb and flow.

The strand invites spectators to experience the alternating rise
and withdrawal of the waters; the availability of this expanse
gives rise to the contrasting fantasies of engulfment and drying
up. The sea is then perceived essentially as a mirror on whose
surface one can witness the marriage of water and air. Were it
not for a few divers' accounts, only the skin of the hydra would
be known.[77] Only through imagination could artists describe the
depths of the abyss. This blindness increases the importance of
the strand as the boundary along which the supposed fecundity
of the deep is revealed; this stage swarming with life is rendered
all the more fascinating by the invisible mysteries whose existence
in the unfathomable depths it seems to suggest. The obscene
bottom of the sea, miraculously exposed and temporarily available for the scientist's observation and the artist's vision, only

belatedly managed to capture the attention of both figures at once. This attention was growing at the same time as the vogue for Romanticism; the development across social lines of the contemplation, representation, and exploration of the strand constitutes one of the major phenomena dealt with in this book. The occasion on which seaweed, shells, and zoophytes were gathered was also a social encounter; the elites, eager to discover what had been a void, came to the hardened sand to meet astonished fishermen.

One of Goethe's experiences in Venice on 9 October 1786 reflects this new fascination. The previous day, the author of *Faust* had endeavoured, so he explains, to reconcile the two images of the lagoon, majestic and humble in turn, according to the rhythm of the tide. He adds:

> I come back once again to the sea in my account. There I observed the customs of shellfish, limpets, and crabs today, and I really enjoyed it ... At first, the teeming mass does not know what is happening and keeps hoping that the salty waves will come back; but they remain absent, the sun quickly burns and dries the place out, and then the retreat begins. On that occasion, crabs look for their prey ... I never saw any crab getting what it wanted, although I observed the retreat of this swarming throng for hours. [78]

Here, curiosity and even wonder are caused not so much by the spectacle of the sea or its vastness, but by the line of contact that reveals another animate world. [79]

Actually, such behaviour had become quite commonplace in England by then. As early as the 1760s, multitudes of tourists came to Margate to gather seaweed and shellfish. [80] The custom of collecting shells had already trickled down the social ladder. But it would still be some time – until 1815, to be precise – before scientists actually appeared on the strand in order to observe the living species that dwelled there. This delay is astonishing, as Audouin and Milne-Edwards, the pioneers in this adventure, pointedly observed in 1832. They were responsible for introducing on the Continent the model of the scientific expedition whose purpose was the systematic study of the flora and fauna exposed at low tide. Milne-Edwards speaks of a new joy that owes little to the pleasures of scholarly geologists. [81] With good reason, he stresses the prodigious mysteries of seashore fauna. As his disciple Armand de Quatrefages de Bréau would write in 1844: 'It looks as if Nature creates the most apparently insoluble problems here, so that it can play around with the difficulties, sometimes overcoming them directly, sometimes evading them by taking the most unexpected detours and making the most wonderful combinations.'[82]

Where the rock and the wave meet, there is a proliferation of zoophytes and amphibians. On the sandy surface, the indistinct borders between the kingdoms seem to reflect those between the elements. As painters were discovering the luminous confusion of the beach, the *Revue des deux mondes* was opening its columns generously to the science of the coastlines. To the novel attraction of things concerning the sea was added a new curiosity on the part of the public about the inhabitants of the area laid bare by the ebbing tide.

Scientists, who were no longer interested in compiling catalogue-like knowledge, gave up the study of fish for that of molluscs, annelida, crustaceans, and zoophytes. Suddenly they wanted 'to visit the places where these species dwell, and to observe them in their living state',[83] in order to undertake a more in-depth examination of their structure, to study their functions, and to observe their peculiarities and customs. This meant that they could increase the number of experiments they performed and dissect living animals. While staying with fishermen in the neighbourhood, Audouin and Milne-Edwards would explore the same shore for many a long week, breaking up their stay regularly with short trips.

The coasts where the highest tides occurred, especially those of the bay of Mont-Saint-Michel, were a natural focus for this scientific labour.[84] The Chausey Islands, where the two scholars stayed in July 1828, seemed to be a paradise for zoologists eager to dig in the sand and silt of the strand. Audouin and Milne-Edwards set up fish tanks and large basins on the beach. In the course of short trips to the archipelago, they collected over 500 different species. They overturned big stones, dug in the sand, fished with little drag-nets, and crossed 'vast shores' at low tide in order to swell their collection from reef to reef. On the beach, they broke open the shells of many a lobster. Sometimes, they would take a boat and dredge 'for whole days, depths that had probably never been explored'.[85] Their greatest pleasure, however, was visiting the rocky crevices in search of ascidia, 'for it is with them that the movement of life seems to have started'.

A few years later, Armand de Quatrefages began to combine the study of the swarming primeval or degraded forms to be found on the sandy beaches of Bréhat with ethnological investigation. 'On the very day after my arrival', he wrote,

I went up and down the shore that had been left dry by the tide, with my tin-plate box slung across my shoulder, my pockets full of tubes and bottles, and my wide iron trowel in my hand ... I discovered a few spots where the sand and the silt, mixed in the right proportions, promised ample yields ... Of course in order to pursue (the molluscs) into their retreats, one needs the skill of a real quarrier. I foresaw strenuous work ahead, but I looked forward to my reward: the difficulty did not frighten me.[86]

This twofold scientific quest, recorded in notebooks, notes, drawings, and sketches of all kinds, gives rise to romantic meditation and serves as a springboard for it; it deepens its resonance. Fourteen years before Michelet retired to Saint-Georges-de-Didonne, Armand de Quatrefages delighted in both his joyous quest and the nostalgia that the sights and sounds of the empty shore produced in him.

During the 1830s, as painters, writers, and the general public were showing an increasing interest in the sea-shore, Charles Lyell, the founder of modern geology, was more and more impressed by the erosive power of the sea. He even went so far as to attribute the main features of the valleys' shapes to its action.[87] From gazing at the power of the waves, he came to overestimate the ocean's role in this process. Lyell saw the sea as the symbol of continuous action, as indefatigable proof of the veracity of actualism, which in this case emerged decidedly triumphant.

This should underscore how artificial it is to exclude the scientific quest from the field of aesthetics. For example, historians long ago detected the Romantic character of the emergence of geology. Like Lyell sitting by the sea, scientists facing the waves would contemplate the repetition of the same primeval and eternal patterns. At this point of contact, where the ocean seems to live by human time, geologists could henceforth visualize better than anywhere else the transient nature of the relief, whose perceived precariousness was confirmed by a look at the ephemeral surface of the strand. Otherwise, like antiquarians of the world, less well-informed geologists would scrutinize the traces of sublime events such as floods, earthquakes, or landslides that had disrupted an earth now indifferent to human history.

Despite the multiplicity of theories, geology provided a stock of concepts, images, and affects that breathed new life into the interpretation of human history, as it dissociated this history from that of the earth. Subtle exchanges took place between politics and science. The French Revolution gave a new boost to cata-strophism by promoting the vision of the world's history as a series of dramatic episodes. Conversely, modern geology imposed a sense of ephemerality and an awareness of the precariousness of things. For the French elites during the reign of the Citizen King,[88] haunted by the longing to return to life as it was before the cataclysm, geological observation easily drifted into nostalgic meditation. Telluric ruins seemed to be a metaphor for the social order, shaken and inescapably undermined by the onslaught of time. On the island of Bréhat, Armand de Quatrefages enjoyed contemplating the pounding of the ocean, whose waves would alternately lift up the powerful rock called 'The Peacock' and then let it drop heavily on to its immovable abutments. For the scientist who actually experienced the slow wearing away of the moving reef and could foresee its eventual collapse, the end of the ancient orders was written on the sands.[89]

5

The Freshness of Wonder

THE SUBLIME CLIFFS OF THE SEA'S
GOTHIC BLACKNESS

Any vast object, wrote Saint-Évremond in 1685, is incompat-
ible with beauty, because it inspires horror. Great expanses of
countryside, immense forests, plains that are too wide, and
boundless gardens are awful; in these unpleasant abodes, 'the
eye drifts and loses itself'.[1] The quest for beauty requires that
such excesses be contained. Artists must prevent the mind from
wandering; just as God was able to contain the abyss, so they
must set limits to vision.[2] This makes it easier to understand
the function of the biblical image, which allowed observers to
contemplate the power of the ocean, and only then, finally, to
remember that it was contained.

In France, the criticism of *vastitas* was part and parcel of the
debate between the ancients and the moderns. The moderns were
especially reluctant to evoke the forces of primitive nature; they
denied that there was any beauty in the desert, the forest, or
the sea. In England, in the course of the debates arising over
concepts of beauty and the sublime,[3] the same reluctance was
apparent. In 1713, Thomas Tickell made a similar criticism of
the evocation of the seascape in the pastoral. He deplored the
fact that since Sannazaro,[4] some poets should have preferred
'arid beaches and the boundless ocean,[5] to the green meadows
of shady dales. To the repulsion felt by beauty-lovers at the
monotony of the waves was added a disgust with sand. The
yellow, arid beach beside the boundless ocean, prostrate beneath
the sun and swept by the wind, stood out as the antithesis to

the 'lovely green' which seemed to offer a peaceful refuge. The shore, linear by nature, stood in contrast to the image of the circle and the concentric pattern that suggests the cradle.[6] No harvest can come from the sand; it prevents laughing streams from developing their course. Along the desolate shore, the pleasant song of birds gives way to the sea-gull's harsh cry, which strikes terror into the sensitive soul. Remember that the beach, absent from the Garden of Eden, as it was from any re-presentations of the golden age, was the negation of harmony. Bristling at some points with reefs and uneven rocks, it testified, as did the horrible mountains, to the invasion of diluvian chaos. It represented a yawning chasm previewing the potential return of divine wrath.

Most poets despised the sea and its shores.

> The muse long sought the woods, and mossy caves,
> Despis'd the Seas, and fear'd the rolling waves,

William Diaper observed in 1712.[7] At this time, however, the classical system of appreciation was beginning to shift. Signs of the change are to be found in the accounts of the emotions experienced by Gilbert Burnet and John Dennis as they were crossing the Alps, the former in 1685, the latter in 1688. The two men, inspired by physico-theology, were obviously repelled by the horrible, chaotic scenes of the earth's *pudenda*. They discounted these rifts of nature which it was poetry's specific mission to fill.[8] But in spite of themselves, they already revealed how sensitive they were to the aesthetics of the sublime. They enjoyed feelings of amazement, 'exquisite horror', or 'terrible joy'. 'At the very time as I was experiencing infinite pleasure, I was shivering,'[9] confessed John Dennis. In spite of themselves, the travellers discovered the power of immensity. Their emotion was in keeping with that of authors of accounts of cosmic voyages, then in fashion; in their own way, these authors took delight in the splendour of infinite space. But there should be no mistake: in Burnet's mind, as in Dennis's, the 'unheard-of transports', a mixture of amazement, horror, joy, and desire, must contribute to the exaltation of divine greatness.[10]

The emotion generated by the sight of the sea, as described in *On the Sublime* by pseudo-Longinus and translated in 1674 by Boileau, was to free itself only slowly from the patterns imposed by natural theology. In this respect, it is worth pausing to consider a foundational text written by Addison in 1712 in the

Spectator.[11] This periodical then had a circulation of three or four thousand. Its public, among which Whigs outnumbered Tories and Londoners outnumbered provincial readers, was a heterogeneous one. The *Spectator* acted as a cultural unifying element; 'socialites and merchants, people from Westminster and from the City, regular visitors of salons and cafés',[12] not to mention a great number of women, all joined together in reading the periodical. Nevertheless, *Spectator* readers comprised a public that was limited by the cultural requirements for understanding a text intended to 'give value to the image of a civilization based on cultured leisure'.

> Of all objects that I have ever seen, there is none which affects my Imagination so much as the Sea or Ocean. I cannot see the heavings of this prodigious Bulk of Waters, even in a Calm, without a very pleasing astonishment; but when it is worked up in a Tempest, so that the Horizon on every side is nothing but foaming billows and floating mountains, it is impossible to describe the agreeable horror that rises from such a prospect. A troubled ocean, to a man who sails upon it, is, I think, the biggest object that he can see in motion, and consequently gives his Imagination one of the highest kinds of Pleasure that can arise from greatness. I must confess, it is impossible for me to survey this world of fluid Matter, without thinking on the Hand that first poured it out, and made a proper Channel for its Reception. Such an object naturally raises in my Thoughts the Idea of an Almighty Being, and convinces me of his Existence as much as metaphysical Demonstration. The Imagination prompts the Understanding, and, by the greatness of the Sensible Object, produces in it the Idea of a Being who is neither circumscribed by Time nor Space.
>
> As I have made several Voyages upon the Sea, I have often been tossed by storms, and on that occasion have frequently reflected on the Descriptions of them in ancient Poets. I remember *Longinus* highly recommends one in Homer, because the Poet has not amused himself with little Fancies upon the occasion, as Authors of an inferior Genius, whom he mentions, had done, but because he has gathered together those circumstances which are the most apt to terrify the Imagination, and which really happen in the raging of a Tempest. It is for the same reason that I prefer the following Description of a Ship in a Storm, which the Psalmist has made, before any other I have ever met with.

Like Shaftesbury and John Dennis, Addison primarily discerns the traces of God in immensity. The ocean sets forth the sensory proof of the existence and omnipotence of a single God. This explains the author's implicit contempt[13] for the Virgilian vision of the sea and his admiration for the Psalmist's text. The sense of the sublime here arises from actual experience; it is the 'agreeable amazement' produced by the waters' calm and the 'exquisite horror' of the storm, both experienced in the course of long crossings, and not the beauty of a literary work, that justify the hymn to the ocean. Admiration arises in the process of interpreting the object; sensation and the workings of the mind become the main focus. Here can be discerned the 'effect/affect of the sublime', the 'stunning stupefaction', and the bewilderment caused by the monstrous Medusa or her equivalent, the storm: as it breaks loose with a sudden din, those caught up in it look anxiously for its cause. This is an indescribable emotion, for those willing to consider the sublime as 'the unrepresentable aspect of representation'.[14]

The text implies the primacy of vision, and gives decisive importance to a spectator's attitude.[15] Unlike John Dennis, for whom enthusiasm derives from meditation, Addison places the emphasis on 'primary pleasures' that introduce those of the imagination. He gives priority to the contemplation of the present object; this reversal in the hierarchy of emotions leads him to praise the work of seascape-painters and to express the wish that a form of descriptive poetry might emerge which would also sing the praises of the sublime sight of the ocean.

In 1726[16] the cultured public discovered the three tempests in Thomson's *Winter*. The author was a Scotsman who claimed, in his youth, often to have walked along the Scottish shores. This explains the poem's northern accent, which is important for our subject. Admittedly, the significance of this publication should not be overestimated. Between 1712 and 1726, as will be seen later, a delight in the sea-shore insinuated its way into English literature.[17] It was Thomson, however, who first managed to modify the classical artistic model of the storm; his account departs from the Virgilian code. Moreover, the author of *The Seasons*, whose influence was to be felt throughout Europe for nearly a century, was able to make his readers see the surprising landscape of the natural coastline.

This calls for a digression. In the England of the 1688 Glorious Revolution, the conquest of the oceans comes to the fore. The lack of sensitivity to the spectacle of the ocean seems to be contradicted by the rising prestige of navigation. Whig ethics imposed the image of Nature as something to be conquered; man must compel her, with all his might, to surrender her treasures.

In taking this direction, human involvement encouraged observers to exalt the vision of the harbour, the activity prevailing on board ship, and the sailors' victory; it suggested the dream of a golden age expanding to include the whole world; and it offered an invitation to praise the ocean for bringing together the continents and the men inhabiting them. Especially after 1720, the fashion for accounts of distant travels developed within this perspective. But 'everything in a landscape that invites conquest, acquisition, appropriation, predation, and exploitation, prevents us from perceiving it aesthetically'.[18] Robinson Crusoe does not admire the splendour of his island, and he remains insensitive to the beauty of its shores. In order to survive, he must become part of the picture. The urgency of taking practical measures prevents him from obtaining the distance necessary to appreciate the spectacle of nature. For Robinson, storms are but a succession of perilous events, and they leave no room for aesthetic enjoyment.

But this Promethean battle breathes to a new rhythm. It harks back to times of retreat and seclusion; it creates pauses for rest intended to allow contemplation of Nature's primitive forces. Paradoxically, it invites the observer to delight in any setting that demonstrates that Nature has sufficient force to resist the pressure of civilization. From this desire for a compensatory tempo, the sublimeness of the ocean is born. Remember that human activities leave no trace upon the sea. As a barren landscape that mankind can neither arrange nor endow with moral significance, the immensity of the waters is the antithesis of the garden.[19] It is in this light that Thomson's text should be read.

Of course this work is still marked by the image of the Flood; it reveals the influence of the Psalmist and of Milton. But a *dizzying sense of boundlessness* is definitely what animates the poem. Thomson appreciates wide rivers and the luxuriance of tropical vegetation. Terrifying meteors sweep through the landscape in *Winter*. The reader's attention is focused on the unleashing of elemental forces, and the landscape has become dynamic, stripped of all human presence. Winds, flashing lightning, and waves join forces. As these hostile forces break loose, mankind feels threatened; and he forgets the Creator. The spectacle of unrestrained Nature gives rise to delight:

Congenial horrors, hail!

The shift is disorienting. The tempest no longer appears as the manifestation of God's wrath; it becomes the impenetrable movement of the unknown. Nature's indifference culminates in the evocation of the Arctic landscape. The description of ice-bound seas, echoing with sinister creaking, and of the desolate shores of the far North soon became attuned to the growing craze for accounts of travels to Lapland. This reinforced the image of a ruthless Nature. The ocean unleashed brought an end to the complicity between man and an Earth governed by God's hand. As a result, the spectator's position was drastically altered. Until then, the pious observer had one eye fixed upon the limit, the reassuring divine boundary. Henceforth, his glance would be cast upon the horizon as he endeavoured to size up the incommensurable. The institution of such a model of appreciation was a historic event: the changes had begun that would lead to the Romantic model of contemplation.

Already, a close link was developing between state of mind and landscape; a new harmony was established between the cataclysms in *Winter* and the poet's melancholy. Faced with the immensity of an ocean whose limits man is incapable of grasping, whose breadth he cannot conceive, the spectator experiences the emotion engendered by the sublime spectacle of Nature.

Henceforth, the aesthetics of the sublime began to spread, and very quickly to become commonplace. In 1731, David Mallet praised 'the terrible solitudes of the shores';[20] already in 1726, the walker presented by Aaron Hill in *The Happy Man* stated that he preferred the boundless horizons visible from the sea-shore to rustic landscapes.[21] At the same time, the prestige of Salvator Rosa was growing among travellers making the grand tour as the work of this Neapolitan artist was reinterpreted from the point of view of the sublime. After 1762, Ossianism strengthened his aura even further. This painter of the sea-shore found himself transformed, not without exaggeration,[22] into a friend of brigands, if not a bandit himself, and a freedom-lover. The vogue for his paintings and the moral echo detected in them helped to extend a taste for rocky shores bathed in chiaroscuro and bordered with stormy mountains and wild chasms. Subterranean passages carved out of the hillsides and the presumed cruelty of bandits gathering on deserted beaches heightened the viewer's feeling of terror. Piers stretching into the water and lighthouses that spoke of shipwrecks both fitted into this pathos of the sea-shore.[23]

All that remained was to codify an emotional strategy that had been slowly developing since Burnet's voyage. Edmund Burke applied himself to the task in 1757, as the crowds were beginning to rush to Brighton in order to encounter the violence of the waves. That year, he set out the moving forces of the sublime, and established the aesthetic value of the emotion produced by contemplating the infinite waters. Burke turned Saint-Évremond's maxims inside out: 'Great size is a powerful cause of the sublime.'[24] The sight of the incommensurable forces man to experience his finite nature; it arouses passions in his soul in which the aesthetics of the sublime take root.

> A level plain of a vast extent on land, is certainly no mean idea; the prospect of such a plain may be as extensive as a prospect of the ocean: but can it ever fill the mind with anything so great as the ocean itself?[25]

Towards the end of the eighteenth century, at the time when Cook and Bougainville were making their travels, the ocean would call forth the image of a 'vast expanse' that was indifferent to human time, like the desert; a place of sublime *vacuity*[26] whose imagined depth was modelled on the very perpendicular sides of the mountains that often bordered it.

Yet it is not only the immensity and negativity of the ocean, what Burke calls a 'privation'[27], that make it sublime, but also its power and its energy.[28] Though the observer on the shore is a mere spectator, the storm inspires terror, especially when unleashed in the dark of night. Burke scrupulously details the passions aroused by the sight of this vast, terrifying, dark event. First comes astonishment, mental panic, and the momentary amazement of a soul so overwhelmed by the dreadful object that it becomes incapable of reasoning;[29] it is simultaneously an irresistible force that suspends the viewer's actions and a sudden apparition that fills him with longing to live in the present. This suddenness breaks the monotonous chain of habit, and creates a break in time that sweeps away visual boredom, one of the major fears of that time. Right down into the observer's body, the emotion arising from the sublime scene causes the experience of the continuum of natural phenomena to coincide with life rhythms. This emotion is like active matter making its way into the psyche, and it runs counter to reverie.[30]

The emotion required to enjoy the sublime spectacle prevents any rest, especially for the eyes. The infinite space of the sea is tiring for vision. In order to gaze at a vast object, 'the eye must traverse the vast space of such bodies with great quickness; and consequently the fine nerves and muscles destined to the motion

of that part must be very much strained; and their great sensibility must make them highly affected by this straining'.[31] The
eye that contemplates an extensive, uniform object 'does not
readily arrive at its bounds; it has no rest whilst it contemplates
them'.[32] This labour has, however, a therapeutic value. Those
who are subject to melancholy are advised to indulge not only
in fresh air and exercise but also in the emotions procured by
sublime sights. They are supposed to train 'the coarse muscular
parts of the constitution',[33] but also to 'shake' and 'exercise' the
most delicate ones. Aesthetics, ethics, and therapeutics, here
artificially distinguished, form a consistent chain. 'Travel', 'exercise', and 'delight' arising from the perception of a sublime
object are merely different tactics in the same strategy.

This same consistency explains the rise in the appeal of the
seaside. The shore easily provides the 'sort of tranquillity tinged
with terror'[34] that defines the aesthetics of the sublime. Walking
along the edge of the abyss on the solid rock of the cliff, facing
the rush of the wind and a feeling of dizziness (without real
danger), venturing into dark sea caves under a cicerone's guidance,
climbing up and then rushing down the slopes of embankments,
succumbing to the impact of the waves on a soft, smooth, sandy
beach, or running the imaginary risk of being ambushed by
bandits and pirates, all give rise to emotions in the soul that
create sublime moments. Unlike the cool valley which had once
been the image of the golden age, the sea-shore allows an accumulation of simultaneous emotions. The infinite view, the emptiness of the horizon, and the proximity of the abyss all emphasize
the horror of the void, and lay the foundations for fantasies of
being swallowed up. This emotion is further heightened by the
contemplation of traces of earlier cataclysms or the pictorial
images haunted by the pathos of the sea that are rooted in memory.
Thus the taste for vast shores gradually asserts itself, a taste for
straight beaches from which the eye can leap to the ocean's
solitude without being disturbed by the curve of the shore.

Seascape painting had been rehashing the theme of the tempest
for quite some time. All told, Claude Lorrain's calm sea-shores
were actually an exception. In Rome at the beginning of the
seventeenth century, his master, Agostino Tassi, or the Genoese
Borzone enjoyed painting the fury of the sea. In order to satisfy
their clientele's taste, a series of Italianizing Dutch painters, such
as Bonaventure and John Peters and especially Pieter Mulier,
known as 'La Tempesta', turned out a great number of paintings
depicting storms, squalls, and shipwrecks, and their work was

much appreciated by Italian art lovers. The fashion quickly spread to Provence and the Rhône Valley. When Anthime-Denis Cohon, Bishop of Nîmes, died in 1670, he had several storm scenes in his collection.[35] It is through such paintings that the young Joseph Vernet discovered the fury of the sea. In Rome, Adrien Manglard taught him the already stereotyped technique of this stormy style.[36]

Travellers on the grand tour discovered with pleasure the pathos of the sea. Later, both French and English clients would commission storm scenes from Vernet. In short, although the accessories that adorn these paintings are taken from the Tyrrhenian coast, and the open sea, in the Dutch manner, was rarely depicted – at most a harbour, an estuary, or a bay – the longing for the sight of painted tempests both indicated and stimulated this shift in sensibilities.

Seascape painting was what introduced many amateurs, especially in France, to sublime scenes of the sea. In those circles, images of the coast remained largely independent of the contemplation of nature. Reading, conversations, travel accounts, and private collections provided background knowledge which, for many, would remain their sole experience of the sea-shore. Beginning in 1740, visitors came to the French Salon to shudder as they gazed at Vernet's paintings, and later, those of Loutherbourg. In this way, a means of appreciating the sea was constructed that was detached from any experience of nature. Otherwise, in the tradition of classical aesthetics, the view of the coast served merely to verify the congruence with pictorial images. Diderot did not discover the sea until 1773, during his journey to Holland, after commenting on it for many years at the salons. Marmontel, who was moved by the beauty of painted seascapes, admitted in 1760 that he was disappointed when he saw the real thing.[37]

At the beginning of the 1760s, the arrival of Ossianic literature heightened the fascination exerted by the North. Already, Pelloutier's *Histoire des Celtes* [History of the Celts] (1741), and even more so the magic of the *Edda* as revealed by Mallet in 1755–6 in his *Antiquités Scandinaves* [Scandinavian Antiquities],[38] had popularized the mythologies of northern peoples. MacPherson's text and the abundant polemical literature to which it gave rise compelled the West to adopt a new system of appreciation of landscape. The Caledonian shore was radically different from the Arcadian scene. In the West, images of the sea were completely renewed.

This revolution in people's imaginations was occurring in Great Britain just as crowds of bathers were rushing to the Sussex coast. It did not reveal its full impact on the Continent until the beginning of the nineteenth century, before declining belatedly during the last decade of the reign of Louis-Philippe. It is important to remember that this time lag also applies to the spread of the practice of taking seaside holidays.

Ossianism contributed strongly to the rise of the appeal of the seaside, and its influence operated in favour of northern beaches.[39] At the heart of a majestic and gloomy hyperborean nature, the aerial point of view, which generated an affective sense of space, proved to be particularly striking. The abundant fog and a low-hanging overcast sky fostered reverie.[40] Along the shores, at the edge of the infinite, solitary heath haunted by sea-birds with their desolate cries, the elements mixed together in a state of primitive confusion. The mood of the sea had an obsessive influence on the landscape extolled by northern bards. The Ossianic tempest, which radically renewed the classical palette, was to be viewed from the shore. It offered the scene of the confrontation between the elements. Dark or green waves fringed with foam launched an assault on the dark rocks of Morven. Though their drive was fragmented by a string of islands and shoals, the northern waves inspired a particularly intense terror; near the Poles, Nature had preserved her primitive force. The violent waves came and beat against the shore fringed with resounding reefs and basalt grottoes in which squalls were swallowed up. In the maze of islands of *The Tempest*, the music of the hurricane and the whistling of the wind through the mossy arches of the ruins overlooking the headlands were attuned to the rhythms of the bard's harp.

This nature, with its cold, fruitful waters, offers the majestic backdrop against which the hero's tale is told and a drama filled with nostalgia and tears unfolds. This virgin coastline was magical because it was, paradoxically, fertile and desolate, and it offered a new means of experiencing the link between the past and the present. Here, the delight in bygone eras could be found only through their evocation or resurrection.[41] Unlike the coast of antiquity, still visible on the Mediterranean shores, the Ossianic landscape did not lend itself to pilgrimages. There was no clear itinerary to follow on the traces of a culture; nor was it easy to gauge the gap between the real present and a vanished past. Despite its few unavoidable sites, the Caledonian coast was resistant to archaeological ambitions; it was more dependent on the magic of the world's fundamental elements.

In 1768, five years after MacPherson's poem was published, Beattie's *Minstrel*, of which Chateaubriand translated the first canto, proposed a new model for the solitary walk along windswept shores. Before young Edwin's eyes, the storm is unleashed, followed by the calming of the waters at noon, with primitive scansion.

> Thence musing onward to the sounding shore,
> The lone enthusiast oft would take his way,
> List'ning with pleasing dread to the deep roar
> Of the wide-weltering waves ...

The terror and beauty of the spectacle merge in his soul. Responding 'To the sprightly pipe',[42] Edwin listens to the winds and eagerly contemplates

> ... Whate'er of beautiful, or new,
> Sublime, or dreadful, in earth, sea, or sky.[43]

The new models of appreciation of the coastal landscape determine the shape of the Caledonian voyage, whose fashionable period touches the chronological limits of this book. As the leisured classes were moving away from Bath towards the luminous beaches of Sussex, the solitary shores of Scotland and its islands, especially the Hebrides, were attracting the migratory attention of travellers obsessed with Ossian and the longing for the North. The defeat at the hands of the Americans encouraged a fashion introduced by Dr Johnson, Boswell, and especially Thomas Pennant.[44] The war had highlighted the need to pay attention to the northern regions of the British Isles; in addition to providing an invaluable route for ships wishing to escape from the enemy and a rich pool of sailors to draw on,[45] they now emerged as areas full of economic potential and poetry. Nearly forty years after Culloden (1746), the time had come to value and develop these shores that seemed destined to play an increasingly important role. Between 1792 and 1815, grand tour travellers forced to abandon Continental itineraries took advantage of this vogue which became part of the discovery of their home country.[46]

In Restoration France, émigrés belatedly introduced the wish to undertake the journey to Scotland. Hosts of Continentals, steeped in Ossianism and fascinated by Walter Scott, travelled through the heart of this fashionable region whose modernity they discovered with surprise.[47] It is important to stress this point: the northern journey, whether made in Scandinavia, Denmark, or the British Isles, soon acquired the reputation of a journey through well-governed countries whose prosperity was increasing. In this domain, too, it developed into the antithesis of the journey to Rome. La Tocnaye's peregrinations in Norway and Scotland and Leopold von Buch's visit to the fjords between 1806 and 1808, as well as the trip made later by young Adolphe Blanqui, reveal the complexity of a single emotional quest combining nostalgia for northern mythologies with a fascination for modernity.

The Caledonian voyage had its own ritual. The boat trip into the labyrinth of the black rocks of Morven, Iona, or Staffa was thought of as an initiatory rite. It was the traveller's duty to encounter the storm that interrupted the bargemen's songs and gaiety. The excesses of aquatic nature simply helped him to become more attuned to the beauty of the admirable Gothic character of the sea. This coastal scene was in fact marked by the prevailing quest for mineral curiosities, itemized as early as 1716 by Martin Martin.[48] The colour and shape of the rocks, the black blocks, and the colonnades and arches visible from the sea attracted travellers' attention before they set out to face the perils of the visit.

What stands out in accounts of the voyage is the description of the rocky promontory, which would become the scene of Romantic confrontation. The cliff offered the pleasure of imagining a great fall. 'If one's foot happened to slip,' mused the earnest Dr Johnson.[49] A traveller, especially if he was a Tory, would not fail at that point to recall the vertiginous Dover cliffs, from the top of which King Lear unsuccessfully attempted to throw himself down. As he left the English shores, Nodier summarized his emotions in geological terms: 'I am losing sight of this land on whose two extremities genius has imprinted two equally imposing seals: Ossian's admirable poetry about the black rocks of Morven, and Shakespeare's admirable poetry about the white cliffs of Dover.'[50]

In the book he devoted to the northern part of the planet, Thomas Pennant, one of the originators of this new sensibility, describes the great Flamborough promontory, one of the most famous natural monuments of sea-shore architecture. He offers an emotional palette to the novice traveller that relies on a series of sensations detailed by Burke and is in keeping with the fascination with underground labyrinths of the Gothic novel.

Pennant paints a picture of the traveller who discovers the majestic site:

> The cliffs are of a tremendous height, and amazing grandeur; beneath are several vast caverns, some closed at the end, others pervious, formed with a natural arch, giving a romantic passage to the boat, different from that we entered. In some places, the rocks are insulated, are of a pyramidal figure, and soar up to a vast height; the bases of most are solid, but in some pierced through, and arched; the colour of all these rocks is white, from the dung of the innumerable flocks of migratory birds, which quite cover the face of them, filling every little projection, every hole that will give them leave to rest; multitudes were swimming about, others swarmed in the air, and almost stunned them with the variety of their croaks and screams.[51]

Fingal's cave on the isle of Staffa remains, however, the masterpiece of this Gothic aspect of the sea. It offers a reflection in depth of the ordinary unsettled character of the rock and the castle; it adds wonder to the fear aroused by the smuggler's cave or by the mysterious tower haunted by shadowy figures whose reputed ferocity corresponds to the cruelty of the sea. Along the northern coasts, the Duke of Argyll's castle, standing over Inverary since 1740, and the inevitable visit to Iona's ruins strengthen the impact of the Gothic revival on vision.

For a long time, travellers visiting Staffa claimed to be discoverers: Joseph Banks, the Scandinavian bishop Uno Troil,[52] Thomas Pennant, and later Johnson and Boswell successively played at being pioneers, as did Faujas de Saint-Fond thirty years later. In contrast, by the beginning of the nineteenth century, the emotional itinerary of the sea's cathedral was definitively codified.[53] The intertwining approaches to the island and the cave made them difficult of access, and allowed entry only through small apertures, which incited womb-like reverie. As they were carried by the surge of the sea into the cave without being able to control the motion of the boat, travellers liked to fantasize about the risks of being swallowed up.

Inside, a cascade of emotions awaited them. Tourists were overwhelmed by the width of the cave and the height of the vault, barely suggested from without. Darkness, accentuated by the colour of the rock, compelled visitors to focus on their auditory sensations. The surge of the sea, like a 'water-dwelling ram', produced harmonious sounds in the basalt columns at the bottom of the cave.[54] As the eye began to discern its surroundings, it would slide admiringly along the columns that gave the architecture its rhythm before coming to rest on the horizontal line of the ocean that stood out against the brightness of the Gothic arch. The pleasure of surprise would blend into religious emotion. As they recovered from their amazement, some tourists endeavoured to reach the furthest point,[55] venturing down to the bottom of the dark cave hollowed out of the island which was often the last stage of their journey. Others satisfied their obsession for taking measurements.

The Hebrides lent themselves admirably to the emotional strategy of the sublime. There, Nature, terrifying and majestic, was decked with the vestiges of cruel times. The islands were inhabited by men who were, at first sight, alarming. But travellers soon realized that this was only a semblance of reality. During the travelling season, the string of reefs broke the force of the storms, and by then, the hospitable inhabitants had clearly given up their savage behaviour; the bloody strife among clans had ceased, and the caves, which had long since been explored, had lost their initial mystery.

By the 1820s, the days of solitary travel or adventurous trips in small parties were over. In Scotland, the practice of the 'falsification of the art of travelling' was already reaching the end of its decline.[56] The traveller who was also discoverer, inventor, and initiator, who mapped out and related his progress as he wrote about it, was vanishing. The first-generation tourist was proliferating, who 'degrades by repetition – that is to say, by convention – adventure into an excursion, the archetype into a stereotype, the model into a mass experience, production into consumption, and the foundation account into endless anecdotes'. As the mere semblance of a traveller, sheltered from the unexpected, he served as a prelude to the mass adoption of these practices, which represented the end-point of the process.[57]

This precocious *touristization* of the Caledonian journey, whose aesthetic imperative took shape clearly as early as 1804 in Cruttwell's guidebook,[58] warrants further examination. Edouard de Montulé and Adolphe Blanqui,[59] the latter obsessed by Girodet's virgins, undertook the ritual voyage to Staffa in caravans. Ducos demonstrates the mechanisms by which such experiences became standardized.[60] In 1826 the excursion was organized at regular times. The long trip already resembled Monsieur Perrichon's journey.[61] After the ritual evocation of the *manes* of Fingal and Ossian, the group set out, full of expectations. 'Staffa and Iona, what is known about them and what everyone expects from them are the things uppermost in everybody's mind, absorbing all faculties, firing all conversations. Imagination lends this journey all the illusions of its enchantment.[62] The silhouette of the 'melodious island' raises joyful cries from the group. Then the epic begins: three row-boats take the travellers to the shore among the basalt columns. 'Everyone rushes to this surface and from one column to another.' Then the caravan stops in order to admire 'the magnificent view'. Very soon, the exploration resumes:

> The slope has become steep and arduous. The most daring come, go, climb up, and come back down alone, like scouts. Many help one another and gather together at intervals. The gaze, gestures, and exclamations of some express the surprise they experience at such a strange new sight. Others uproot marine plants from the crevices in which they are growing or gather shells, pebbles, and various fragments. My travelling companion has taken off his boots because their thick soles prevent him from getting a solid footing. We just heard a cry of fear: some clumsy fellow was about to fall down but was held back. Bursts of laughter followed this brief alarm. What a lot of lively, picturesque scenes![63]

The caravan itself became part of the sublime scene, until it sank into an atmosphere of 'meditation' as the group approached the 'sanctuary', the 'temple of nature'.[64] The freshness of wonder, physical involvement, and exaltation were characteristic of the dawn of group tourism along the ocean's coast.

On the Continent, the fashion for travel to Brittany,[65] long a nascent one, added very little to this Caledonian model. The voyages of Count de Guibert and especially of Cambry are, however, of some interest. The former, an inspector of institutions for invalids, arrived at Brest in 1778; Chateaubriand was barely nine years old. As he neared the town, Guibert observed, 'My eyes searched and scoured the region to discover the harbour. The sight of the sea always has an effect on me: it uplifts my mind, saddens it, and finally, fills it; but never with pleasant feelings. The outcome is always that I fall into vagueness, darkness, a sense of infiniteness. It is like the sight of the sky and the thought of eternity.'[66]

This premonitory range[67] of emotions is of secondary interest compared to that aroused by the emotional mapping of the sea-shore undertaken by Jacques Cambry in 1794 and 1795, when he was entrusted by the *Directoire* with the task of inventorying the monuments that had escaped vandalism. His voluminous report is also an invitation to wander along the Armorican coasts, whose absence from literature he stresses: 'Inhabitants of Paris, you whom I have met as you sought great scenes and pleasures in the Apennines and the Alps, flee these much frequented countries: *come and wander on our shores* that no modern author has described and no poet exalted. I promise you will find grand scenes and new sensations.'[68] More than a third of a century before young Michelet, Cambry became carried away on the island of Batz as he encountered the infinite waters.

> What an immense spectacle ... These waves that unfurl and break into foam on the sea's extended shores ... these capes, these headlands; the dull, majestic sound of the waves, the sky across which the sea-gulls fly, the thunder that resounds in the deep sea-serpent's cave; nature's silence between the ebb and the flow, the strange cries of so many sea-birds. I cannot describe the feeling, the excitement that hinges on the majesty of this scene, the recollections it has aroused, and the boundless expanse it reveals.[69]

At the foot of the abbey on Saint-Mathieu Point, he dreams in the manner of Ossian's disciples and gives way to exhortation. 'There, on this wild rock, when the sun sets in the west, when the sea rises and roars, heralding a storm: come, sublime minds, profound philosophers, strong souls, melancholy hearts, impassioned poets, and meditate in silence.'[70] In the course of his journey, he refines his emotional tactics: 'I waited for a storm to come before going to Penmarch'; and his reader's reward is a sublime piece that expires with a secularized evocation of the limits set to the sea:

> These distinct, black rocks extend to the edges of the horizon; thick clouds of spray swirl up, and sky and sea merge. Through a dark mist, all you can see are enormous globes of foam. They rise up, break, and leap into the air with a dreadful sound; you think you can feel the earth quaking. You flee instinctively; all your faculties are seized by a stunned feeling, fear, and inexplicable agitation. The rising waves threaten to swallow everything up; you feel safe only when you see them glide across the shore and expire at your feet, subdued by the laws of nature and invincible necessity.[71]

And his journey continues with a grandiloquent hymn to the beauty of the point at Raz, magnificent at sunset, when Ouessant, the bay of Audierne, Penmarch Point, and the 'vast sea, made rough by the evening winds, form a boundless sight that is linked only with the sky, the universe, and eternity'.[72]

As Pennant had done in Scotland and Mallet in Scandinavia, Cambry in the course of his journey outlined a virtually definitive itinerary of sublime sites on the Breton coast. All that remained for the guides who would subsequently endeavour to found the imperative aesthetics of these natural attractions was to comment on his work. [73]

A BELATED AWARENESS OF MARINE PICTURESQUENESS

During the first quarter of the eighteenth century, even before the publication of Thomson's *Seasons*, a loco-descriptive poetry began to appear in England, intended for a public familiar with Virgil as well as the Bible and the Book of Common Prayer. [74] Guided by didactic ambitions and inspired by the model of the *Georgics*, the poets whose work revealed the emergence of this trend were also influenced by physico-theology. Actually, it was not contemplation in the bosom of nature that was new, but the ways of interpreting the landscape. The emergence of the topographical poem coincided with the sensualist theory of imagination. By focusing attention on the spectacle of nature, they renewed the taxonomy of emotions aroused by the rural retreat, whose social base was spreading. There was an increasing production of estate poems, journey poems, and above all hill poems, all of which extolled simultaneously the new mechanisms of visual pleasure and a physical awareness developing from the practice of rural sports.

The hill, often dubbed a mountain, became a favourite scene for contemplation. It gradually took the place of the cool dales found in classical pastorals, and made readers forget the melancholy solitude of the woods. Praising this site gave rise to a new pleasure in writing. Poets would speak of climbing up a hill in order to reach the panorama that gave focus to their walks, stimulated their endeavour, and allowed them well-earned rest. There was a growing desire to enjoy panoramic views and to capture the landscape thus discovered in a descriptive tableau.

The debate continued for some time about the sources of this fabulous broadening of vision. Already in the fourteenth century[75] Petrarch enjoyed going up the slopes of the Ventoux to contemplate the panorama that unfolded at his feet. The Church and the mighty of the Renaissance world appreciated the dominant positions that were actually so many rallying points, symbolizing their ascendancy over the world. The Italian *vedute* had learned to take a comprehensive view of their cities, and for ages tourists had rushed to take in the Bay of Naples from the terraces overlooking the city. As he approaches the earth, Milton's Satan hovers in the cosmos, discovering the panoramic view it offers. Whatever its roots, the 'prospect view' offered a pleasure, combined with walking and the ideal day, that gave rise to a new way of seeing. Taking in the panorama with a sweeping glance, evaluating its variety, recognizing the hand of God in the diversity of space, letting the eye slide from the distant horizon down to the foreground in a sort of travelling shot, and learning to expand the depth of field of one's vision were so many new delights for enthusiasts.

This collective attraction for views confined within the limits of a picture, this 'picturesque revolution', took place in the heart of the countryside in southern and central England it applied originally to limited prospects, set in perspective; the new sensibility led to a celebration of large estates, their meadows, and their forests; it provided fuel for the verdant reverie that was to give the English landscape garden its full splendour. This revolution was unaware of the ocean and its shores: immensity cannot be confined within a picture, and descriptive poetry was repelled by the monotony of the sea. In 1785, Jacques Delille, though fond of walking along the Channel coast, still suffered from the weariness – indeed, the disgust – which he claimed was aroused in travellers when they contemplated the waters' immensity.[76]

Gaston Bachelard discerns other motives for this repulsion.[77] Unlike water for everyday use, which comes down from the sky and can be heard singing in the spring or the tap and which refreshes with its tumbling sound alone, sea water is inhuman and unsightly, burdened with salt that cuts short any reverie: it is unable to quench our thirst. The rest of nature cannot be reflected in the waters of the ocean as it can in those of a lake or a river.

It is important, however, not to overemphasize this combination of blindness and revulsion. As was seen earlier, between 1700 and 1726 descriptions of the sea-shore began to make their way. While tourists rushed to Italy to admire the scenes depicted by Claude Lorrain, the coast, reminiscent of the vestiges of the Flood, was still banned from the description of the Garden of Eden, but began to find its way into panoramas.

Among the signs of this incipient event, the 1715 poem by Alexander Pennecuick the Younger, *Morning Walk to Arthur's Seat*, foreshadows the fad for panoramic views of Edinburgh, the new Naples destined to become the Mecca of the West in the collective quest for the beauties of nature. In the distance,

> The Sea's smooth face calm as the cradled Infant,
> When lull'd asleep with Nurses' dreary Songs.
> Ships in Leith's happy Bosom spread their Sails;
> Which sported with the Winds of gaudie Pomp.
> Morning Tide had early rouz'd the Sailers:
> With labour'd Stroaks they reach'd the Southern Shore.[78]

Four years earlier, at the other end of Great Britain, Henry Needler sent one of his friends a poem about the sea, composed in Portsmouth; it was one of the first portraits of the picturesque charm of the Channel landscape.

> A smooth unwrinkled plain accosts the Eye
> Which seems to meet and reach the bending sky;
> One Even, Uniform, Unvari'd Scene
> On ev'ry Side extends its wat'ry Green,
> A spacious Field, which leaves the Sight behind
> By Nature to a nearer Bound confin'd.[79]

Having traced this genealogy of the delight in the picturesque quality of the coast, it also seems necessary to consider John Gay's walks along the Scottish shores. In 1730, he demonstrated an entirely new visual precision, this time without any symbolic purpose, as he described the sun setting over the sea, then the sparkling reflection of the moon and the stars on the moving waves.[80]

There is another kind of writing about the sea, based completely on imagination this time: it is no longer an exaltation of the picturesqueness of the shores,[81] but rather a transfer of the Arcadian code of delight to the imaginary world of the underwater depths. This is what William Diaper undertakes in his *Nereides*.[82] The author dreams of a golden age buried beneath the waves, and he unfolds an underwater Arcadian panorama before his readers. This slow-motion space is lit by soft light that filters through the green waters. It is a mirage of plains and marine vegetation scattered with coral groves, and storms can disturb neither its serenity nor its crystalline limpidity. It is a glaucous refuge for original innocence.

In 1754, at the dawn of the cultural revolution that was taking place in the West,[83] Fielding, who was gravely ill, could not suppress the emotion he felt at the beauty of the coastal landscapes he could sight from a steep shore or the deck of a moving boat. The picture in motion that he painted marked the beginning of a sensibility to the picturesqueness of the sea which was fanned by the yachting fashion and the growth of regattas. Fielding hoped that another Alexander Pope would appear who could exalt the beauty of the marine prospect at Ryde, but he admitted that he was even more sensitive to the beauty of the open sea: 'For my own part, I confess myself so entirely fond of a sea prospect, that I think nothing on the land can equal it.'[84]

In the same year, and again in 1764, Thomas Gray made several trips along the Spithead coasts and the bay of Southampton which offered a precocious illustration of the practice of picturesque travel. This lover of mountains had praised the charm of the river Wye before Gilpin, and he now introduced his readers to what was picturesque in the sea-shore. He appreciated their profile, he observed the numerous colour contrasts, the lustre of the sails, and the majesty of the cliffs of Dover. In a word, 'Gray likes the marine landscape to the extent that it offers him the picture he expected.'[85] The intertwining of the countryside and the sea is what produces the variety of the picture and gives pleasure to the eye; Gray remained insensitive to the vision of naked immensity or to the dizziness of the cliffs.[86]

These scattered images must not, however, overshadow the repulsion expressed in descriptive literature, which remained reticent as regards the vision of the shore for some time.[87] This would no longer be the case at the end of the century, when the hour for the aesthetics of picturesqueness, codified by William Gilpin, finally arrived. From his childhood on, the future vicar of Boldre enjoyed the sight of the Sussex coasts. His father, who was an artist-topographer himself, influenced by Alexander Cozens and probably by Paul Sandby,[88] was captivated by the spell of the cliff scenery. In 1767, Gilpin went to visit the Isle of Wight, which was destined to become the most highly reputed setting for the emotions procured by the picturesque seaside.

At this stage, it is necessary to review briefly the fundamental characteristics of the picturesque voyage for which Gilpin refined the model.[89] The Vicar was not a grand tour man; over the years, he developed an approach that was well adapted to the retreat within national boundaries made inevitable by the war. The picturesque voyage, as understood by Gilpin, lasts at most four weeks. Preferably it takes place in summer. It satisfies a twofold purpose, aesthetic and moral. Like excursions, which are shorter, the voyage requires an ascesis. Haunted by the fear of perverting innocence, however wild it may be, and by the feeling that urban civilization is vain, tourists must lead a thoroughly simple life. They are to avoid the dissipation that characterizes 'high society'; they are to tire their bodies and content themselves with frugal meals. Thus they learn that their real natural needs are limited, and that a careless traveller may easily disrupt the customs he is observing, thereby making the landscape he would admire less picturesque.[90] Tourists are also to record their emotions in little notebooks. This little memory device makes it possible to publish a picturesque travel account, which is the ultimate aim of the undertaking. The success of such a work requires accord between text and images. Only this harmony will arouse the reader's desire to follow in the traveller's footsteps.

This experience, which concludes with an invitation to travel, also takes the form of a prayer. In keeping with natural theology, Gilpin transforms the walk into a spiritual exercise. Pursuing his aim, which is to edify, the Vicar uses a sentimental survey that had been fashionable since the beginning of the century.[91] Glorifying the Almighty does not, however, prevent him from displaying a solid classical education. Here again, numerous references to the works of Ovid and Virgil, as well as those of Milton, Claude, and Salvator Rosa, are combined with biblical allusions.

The picturesque journey is a quest, and it implies a refined control of the chain of emotions. Its purpose is to reconcile the ecstasy provided by the general impression with a scrupulous analysis of details. This learned organization of desire and delight has its own code. In the first place, it requires constant attention to temporal rhythms: the alternation of the four seasons, a day divided into four parts, the succession of the four ages of life. These are interlocking series, endlessly recapitulated, which surround the whole picturesque genre. Even more significant, though, is the search for the propitious moment: this is a valorization of the fleeting experience of a synchronicity between the world and the self, and the aim of the travel account is to revive this experience.[92] 'In the second half of the eighteenth century', writes Edouard Guitton, 'everything leads poetry to provide snapshot images in order to comply with vital rhythms.'[93]

The picturesque journey also requires a model for appreciation of place.[94] It dictates the choice of viewpoint, favouring the one which most easily frames the natural spectacle in a picture. Taken on a vast scale, nature remains incomprehensible to tourists; they are obliged to choose among these images, and partitioning an environment is perceived as a visual performance.

Picturesqueness is part of a taste for limits; it emerges out of the search for the microcosm. Remember, it expresses a sensibility born in the English countryside, which was to come into its own in the meadows of central Germany or Normandy.

A tourist's choices are governed by his culture. The picturesque journey is an ongoing effort to project artistic memories on to the spectacle of nature. In the second half of the century, mankind no longer restricted his search for pictures to the garden alone. Along the River Wye or on the cliffs of the Isle of Wight, the artist and the spectator gradually began to merge into one. Needless to say, this approach is part of a strategy of distinctions; it implies a vocabulary of exclusion, and demands a culture without which, Richard Payne Knight was soon to claim, it is impossible truly to appreciate a landscape. Distinguishing the picturesque element within a natural scene is based on notions of composition and effect. It makes use of viewing skills that travellers had long trained in gardens. Sometimes it even requires the use of instruments such as Claude's mirror, which facilitates the assessment of a picture's quality. The sensibility to picturesqueness, like the classical appreciation of Neapolitan beauty before it, was based on a form of complicity: the text produced a public that was socially restricted for the time being; the culture of this public allowed it to use stylistic short-cuts and more sparing descriptions through the use of references.

By the same token, this model of appreciation and delight prescribes places, postures, and attitudes, all of which are inspired by those advocated in the normative literature intended for artists still in training. Joseph Vernet, for instance, carefully explained how to acquire a sense for the right moment, and he taught how artists should place themselves as they lay in wait for a picture and how they should move. After him, Valenciennes explained how to appreciate a panorama: 'Nature', he stated, must be 'taken in in a single glance; the viewer's head must remain still ... If you were to turn your head, the viewpoint would change.'[95]

A detailed examination of the code developed by Gilpin is not part of the present project.[96] However, it is necessary to list rapidly its major prescriptions. It is important, for instance, to stress the attention he pays to composition, layout, the sequence of fore-, middle-, and background, and the use of light and shadow. For Gilpin, the picturesque scene excludes any discontinuous movement occurring near the spectator. The imperative for variety and the desire to breathe life into the landscape determine the way the eye partitions a scene. Mankind's only function is to incorporate these two qualities. This explains the interest travellers found in ploughmen, harvesters, or bargemen who moved about in the distance, and later, in bright costumes and in picturesque groups in festive urban scenes. Gilpin's sensibility leads him to prefer roughness to the softness of polished objects, the bark of an English oak to the brilliance of marble. It is important, however, not to compartmentalize the various systems of appreciation too rigidly; Gilpin was fascinated by Shakespeare, Milton, Thomson, and Salvator Rosa, and so devoted himself solidly to the sense of the sublime; he, too, fell under the influence of the Gothic revival.

What remains most important, however, is a longing and an endlessly renewed quest for the ideal viewpoint and the search for images of the moment. The picturesque journey is an endless pursuit of spectacle and the pleasure it provides.[97] The traveller lives 'in the hope of continually seeing new sites rise before his eyes'.[98] This perpetual quest for novelty is what 'keeps the soul in a state of continual, pleasant expectation'. This is no doubt the point at which the fundamental break can be identified between this model and the classical system of landscape appreciation.[99]

At this point, a passionate interest in views began to break loose in Western Europe. A network grew up focused on a few sites which were natural sanctuaries, seen as compulsory stops for travellers pursuing the aesthetic imperative that rapidly emerged.[100] Its novelty was due to the fact that this time, the network no longer included the hills of Genoa, the summit of Mount Vesuvius, or the vista over Marseilles,[101] but rather, northern resorts, especially Calton Hill and Arthur's Seat,[102] as well as Shakespeare's rock at Dover,[103] and later Mont-Sainte-Catherine near Rouen,[104] the sight of the Sound[105] contemplated from Elsinore, and much later, Mont-Saint-Michel seen from its shores. Whole regions – Wales, the white cliffs of the Channel, the sites on the Isle of Wight, or the coast stretching from Le Havre to Le Tréport – emerged as ideal hunting-grounds for lovers of the picturesque.

The picturesque sensibility quickly reached beyond the circles of the English upper nobility and gentry, whose daughters were particularly devoted to sketching landscapes and copying paintings of Mediterranean scenes brought back from the grand tour by the young men of their families. The already well-worn practice of aristocrats and clergymen eager for opportunities for contemplation, which consisted in climbing a hill to overlook the natural scene, was gradually reinterpreted even as it spread. The utilitarian aim, which accorded with both middle-class mentality and the spirit of the 1688 Glorious Revolution, reinforced what had been an elitist pleasure. Soon, anyone and everyone was rushing to the slopes of Calton Hill or Mont-Sainte-Catherine. The picturesque view became the focus of a complex network of sensations, memories, and knowledge. Curiosities also found their place there. Contemplation, which requires solitude, was gradually combined with or replaced by the didactic function of the view. On these picturesque summits, sites of overlapping temporalities, the present was not expelled by evocations of the past. Before the tourist's eyes, telluric ruins and remains of ancient times often coexisted with bustling industrial and commercial activity.[106] In order fully to enjoy a view, versatile knowledge and a variety of curiosities were now required. Contributions to the endeavour were expected from geology, geography, botany, history, economics, aesthetics, and ethnology. This practice corresponded to the ways of travelling dictated by guidebooks whose surprising voracity intrigues modern tourists.[107]

Once he reached the viewpoint, the happy traveller could savour emotions of surprise and wonder and perhaps even discovery, satisfy a desire to accumulate and store away knowledge, and enjoy the pleasure of naming and listing the places and things seen. And, of course, he had the satisfaction of participating in the celebration of a picture that had been framed by the undeniable standards of many prestigious artists. By the end of the eighteenth century in England and the beginning of the Restoration in Bourbon France, the picturesque traveller was giving way to the 'lover of beautiful sites', who had no guidebook but had been supplied by an obliging coachman with a list of views not to be missed.[108]

The sea-shore became caught up in this picturesque web. More and more, it comprised one of the elements that introduced indispensable variety. At Edinburgh, Dover, Rouen, or Copenhagen, fragments of the sea's immensity could be included in the picture thanks to the bay's outline or curve or the profile of the shore. Obviously the picturesque code favoured jagged coastlines, because they increased the number of planes which the eye could analyse and introduced the factor of interpenetration between the elements, which ensured that variety was maintained even in the remote background. 'Assuredly, nothing is more sublime than the ocean, but it would not be very picturesque if there was nothing to accompany it,'[109] observed Gilpin, who took particular delight in the motion of sailboats which enlivened the scene. By diversifying the emotional palette, this variety of codes could, if necessary, add to the viewer's enjoyment. Bathers at Brighton or Weymouth appreciated the long, straight, sandy beaches which offered the fulness of aquatic emotion. At the same time, the proximity of jagged coasts like those of the Isle of Wight offered opportunities for excursions, once travellers had tasted the pleasure of the waves.[110]

Gilpin's travels, especially the 1774 Southern tour,[111] are sprinkled with such pictures in which headlands and a portion of a bay are part of the composition. However, it is important to recognize that the way in which the Vicar of Boldre appreciates the sea-shore is generally oriented by the aesthetic primacy of the banks of lakes.[112] Though he is moved by the sight of a storm and the pathos of the sea-shore, Gilpin nevertheless admits that the picture he finds most beautiful is one that combines the sublime and the picturesque, after the manner of a seascape by Van de Velde the Younger; the sight of a segment of a bay stirred by the swelling waves of the storm 'prevails over the most beautiful scene the lakes can offer'.[113]

In the last years of the eighteenth century, the journey along the Downs, together with the voyage to Magna Graecia, recently discovered by artists, comprised the ideal experience of sea-shore picturesqueness. This path of delight became a decisive component in the pleasures provided by seaside resorts. Having a view of the sea was now a major factor for anyone seeking to praise the qualities of a seaside resort, a villa, or an estate. In this connection, the role of topographic artists cannot be overestimated. Paul Sandby, who was touched by this type of scenery as early as 1745, and later Alexander Cozens and then William Daniell, the tireless traveller along the sea-shores, all contributed heavily to preparing viewers to appreciate these white coastal landscapes. The crossing from Dover to Calais, a virtually compulsory route for grand tour enthusiasts and for visitors from the Continent, also paved the way for the rise of this new fashion. Finally, the prestige of the white cliffs was well suited to the growing vogue for seaside holidays on the beaches of southern England.

The Isle of Wight, a microcosm carved out of the thick whiteness of these luminous coasts, offered travellers the whole range of picturesque attractions built up there layer upon layer. John Hassell's *Tour of the Isle of Wight*,[114] published in 1790, provides an effective vehicle for analysing the forces that drove this new enjoyment of visiting. This journey is openly guided by the longing for coastal beauty. This new purpose encourages sailing along the coast and sea excursions, which are the only means of obtaining all possible views off shore of the coastline. Hassell knows nothing of the emotional tactics involved in the initiatory journey into the Hebrides' insular maze; he does not share the coenaesthetic sensibility that Townley revealed when he took refuge on the Isle of Man; his sojourn is not modelled on a series of Robinson Crusoe-like adventures. The peregrination which this traveller and his companions undertake is merely an endless pursuit of aesthetic surprises which they hope to see repeated *ad infinitum*. This quest results in close management of emotional resources. Hassell and his friends have trained their eyes to take in an aerial perspective, and so they watch for changes in the atmosphere and await the effects of a thunderstorm with impatient delight. They assign a picturesque rating to each view, referring to Claude Lorrain, Vernet, or Thomson. Where appropriate, they point out the flaws in the landscape, as they would in a painting, paying careful attention to the use of planes, contrast, special effects, and variety.

These travellers plunge physically into this hunt for a good view. Constantly in search of a fine viewpoint, they change tack or cast anchor when necessary, and climb up dunes or even cliffs. They are less interested in caves and crevices than the adepts of the Caledonian journey; but they untiringly reiterate descriptions of the beauty they find in those bits of the sea that fit into their pictures. In this strategy, the picturesque journey is far removed from the reverie of a walker along the sea-shore, which presupposes that his soul has been emptied by yielding to the pure background noise of the waves.

Hassell and his companions are also ignorant of the perils of the sea. For them, thunderstorms and tempests are nothing more than pictorial elements. They cross a bay as they would a lake. Hassell writes the way a topographer who pays little heed to the pathos of a storm draws, using his pencil to put order into the coastal rocks, tame the profile of the coastline, and make it all fit into the framework of his picture. In short, structured this way, the picturesque appreciation of the shore is radically opposed to the quest for the coastal sublime.

Sir Henry C. Englefield stayed on the Isle of Wight during the summers of 1799, 1800, and 1801, 'visiting repeatedly almost every part which contained anything worthy of notice, making copious notes and numerous sketches on the spot, and measuring many heights with the barometer'.[115] This traveller was a scholar who had undertaken an exhaustive journey. In his book, an ambiguous process leads to a mingling of aesthetic appreciation and scientific research. Thomas Webster's numerous illustrations alternate between chaotic rocks, solitary creeks, and vertiginous cliffs of dazzling whiteness, whose verticality is deliberately accentuated. They feature no walkers collecting shells at low tide, no boats run aground, and no storms. The impression is one of solitude: in the absence of man, the viewer's vision is filled by the cliffs and the fallen rocky blocks.

The text follows the same pattern as the images: it seeks to provide a systematic catalogue of views, a disciplined trace, and a taxonomy of the island's coastal landscape that the visitor/reader is expected to rediscover from the same viewpoint. The extreme precision of the account and the accumulation of prospects within such a limited space heighten the attraction of this insular quest. Englefield points out tiny paths visitors should take; he designates resting points from which one absolutely must view the panorama; he lists the views which tourists absolutely must not bypass. To cite just one example: 'But to enjoy in all its glory the complete view of the tract, which in its detail has presented so many separate beauties, *we must ascend* the chalk range which rises immediately from the woods of Nunwell ...'.[116]

Englefield makes a list in which he rates the finest landscapes and points out the most beautiful clusters of rocks, often unknown to tourists; he specifies the point of 'high perfection'. His book concludes like a hymn in praise of chalk. Such 'picturesque voyages' offer a way of speaking about the landscape. They inherit this role from the critical literature of the early eighteenth century, which taught men of taste how to discourse on the paintings of Claude Lorrain.

In Englefield's book, the collection of landscapes is part of the same longing for accumulation as the gathering of rock samples, shells, or plants. These various quests reinforce one another: grasping a landscape complements, explains, and sometimes encourages other sorts of collecting.

This new art of the poetic voyage made its way across the whole of Europe. Even before the French Revolution, some French tourists discovered with surprise this learned way of appreciating the picturesque landscapes of a chalky coastline.[117] Later, the presence of émigrés furthered the spread of this new way of experiencing nature. The success of Ann Radcliffe's novels, which describe the desolate shores of the Adriatic coast and the wild splendour of the Apennines with the same appreciation, increased the prestige of what had become virtually a popular fashion. Already in 1795, when Cambry published his account of his journey in Finistère, Noël de la Morinière marked the picturesque path along the cliffs between Le Havre and Étretat. He wrote a thorough guidebook including the shores and views, broken up by meditations that reveal a new receptiveness in France to the excitement provided by the view of the seashore.[118]

From that point on, the picturesque code became commonplace, losing its value as it fostered a copy-cat attitude. Benches sprang up to encourage admirative contemplation, viewpoints were fitted with belvederes and orientation tables, and the visit to the lighthouse became part of the ritual of any excursion. At the same time, in town, optical shows and, shortly thereafter, panoramas appeared to satisfy the curiosity of a constantly widening public. The crowd of tourists soon became nervous slaves of how, as well as what, they were seeing. Travellers, or rather tourists, forced themselves to dissect their impressions in order to heighten or better control their emotions.[119]

The picturesque vision of the English coasts had reached such a degree of triteness by 1817 that Jane Austen, who was otherwise unable to resist the new fashion,[120] set out to make a mockery of it. While tourists in caravans on Staffa were savouring stereotypical emotions, the derisory characters of *Sanditon* on the grotesque beach imagined by the novelist filled their conversations with clichés about the picturesque beauty of the sea, clichés which the beaux, those ridiculous suitors, wield like blunt-edged swords.

THE REPULSIVENESS OF HOT SAND AND NEW FEELINGS OF CLARITY

During the last two decades of the eighteenth century, artists and antiquarians discovered the southern coast of the kingdom of Naples. The time had come at last to explore aesthetically the Magna Graecia recently unveiled by Winckelmann. As a result, praises of the picturesqueness of the Sicilian shores finally counterbalanced the repulsion that Italian beaches had long inspired. This repulsion is worth pondering, as it developed in such negative contrast to the fashion that was making the English rush to the Sussex coasts.

While scholars were beginning to praise the healthful virtues of the perfect balance among the elements to be found along northern shores, swept by the winds and washed by the tides, medical science was using already old scientific beliefs to refine and give consistency to the devaluation of Mediterranean beaches. A vehement discourse, based on the deadly character of the malaria-infested Italian coasts, developed to discredit these beaches of burning sands washed by a lukewarm sea. This learned discourse both presupposed and reinforced the dominant system of appreciation. It is important to analyse carefully the theoretical justification for the repulsion that these hot and sunny beaches inspired, while bearing in mind that at the same time, the desire was developing to expose one's body to the invigorating waves of the northern seas. Scarborough and Brighton emerged as unexpected forms of recourse against the evils of civilization, whereas the excessively old Tyrrhenian shores had worn out their vivifying strength over the centuries to such an extent that they came to stand out as images of Hell. Unhealthy realms of immorality and degeneration, they symbolized the traveller's discomfort.

The fact is that on these shores, at this nodal point where the four elements come together, a balance proved impossible to obtain. From Lancisi on,[121] scholars had endeavoured to analyse this disharmony and to discover the secret of its 'bad air'. This effort culminated between 1787 and 1797 in Thouvenel's undertaking, which sought to produce 'an exact analysis of the atmosphere of [Italian] *beaches*'.[122]

The fundamental intuition of this climatology which underlay the medical understanding of coastal regions was rooted in neo-Hippocratism. According to scholars, the main characteristics of the climate of a particular place could be deduced from the shape taken by the perpetual conflict between the elements.[123] And the beach was the supreme stage on which this struggle could be watched as it unfolded. This *limes* was often the point at which atmospheric troubles arose that would then affect the continents. Neo-Hippocratic climatology was constructed of many layers, and the sea-shore was considered to be a useful laboratory allowing the observation of major meteorological phenomena on a reduced scale. This explains the importance attached to these unassuming spots, whether salubrious shores or deadly beaches.

Newly refined in this manner, this climatology refused any unequivocal explanations, preferring instead a complex analysis that appealed simultaneously to what we would now classify as chemistry, physics, biology, and geology. Thus the causes which Lancisi had identified earlier, such as the bad smell and the sea's mephitic exhalations, were certainly taken into account. Likewise, Thouvenel, applying the analytical resources of modern chemistry, dissected the various saline, ammoniacal, muriatic, bituminous, and compound emanations that prevailed on these beaches. This was not thought to provide sufficient explanation for their insalubrity, however. Similarly, no one believed that the presence of stagnant waters in coastal swamps, though full of vegetable debris, poisonous plants, and animal carcasses mingling their horrible corruption stimulated by the blast of burning winds, could alone account for the source of this bad air.

In order to disentangle the complexity of mephitism on the beaches bordering the Mediterranean, it was important to consider each of the elements, and not just the quality of the waters. Summer drought prevented the atmosphere from being adequately cleansed in these areas that were unduly exposed to the sun's heat. The breezes or inland winds that arose at night caused low, whitish fog to accumulate along the shore, and its thick, damp, miasmatic air represented in and of itself a terrible danger. The midday air was made suffocating by the burning winds blowing in from the sea, by the sun which was visible above the horizon, and by its heat reverberating off the sand and the water.[124] The worst was not so much that the fluid then reached a high temperature, but rather that it became heavier and therefore less elastic and more subject to fermentation. Often, the unhealthy effect of this stifling air was reinforced by the circular or semi-elliptical wall formed by the Alps or the Apennines.

This 'overabundance of fire'[125] in the atmosphere upset the desired balance; in the noonday heat, coastal mists that had not yet dispersed were 'sublimated by the action of the sun', and released a 'putrefying gaseous exhalation' which was a potential cause of putrid fevers – unless volcanic vapour happened to adjust this bad air. Indeed, it is important to realize that the most frightening sight on these coasts was not the volcano, but rather that of an enclosed swamp simmering in the sunlight.

Moreover, the Tyrrhenian beaches were highly exposed to the two hot winds that develop on the African coasts: the sirocco and the libeccio. Consequently, they were 'frequently invaded by miasmas and foreign winds'.[126]

Overall, the atmosphere of these sea-shores was subject to 'strong and frequent vicissitudes': 'storms and raging rains', 'fetid fogs', and 'dry vapours'. The meteoric succession of these changes made these shores a portrait of inconsistency.[127]

The earth, too, exerted an influence on the atmosphere of these insalubrious beaches. In addition to the 'putrefying mofette', there often existed a 'stifling mofette' caused by the presence of 'subterranean fossils subject to fermentation'.[128] As he walked along these accursed shores, Roland de la Platière stressed the intensity of the mineral vapour emanating from them.[129] Between the two kinds of mephitism, one 'putrefying', the other 'stifling', mixtures could occur that increased their danger.

The quality of the parent rock, far from correcting the bad air, sometimes only aggravated its insalubrity. Thouvenel considered, for instance, that clay and marls gave rise to a slimy, damp soil which tended to generate, or at any rate foster, mephitic miasmas.

Morphology also had a role to play in the deployment of this fascinating phenomenology. The sand of beaches and dunes, river deposits, and the sea itself were so many obstacles that prevented the rivers from cleansing themselves. They caused puddles, swamps, and even putrifying ponds to form. The 'rising walls' in which the Italian sea-shores were 'enclosed'[130] confined the electricity produced in low, damp regions. It was there, according to Thouvenel, that terrible 'vaporous, electric' fogs, such as the one that infected Europe in 1783, were fomented.

Mankind also contributed to the formation of bad air. Land left uncultivated facilitated the spread of entangled undergrowth, in which animals teemed, animal carcasses rotted, and their remains fermented. These deserted shores, these repulsive strips of overheated sand, were the antithesis of the pleasant, healthy, manicured fields that reassuringly demonstrated that mankind had the power to overcome the forces of chaos. Just as these enchanting spots gladdened travellers' hearts, so the Tyrrhenian beaches saddened them. There, observed Roland, the bad air exfoliated trees, dried out their bark, corroded the walls of houses, and undermined the inhabitants' health.

Admittedly, the degree of insalubrity varied from one area to another. This suggested the necessity of taking exact measurements in order to map the distribution of bad air. Thouvenel first of all specified a 'pestilential degree' so strong that nobody could resist its disease-causing powers, such that merely journeying to such countries comprised a terrible health risk: according to him, 'all truly swampy *and maritime* areas'[131] belonged in this category. Along well-drained beaches, children and adults could survive, but they would be reduced to the state of living ghosts. Only the mountains and hills were habitable, by the country's natives at any rate.

The Adriatic coasts seemed less unhealthy than the shores of the Tyrrhenian Sea. The rising sun, which the neo-Hippocratic tradition considered to be its only beneficial phase, shone there early in the morning, and dispelled the harmful vapours of the night. These beaches, less narrowly hemmed in by mountains, were more easily swept by winds that cleansed the atmosphere. Generally speaking, capes and headlands were considered more salubrious than gulfs and badly ventilated coves.

The season for bad air extended from the summer solstice to the autumn equinox. It therefore corresponded to the beginning of the bathing season, which was bringing fame and fortune to northern shores. The degree of peril varied according to the time of day: it was especially strong just before sunrise and immediately after sunset, but the middle of the day, as the sun approached its zenith, was also dangerous. The only safe hours were those in which travellers could enjoy the morning and evening sun.

In that age, when scholars and travellers felt that they were invested with an edificatory mission, each description and each analysis of the bad air along Italian coasts was accompanied by a programme for reform. The proposals generally consisted in draining or filling in the swamps, clearing undergrowth, planting the dunes or changing the orientation of the houses. Roland, a future French Minister of the Interior, advocated building artificial ramparts and, above all, redistributing land to small landholders. In this case, a plan for social reform was guided by health concerns.

The bad air of these beaches also had harmful consequences for human morale. In this respect, it is important to stress that appreciating landscape is not only an aesthetic process, and is not a monopoly of the artist; it also concerns the health of the soul, and a traveller's reactions were found to be consistent with the nature of the climate. As it was then understood, the definition of climate applied to an area in which morphology, appearance, the prospects offered, and the activities made possible by this climate all combined to influence the morale of visitors and even of mere spectators. It is therefore useless to try to dissociate the history of systems of aesthetic appreciation from this geography built up from medical considerations. Individuals of that period used their coenaesthetic impressions, their diaphragms, and their viscera to appreciate the terrain through which they travelled. Each type of landscape had a specific effect on the traveller; it could be harmful or beneficial, but like the medicinal value of the place, it could not be neutral.

Admittedly, the eighteenth-century tourist had lost the sense of a direct link between mankind and the macrocosm; for him, nature had become a show. All the same, these travellers had not yet acquired the panoramic detachment which is too often attributed to them. They believed and felt that the lands they visited offered beneficial effects or represented dangers, which could cure or kill them. By the same token, descriptions of a journey's aesthetic qualities in travel accounts were subordinated to an overview of the climatic and topographical characteristics of the region in question.

This makes it easier to understand why travellers, physicians, and administrators made such efforts to analyse their 'sensitive affections', and brought them up in the face of the 'depressing spectacle' that unfolded along the Italian coast. By following this coastline, asserted Thouvenel, foreigners first of all ran the risk of 'despondency' − that is to say, 'stupour of the soul' − which is the very opposite of the vivifying effect sought along the Channel coasts. A stay on the Mediterranean coast *attenuated one's energy*, since it undermined physical strength through 'aches and pains' and 'colliquative sweats',[132] both signs of endemic fevers. Of course, by judiciously choosing the time of his journey and taking care to avoid the sun's burning heat, the traveller could guard himself against illness. He could not, however, escape the repercussions of the area on his morale. Physicians were very aware of this, as, when discussing these beaches, they wielded the full range of vocabulary for repulsion. Through a gloomy discourse, they outlined a nightmare whose significance can be more fully appreciated by contrast with the laudatory rhetoric displayed with respect to Brighton's salubrious shore. From Thouvenel's three volumes it emerges that the Tyrrhenian beach during the summer season simultaneously suggests all that is greasy and thick / damp and vaporous / stagnant, foul, and fetid / fermented, corrupt, and rotten / stifling and distressing, in addition to sunburn / a sense of confinement by the topography / and bouts of foreign miasmas. In short, this work unfolds a catalogue of fantasies which, as was shown earlier, exerted a decisive influence at the time.

There is no need, therefore, to go into a detailed analysis of the chorus of complaints that issued from those who visited the Roman countryside, the Tuscan Maremma, and even the beaches of the Riviera. Long before Chateaubriand indulged in the nostalgia that these desolate regions aroused in him in 1804, the list of grievances proved endless. Travellers who had recently learned to read the signs of the march of time in a space still in the process of being formed quickly acquired the habit of assessing the quality of a country's politics by examining the recent traces it had left on the landscape.[133] As early as 1687, Gilbert Burnet described the depressing aspect, the 'stagnant waters' and the 'stinking' air of the countryside that bordered the road from Rome to Civitavecchia, and placed the blame on bad government by the Pope.[134]

In 1739, it was the turn of France's President de Brosses to be distressed and to criticize papal administration, which soon came to represent the flip side of enlightened despotism.[135] At the end of the century, Roland de la Platière found the bad air of Tyrrhenian coasts oppressive. Choiseul-Gouffier was horrified, in the course of his journey to Greece, by the corruption of the beaches at Milo.[136] The diatribe lost none of its intensity in the following century. In 1801, Creuzé de Lesser, admittedly not very generous with his praise, went so far as to denounce the insalubrity of ancient Baiae's sea-shore: in his view, it was nothing but 'a marshy coast, already half eaten away by the sea, and its pallid inhabitants narrowly escape death in the summer ... When one bears in mind that, from the point down to the Maremma in the region of Siena, the whole Mediterranean coast is uninhabitable almost everywhere, beautiful Italy seems a lot less attractive.'[137] Travellers who fell asleep as they crossed the Pontine marshes ran the risk of never waking up again; and what could be said of the detestable air of Ostia and Latium, of which Charles-Victor de Bonstetten[138] had just painted a terrifying picture in his report? This 'horrible countryside' and its 'hideous, pestilential' coast were supremely repulsive.[139]

During the second half of the eighteenth century, the Mediterranean shores overall, and not just the coastal beaches, inspired sadness if not repulsion. Travellers from the North did not appreciate the barren, bare rocks and the rocky brown slopes drenched in sunlight. The shores of Provence and the Riviera aroused admiration only when they were covered with orange- or lemon-trees, or when they offered a picture of pleasant towns surrounded by verdant gardens. In 1754, Fielding described his disappointment upon seeing the arid landscapes of the Tagus estuary, which made him nostalgic for the colours of the English countryside.[140] One after another, Sulzer, Guibert, Millin, Young, and Roland, to name only a few, all described their disappointment in detail. In October 1775, Sulzer, who may have been touched by the 'admirable view' travellers could discover from the cape of Antibes, and attentive to the sight of waves breaking on the rocks at Nice that looked to him like so many waterfalls, preferred the pleasant, cultivated landscapes to the 'the uncultivated coast, bristling with rocks' that separated Menton from Monaco.[141] As Millin crossed the chalky shores of the Toulon region, he wrote: 'Everything contributes to increase the horror of this place, which one could take for the gates of Hell'. On the other hand, the presence of the beautiful, fertile plain surrounding the big port seemed to him like the passage from 'Erebus to Elysium', and the town of Hyères was, in his view, a 'new Hesperide'.[142] This kind of discourse was sometimes accompanied by a diatribe against the heat of the sun, especially when there were no caves or foliage in which to seek shelter from it. Under such conditions, travellers did not fail to break off their walks.[143]

Many a decade would elapse before another system of appreciation would replace this weary way of seeing. Before this could happen, changes had to occur in the way people related to the sun, hot sand, barren ground, and chalky relief – in short, to the elements that today shape the images of the beach and the holiday club. The trap of anachronism lies wide open for careless readers who fail to understand that the rush of tourists to the Mediterranean was initially caused by a cluster of attractions that brought together a now forgotten combination of a nostalgia for ancient times, an interest in archaeological discoveries, a desire to find the landscapes painted by Rosa and Claude Lorrain, hopes of being cured, and the sweet pleasures of holiday-making on the coastal heights or seaside strolls along headlands cooled by breezes from the open sea.

The same negative values stressed by neo-Hippocratism reappear in the field of aesthetics. According to the painter Valenciennes, of the four parts of the day, only midday presents problems to the artist: 'The tiring aspect of washed-out light, the birds' silence, the cicada's strident cry, the weariness of animals of all kinds, man himself covered with sweat',[144] all discouraged lovers of beautiful nature.

If Valenciennes is to be believed, landscapes flattened by sunlight have never given rise to a masterpiece. The artist is compelled by the silence, the immobility of natural objects, the immutable impression, and the emotional vacuity of the scene, to introduce pathos into it by staging a fire or a storm if he hopes to capture the attention of those who look at the painting.

An altogether different kind of delight in the coast emerges from travel accounts of voyages along the shores of Magna Graecia in the aftermath of the journey undertaken by Baron von Riedesel in 1767.[145] The then prevailing neoclassical aesthetics truly blossomed on Greek soil, where an unprecedented model of appreciation for substance emerged: a new emotion arose that contradicted the repulsion evoked by the coasts of central Italy and broke the monotonous series of clichés offered by Vernet's successors.

To all appearances, the practice that developed at the end of the century was an extension of classical tourism. This journey, which often took the form of a scientific expedition financed by a patron, was inspired by a passion for antiquities and enjoyment of archaeological sketching. The ancient texts continued to haunt the imagination of travellers who would scrutinize the landscape with copies of Homer or Fénelon in their hands. On the shores of Sicily, tourists imagined themselves gazing through Ulysses' eyes at the landscape where Nausicaa came to frolic with her friends.

These artists and amateurs complied more or less consciously with the modes of travel prescribed for them in the codes developed by authors of good treatises for their use. They accepted the imperative need to record their impressions, in preparation for recalling them later. Even more than Gilpin, Valenciennes, who was scarcely concerned about praising his Creator, insisted upon these 'studies in remembrance' which were, in his mind, the primary aim of the journey undertaken. A young artist who hoped to maintain a permanent connection between his quest for the picturesque and the preservation of his memories, must record everything. 'Keep a brief journal of your travels,' Valenciennes advises. After a few years, 'take home your portfolios filled with drawings, sketches, and studies, and your tablets full of notes and observations. You will then have to bring all these various materials into play.'[146] During the journey, the aesthete's initial approach was indeed to write, to report or prepare a fragmented account, according to the rhythm of the discovery of natural events. When an artist encountered a motif, he would venture a sketch without even looking at the model, then compare his work with reality, in order to detect any possible omissions. At the first stop, he would hasten to produce sketches based on the 'idea that was still fresh in his head'. In this way he developed a *collection of snapshots* that would enable him, once the journey was over, to complete his pictorial work, or at any rate, to delight in the 'pure and innocent' pleasures provided by recalling these 'varied, regular moments that form the chain of our lives'. For 'it is by contemplating them that the soul blossoms and realizes the full value of existence'.[147]

Valenciennes expected the artist to extend this youthful journey over a period of several years. He saw it as an initiatory rite that would postpone until manhood both the major production, achieved after mature reflection, and the delight of recollection.

But beyond this series of injunctions, the accounts of travels to the 'novel landscapes'[148] of southern Italy reveal an original experience of substance, of the qualities found in water, sand, or rock, and of the brilliance of the heavens. Southern Italy heightened perception. In this respect, the progressive enrichment of Goethe's emotional range is significant as he discovers the Mediterranean during his journey in 1786 and 1787. In Venice, first of all, this tourist experienced, *à la française*, as it were, the emotion aroused by 'seeing the sea'. In the vicinity of Gaeta, he admired one of those landscapes that painters had engraved in his memory. Along the shores of the Bay of Naples, the pleasures of the rituals of sociability introduced him to unexpected delights of the shore: walking along the water, watching handsome adolescents bathing, or conversing on the beach or a terrace all called forth ancient Baiae. As he sailed for Sicily, Goethe discovered the immensity of the sea seen in its midst. But it was at Palermo that he received the revelation of the inexpressible picturesqueness of southern shores and the harmony produced by the magical effect of the air.

> No words can express the hazy clarity that hovered over the coasts as we approached Palermo on a beautiful afternoon. The purity of the contours, the softness of the whole scene, the contrasts of tones, the harmony of sky, sea and land. Whoever has seen this will remember it forever. It is only now that I can understand Claude Lorrain, and I hope I shall be able to summon up from my soul images of this pleasant abode when I get back to the north.[149]

And he concludes: 'Here is the key to everything.' The emotion aroused in him by his discovery sent him back to his references; it earned him an understanding of both the *Odyssey* and Claude Lorrain's paintings.

At the end of a century fascinated by prisms, the literary analysis of aquatic colours reached a hitherto unimagined degree of precision. Forster delighted in watching for optical variations in the ocean waters.[150] Valenciennes, who had made the journey to Sicily and had evaluated the long labours of observation performed by seascape-painters, gave a detailed description of the colours and shades of the Mediterranean in all its states. 'During a storm, the waves of the sea are brown, greenish and foamy'; when the moon shines in a cloudless sky, 'they seem silvery, like long strips of crystal. The part of the water that is in shadow shows only a very strong black-green shade with bluish reflections.'[151] When the air is misty,

> the sea water is grey and takes on the colour of the sky, especially when it is calm; but if it is choppy, it takes on different shades: green, blackish, greenish blue, dark purple mingled with the foamy whiteness of the tips of the waves which disappear as they roll and break over themselves. The waves that come to the shore and touch the beach are foamy and whitish, whereas in the deep areas they most often stand out full of colour on the sides caught by the sun. The reflection of the mountains and the clouds, and the occasional mingling of foreign bodies contribute to the colouring of sea water. That of the ocean on the coasts of France does not in the least resemble that of Holland, and the Mediterranean is of yet another colour.[152]

This extremely precise normative text is intended to be an introduction to seeing. It encourages viewers to pay attention to the shifting nature of the aquatic substance, to watch for fleeting colours, and to grasp the limpidity of the sea as it emerges 'rejuvenated' from a storm. On the basis of this exercise in virtuosity, Valenciennes proposes to draw up a chromatic map of all the seas on earth.

The novelty of this pleasure does not result solely from analysing these colours.[153] The visual complexity of the marine environment allows a panoramic view, a kinetic perception of objects, and a perpendicular look into the limpid depths, all at once. In this respect, only the experience of aeronauts can be compared to that of spectators of this watery mass.

The shores of Magna Graecia – and soon those of Greece – let travellers discover the emotions aroused by the moving transparency of the water and the dizzying view of its depths. Because the eye could so easily penetrate the waters, viewers could develop a form of vertical vision that was all the more exciting since the underwater spectacle kept all its mystery intact. Vivant Denon[154] asserted that he was struck by the limpidity of these waters. Valenciennes invited viewers to study the crystal-clear waters, in the depths of which one could distinguish even the smallest pebbles. Jean Houel, more sensitive than anyone else to the 'transparent rocking of the waves', displayed the full range of his emotions before his readers. From the watery vantage-point of his skiff, he fell under the 'spell' of the cave's limpid waters. His mode of enjoyment, a 'black-out of his senses', distances him from the excitement of the Staffa tourists.

> As they (the caves in the Palermo region) are deep and sheltered by the rocks, the water there remains in a state of calm that makes it absolutely pure. It is so clear that it somehow vanishes before the eye that observes it; you can see and distinguish perfectly the slightest objects at a depth of more than twenty feet ... What makes this situation admirable and unique in the end is the water's extreme transparency; you do not notice it at all; you have the feeling that you are magically suspended in a boat hanging over the objects that you observe. You cannot believe there is a fluid interposed between them and yourself. As a crowning surprise, this very pure water offers almost no resistance to the boat, so that you move from one spot to another with a stroke that is so soft you cannot feel it. This is truly enchanting.[155]

On the Sicilian sea-shores, Houel invented a new way of experiencing the void: on the surface, or rather on the heights of the limpid waters, he performed a virtually aerial journey along the cliffs bordering the depths of the abyss. Like the modern wind-surfer, he felt the dizziness of transparency.

> These rocks plunge into the sea in the same direction, to a tremendous depth. Their whiteness makes them easily discernible in the sea water, which seems very black because it is deep and because, facing north, it is shaded by the rock. And as it is absolutely transparent, under the boat that carries you, near the rock, it reveals an abyss of horrible depth, and the boat seems suspended in its midst next to a smooth, vertical wall that is of no help at all. You have the feeling of being completely isolated and held in the air, as if by magic: you imagine that if the boat were to disappear, you would fall a dreadful distance. The thought of drowning does not occur at first, it only arises upon reflection. One does not experience this same fear on the high sea, because in the absence of objects that make you *see the transparency* of the water, your imagination does not see water, as it does in this spot; here it seeks the foot of the rocks at the bottom of the sea, where the two merge together in the depths of the water's darkness.[156]

Here the fright caused by the angry sea, the crushing agitation of the waves, and the bite of sharp reefs gives way to images of being swallowed up in transparency. The fantasy of drowning yields to that of falling. Houel's experience introduces the Romantic contrast between the storm and aquatic absorption which was to mark Gilliatt's[157] tragic fate as well as that of the Ourque's unfortunate passengers in *L'Homme qui rit*.[158] At a time when Lavoisier was showing by analysis that the bodies comprising different 'fluids' were similar, and was juggling with their possibilities for transmutation, travelling artists had the sense that these 'fluids' were equivalent.

Although this magic is exclusively a characteristic of the ocean, it is appropriate at this stage to discuss the intensity then aroused by the astonishing experience of phosphorescence in the waters. This phenomenon, which has now disappeared, called forth other fantasies about the depths.[159]

At the end of the century, travellers were also familiar with another practice, which consisted of using the deck of a boat to gaze at the heavenly bodies and their reflections in the night-time waters. In 1790 Georg Forster indulged in this aesthetic experience off the coast of Dover.[160] The Mediterranean magnified this range of sensations, as it did that of the bright lights of the Neapolitan night. The mild temperature and the clarity of the atmosphere there engendered a special emotion. Roland, for one, spent hours waiting for the sun to rise over the Tyrrhenian Sea.[161] André Thouin remained on deck off the coast of Livorno for part of the second night of his journey in order to gaze at the 'luminous sparks' that shot forth from the ship's wake; he admired the 'pretty silvery colour' which the moon gave to the waters of the sea.[162] Jean Houel waited for daybreak sitting on the pier of the main Lipari island.

> The sea was lapping against the foot of the pier; the moon that had risen a few hours before covered the waves with a silvery light that sparkled vividly and gave my meal a very picturesque character ... I spent the rest of the night lying on the shore, wrapped in my cloak.[163]

The aridity of the climate, the great number of archipelagos, and the archaic anthropological structures then made it possible along the shores of Magna Graecia, and soon of Greece itself, to have a Robinson Crusoe-like adventure that was wilder, more primitive, and therefore more authentic and more revelatory of substance than the adventures invented by an invalid on the Isle of Man. One example is Jean Houel's experience at Volcanello during his systematic exploration of the volcanoes on the Lipari islands. After examining the crater, this traveller returned to the beach to eat the meal that his bargemen had prepared for him.

> This volcano, this fine, white sand stretching along the shore, the waves that washed against it even though the whole expanse of sea was perfectly calm, the loneliness of the spot, the great thoughts that volcanic mountains always call forth, and the vast expanse of the sea all gave an inexplicable charm to the place, which one always savours with delight. After dinner, I gave myself up, as is the custom in hot countries, to the pleasures of sleep. I had the softest sand for a bed; for a pillow, small sand-bars speckled with the liveliest colours that lithophytes and marine moss can produce. It was a rich piece of embroidery fashioned by the hands of nature. The sail of my little boat served as a curtain and protected me from the sun. [164]

Finding himself washed ashore by a storm on one of the little islands off Malta, Roland, thrilled by the real or imaginary threat of pirates, indulged in a Crusoe-like adventure for three days and four nights. Like Daniel Defoe's hero, he set out at the crack of dawn on his first day to discover his island. He scrambled to the heights to obtain a better view of the island's shape, then journeyed around the shore before returning to the shelter of his boat and his crew of sailors. This playful regression, scrupulously pursued, provided him with the unimaginable enjoyment that comes from the temporary renunciation of the code of civility and hygiene and the sensation of a body tired out by adventure. [165]

Gradually the longing for a new kind of tourism was being invented, far removed from the classical journey to Naples. As the century drew to a close, Jean Houel enjoyed the sea-shores in all possible ways. [166] He savoured views of the sea, happily contemplated the dance of sailboats on the water, waited for the scenes that marked the four parts of the day, and, as an artist, thrilled to the sight of volcanoes along the coast. He let himself be fascinated by the dizzying depths and the life in the clear waters. Above all, he enjoyed the open air, sleeping on the sand amid the seaweed, dining on the shore, and swimming. Never, however, did he forget that the primary aim of his journey was to draw antiquities. He was content in the company of sailors; he observed the fishermen and their families, sensitive

as he was to sea-shore scenes and to the frugal meals he shared
with his simple, kind hosts, both relics of Homeric times.

The fulfilment of this picturesque journey on the shores of
Sicily had its counterpart at the other end of Europe in the
reveries of Bernardin de Saint-Pierre. No one felt the longing for
the shore more intensely than he. For him, this vantage-point was
the supreme observatory for nature's picturesqueness.

> It is not on the mountain top, but at the seaside, not in the boxes
> of a theatre, but in the pit, that the perspectives, the décor, the
> concerts, and the drama of the architect, the painter, the musician,
> or the poet of the universe all converge ...
>
>   It is on the shores of the ocean above all, at the bottom of
> this immense valley that encloses it, that the harmonics of all the
> powers of nature come together.[167]

The awareness of the primacy of this site leads Bernardin to
speak movingly of the happiness of a never ending, wild progress with his family along innocent shores where the 'aquatic
harmonies' are ceaselessly unfolding. Let us end this journey in his
company:

> If my fortune had allowed it, I would have undertaken a journey
> around Europe, and perhaps around the world, that would have
> been more pleasant and more useful than the one people make
> every day to visit the mountains of Switzerland. I would have set
> out by sea, travelling along the coasts, in the manner of savages.
> A light canoe with a lateen sail and a few mattresses would have
> been my conveyance. Two sailors and their wives would be my
> crew. I would not hesitate to set sail with my own wife and
> children; everything would become a source of pleasure and in-
> struction for me. Am I interested in minerals? Cliffs would open
> up their sides to me; at their feet I would find metallic pebbles
> that rivers and current rolled as they wished. Am I fond of plants?
> I could collect some on the sea-shores, carried there by the waves
> from the most distant lands ... Each pull of the oar turns a new
> page in the book of nature for me and reveals new scenery ...
> Meanwhile, the men go their separate ways to hunt or fish; the
> women light the fire and prepare the food.[168]

6

The Ephemeral Journey

The Romantics did not discover the sea. Well before the end of the eighteenth century, the ocean shores had been places of contemplation and delight. Climbing up a dune in search of a view, striding along the top of a cliff, drinking in the sublime sight of a storm, or watching for a perfectly framed seascape in a panorama were already commonplace activities when *Childe Harold* appeared in 1818. As early as the end of the 1750s, crowds came to Brighton to enjoy the pleasures of bathing; in 1776, Jean Houel savoured the freshness and transparency of the Mediterranean waters in which he relaxed in the company of young Sicilians. The sea, viewed as a recourse against melancholy and the harmful effects of unhealthy city life, was assigned a decisive role in the history of the planet by Neptunian scholars. And amateur geologists gathered samples along the shores on which shell-collectors had strolled for centuries.

The Romantics themselves borrowed from previous models. As most of them were heirs of Ossianism, they often found themselves interpreting the picturesque value of a scene like ordinary tourists. In 1828, Dorothy Wordsworth made a thorough visit of the Isle of Man, and followed every detail, one after another, of a ritual of the seaside holiday that had long since been codified. [1]

Romantic creative artists, on the other hand, were the first to propound a coherent discourse about the sea. [2] They powerfully enriched the means of enjoying the beach, and stimulated the longing inspired by this fluctuating boundary. They renewed the meaning, and enlarged the significance of practices that were already solidly established. They provided models of contemplation, or rather of confrontation, that gradually displaced previous prescriptions for the enjoyment of wind, sand, and sea. The Romantics gave a fresh impetus to the way people rode on horseback, walked, or wandered along the sea-shore, or the way they posed on the headland. With a new volubility they were able to describe the excitement of sea bathing and the subtle impressions of the beach.

This is why it is necessary to trace the broad outline[3] of a system of representations, an emotional strategy, and a network of practices developed and proposed by Friedrich von Stolberg,[4] Byron, Shelley, and Chateaubriand at the same time as Jane Austen was ridiculing the triteness of the admiring discourse on the picturesqueness of the sea-shore. The key is to analyse the way in which these models of contemplation offer new itineraries for coastal reverie.

The Romantics made the sea-shore the favourite spot for self-knowledge. Within the perspective of the aesthetics of the sublime recently advanced by Kant, standing on the shore stirred the self in a special way; this stirring arose from the exalting experience of confrontation with the elements. The sea-shore offered a stage on which, more than anywhere else, the actual spectacle of the confrontation between air, water, and land contributed to fostering daydreams about merging with the elemental forces and fantasies of being swallowed up, as it unfolded the mirages of what Ruskin was to call the pathetic fallacy. The ocean's vastness became a metaphor for the individual's fate, and made of the beach a fine line marked by the rhythms of the water, which were in turn driven by the lunar cycle; treading this line became an invitation to reassess one's life periodically.

In every case, it was henceforth the spectator who took the measure of the sea-shore. Individuals no longer came there to admire the limits set by God to the ocean's power. They came in search of themselves, hoping to discover – or better yet, perhaps, to rediscover – who they were. This helps to explain the considerable enrichment in the manner in which sea-shores were experienced. The impressions they offered became something to be captured by all five senses. Moreover, the modes of coenaesthetic appreciation broadened to constitute a major event in the history of sensibility. Unprecedented ways of standing or posturing on the beach or of sitting or lying on the sand were the signs of this deepening of the quest.

THE NEW WEALTH OF EMOTIONS

Romantic painters and writers made the beach into a stage on which the story of the collision of the elements unfolded.[5] For over half a century, they founded new clichés: the sea climbing up to the sky like a huge whirlwind, the outline of the heavenly bodies distorted and dancing in the splashing waves, sea spray and vapour fogging up the air, and wet sand that seems to have lost its telluric consistency. In a reinterpretation of an earlier practice of Dutch artists, Romantic painters lingered over the spectacle of the shore.[6] They paid particular attention to the strand, that territory in which the elements were supremely ephemeral and undefined, and which encouraged pantheistic reverie. The Calais beach at low tide,[7] painted by Turner in 1803, portrays this confusion among sky, sea, and sand. In the series done by Constable at Weymouth thirteen years later, the horizon no longer separates, 'as it did in the classical land-scape, two heterogeneous visual worlds with codified colours and values'.[8] Water, land, and sky began to resonate to one another. This abolished the very divisions of the classical scene.[9]

Romantic writers both used and abused the theme of the boundless ocean.[10] They experienced intensely the Mosaic image of the sea as the substance that preceded all creation and all form.[11] Romantic painters, too, transmit this sense by placing the spectator on a level with the water and lowering the horizon. In many of Bonington's paintings, for instance, the illusion of the sea's infinity coincides with the impression of the immensity of the shore and the sky.

Following the example of Byron when he wrote *Childe Harold*, the Romantics also worked over the theme of the eternal integrity of the ocean, unaffected by historical change. Its power and its energy were all the more fascinating in so far as they seemed undiminished since the beginning of time. The emotion intensified as viewers contemplated the combination of this unimpaired force with the ocean's vacuity and experienced with all their senses the immense, palpable void, endlessly animated with the movement of life. Turner, using dynamic perspective, makes observers feel the sense of power in these empty spaces and their perpetual agitation, using a boat which is merely a pretext for rendering the experience tangible.[12] An imperilled sailing ship reveals the extent of the waves' swell, and lets us feel the strength of the wind and analyse how currents work.

The new attention paid by the Romantics to air, at the time when Lavoisier's ideas on chemistry were spreading, gave new life to forms of contemplation.[13] This 'active nothingness whose role is to awaken us',[14] embodied in clouds and birds, created an association even more paradoxical than that of the ocean between energy, virility, and emptiness. The insistent force of its appeal and its unforeseeable character gave rise to a temptation to flee rapidly towards the supreme refuge.[15] Because it is both the voice and the ear of nature, this 'concentrated nothing'[16] speaks tirelessly to the Romantic soul, with which it seeks a dialogue.

The land arouses interest here through its zones of contact,[17] dangerous strips for sailors, but a comforting surface for swimmers who can feel the bottom. Shortly after Turner and some time before Bonington and Eugène Isabey, Constable enjoyed drawing this point of contact, as a landscape-painter rather than a seascape-artist. As early as 1816, he gave preference to wet sand in his painting of the elements.[18] Viewers of his paintings, their feet in the grey-green water near the last small waves of the English Channel, can feel the unsteady presence of dampened soil.[19]

As had been the case not too long before for admirers of Vernet's storm scenes, Romantic artists were intensely influenced by the theatrical nature of the struggle among the elements aroused by the forces of passion. What remained most important for these artists, however, was the confrontation between man and this energy of concentrated emptiness. A tradition of the Dutch seascape thus came to be reinterpreted in the light of the aesthetics of the sublime.[20]

What is indispensable in this enjoyment of confrontation is a *literary* experience of holding a dialogue with the elements. Childe Harold converses with the sea, and interprets its moaning. Along the loudly echoing shore, the poet in Shelley's work seeks to fill the void in his soul by listening to the ceaseless discourse of the elements.[21] There the ocean breathes; it carries on an uninterrupted dialogue with caves and cliffs that the attentive walker on solitary shores can grasp. In the penal colony, Chactas is moved as he listens to the voice of the waves speaking to him of his America, and he tells them of his miseries.[22]

Caspar David Friedrich's work, the culmination of the Romantic representation of the shore, warrants a long examination.[23] Haunted by a nostalgia for roots that leads him on an endless journey over the long sandy beaches of Greifswald, Friedrich turns the shore into the scene of metaphysical anguish.[24] The painter places the viewer on the edge of the abyss that everyone carries inside himself.[25] Placed in the company of a man who is visible from behind and who stands rooted to the shore at the boundary between land and infinite water, the viewer of the picture experiences a sudden shock of anguish.[26] Though he makes no use of the theatrical nature of the assembled elements, Friedrich projects the most pathetic reading of the seashore. The painted object in this intensely silent art belongs to inner space.

The expansion of the shore furthers the longing for pantheistic merging, the desire to become one with the universe that plagues the Romantics. In this place that breathes in unison with the universe, fantasies take a free rein.

The sight of the boundless sea's emptiness creates a sensation without an object that encourages the imagination to plunge into the waters. The marine monotony is conducive to sleep, and it creates a temptation to let oneself be swallowed up. What paved the way for this dream, then very pregnant, was the unreality of the Ossianic landscape and the undetermined character of the 'Celtic twilight'.[27] Along sandy beaches, there are few ascending symbols; symbols suggesting falling, or more specifically, diving, are easier to find. Black water and the fatal sea of darkness, a blatant analogy with blood, evoke death.[28] In the Romantic's impatient soul, this stimulates nostalgia for primeval times.[29] The power of this invitation to undertake a journey from which no one returns helps to explain the impact of the seashore on the collective consciousness of the times.

The sea as grave fuels dreams of regression. Its association with the mother constitutes a well-known and constant feature in Jungian analysis.[30] The fantasy of being swallowed up by the watery abyss or by the sands reveals a longing for this coenaesthetic return, just as it expresses 'an acceptance of one's temporal condition'.[31] Shelley's *A Vision of the Sea*, published in 1819,[32] gives textual expression to this dream of the body returning to the sea. In this poem about shipwrecks, devouring levels consume one another, until everything is finally absorbed by the sea,[33] that 'primordial and supreme swallower ..., the archetype of the descent into and return to the original sources of happiness'.[34]

This cluster of fantasies determines the guide-lines for Romantic bathing. With these same scenes it is important to include the destiny of the theme of the cloudburst, which combines being sucked up with being swallowed, and even more especially, that of the whirlwind or the maelstrom, which gives a new form to the pathos of the sea.[35] Still more evident in this perspective is the fascination exercised by the rising tide and the temptation of slipping passively away, as presented by Walter Scott's *The Antiquary* long before Victor Hugo's unfortunate Gilliatt gradually fades away.

This is where horror at the viscous immobility of the sea creeps in. The image of what has become heavy, fetid, and excremental, teeming with monsters, coincides with the grip of the swamp on collective imagination, and with that of infectionist theory on medical thought. The dreamer is horrified at the thought of confronting the inexpressible. On the monotonous strand, the imperceptibly rising water asserts 'the insinuating progression of viscosity'.[36] It leads to the new theme of the seductive monster. Near the coast, within reach of the mud and the silt, nightmares take shape, which were manipulated by Crabbe. They were to be the death of Peter Grimes, from gazing at the muddy liquid of the desolate coast and the monsters buried in his solitary soul.

Thus the fascination of the sea-shore and of zones of contact gradually developed which was to give me the idea for this book. Maleficent, pestilential breezes crept in through the interstices and the gap created by the contact between the elements. Through them all the intrusions and the threats that solicit the imagination wormed their way in, while dejections and discharges poured out. The interpenetrations that were to ensure the intense sexualization of the strand became possible. The beach became incorporated in the rich phantasmagoria of borderlands from which perils and magic spells spring.

This is when the certain acknowledgement emerges that the sea, especially in its nocturnal clarity, symbolizes the dark regions of the subconscious, whose constant presence is revealed by coenaesthetic impressions.[37] An analysis of Shelley's work reveals that the poet had a clear 'sense of the correspondence between marine depths and psychological depths'.[38] Jean Bousquet has shown at some length the symbolic significance of the dream of liquidity during the Romantic period.[39] Since Jean-Paul's [Johann Paul Friedrich Richter's] dreaming, the activity of the sleeper's soul has often been seen as an irresistible stream.[40] Madness itself was soon to be categorized using the image of a plunge.

> I love the sea as my soul. Often, it even seems to me that the sea really is my soul. Indeed, just as in the sea there are hidden

aquatic plants that rise to its surface only when they are blossoming and sink back into it when they fade away, so from the depths of my soul there sometimes spring up images of flowers, that spread their fragrance and sparkle and then vanish again – Evelina!

When I go for a stroll at night on the sea-shore, and I listen to the song of the waves that awakens in me all kinds of memories and forebodings, it seems to me that I once stood in such a way on high looking down, but that, seized with dizziness and fright, I then fell down to earth. [41]

These were the words of Heinrich Heine in 1826. The sea sheltered sunken cities whose bells the Norderney seamen claimed to hear ringing on Sundays; at the bottom of the poet's soul, there lay a submerged world that revealed itself only in the mirror of dreams.

On the shore more than anywhere else, the dreamer, swayed by aquatic rhythms, can come and experience the passage of individual time.

The correspondence perceived between the cycle of the tides and the menstrual cycle, between the alternating ebb and flow and the nocturnal rhythm, between the scansion of waves and human heartbeats, encourages a coenaesthetic attitude toward the self, and gives a somatic character to the quest for the self. In 1777, Friedrich von Stolberg expressed this correspondence between bodily sensibility and the rhythm of the sea. [42] The young man in Novalis's *Disciples at Saïs* speaks of a correspondence that develops between the alternation of ebb and flow and that of waking and sleeping. [43] 'The perpetual flow that we hear in silence is the undertow of the sea,' writes Jean Perrin, a subtle analyst of Shelley's work. [44]

In this perspective, it is not so much the surface of the ocean, but the place where the music of the tides is heard, that proves to be fascinating. The spot that best lends itself to perceiving these rhythms is not the high sea, but the beach, where the regular undulations of the water come and break noisily.

Romantic travellers would make periodic visits to the sea-shore. Confronted by the unchanging sea, they could easily experience the flow of individual time. The ocean, unaffected by historical change, proved itself to be an eternal reference. In this respect, however, it is important to observe a degree of nuance and precision. Many scholars then held that since Creation the sea had undergone a weakening of its vitality and its fertility; few were those who felt that its purity was threatened, and no one felt nostalgic for an element that could be imperilled. If the creative power of the sea seemed to some to be in jeopardy, this was considered due more to its removal from its original state than to modern pollution.

Periodically in an empty spot that lent itself to total individual solitude, a walker might find himself confronted once again with something unchanging, with spatial and temporal infinity. In this respect, Chateaubriand represents an exemplary case. The milestones of his life are set along the sea-shore at Saint-Malo. The maritime motif gives a kind of 'subdued vibration' to 'the whole spatio-temporal volume' of his existence.[45] In the framework of this life, as in the episodes of *Childe Harold's* imaginary epic, the shore fixes the memory of departures.

Custine, who moves between the two poles of the Romantic voyage, Magna Graecia (Calabria) and Caledonia, dreams on the shore at Inverness, one of the two extreme points of his journey. This traveller compares the 'nebulous shores' of that 'icy land' to the magnificent coasts of Calabria. 'This recollection', he writes, 'has led me to survey the whole story of my life ... In one glance, I measured the effects that time had wrought upon me ... These thoughts made me feel distinctly, for the first time, that henceforth there would be more regrets than hopes in my life'.[46]

The sight of the open sea completes the impression made by the shore: it contributes to the discovery of the deep inner self. Everything in existence which 'instinctively relates to our origins – desire, wandering, verticality, indifference – can be identified only with this cardinal interval, with this vertiginous, hollow liquid palpitation'.[47] For the Romantic hero, real life is the sea, an intact place of freedom that insulates him from the triviality of the earthly sojourn.

As it issues an invitation to indulge in *nevermore* dreams and encourages observers to change their ways, the shore becomes marked with derivative themes. Stolberg, Byron, and Chateaubriand, or, in the field of fiction, a series of writers from Crabbe (*Peter Grimes*) to Philarète Chasles all portray the feeling of nostalgia for childhood days at the seaside. Those who cannot experience it convey it by anticipating others' reactions, or are moved as they think of their descendants facing the sea's infinity. Michelet, who visited Le Havre on 7 August 1831, confessed, 'I am quite moved at the sight of my little child looking pensive before the sea, a delicate child in whom I have invested my life and whom I shall not be able to protect ...'.[48] Nodier and

Hugo, eager to witness such an encounter, regretted that their daughters were not there when they discovered the beach for the first time. [49]

This emotion was in keeping with medical theory, which was then in the process of changing to laud the beneficial effects of the sea on the health of young children. As the praise of maternal instinct was developing, stimulated by the discourse on feminine nature, and increasing attention was being paid to middle-class childhood, the beach was emerging clearly as the site of extended maternity. Aroused by the proximity of the mothering sea, [50] feminine instincts blossom there which, in a wife, help to protect virtue when it is threatened by a husband's prolonged absence. Finally, the ascendancy of the figure of the child on the sea-shore coincides with the renewal of Robinson Crusoe-like adventures; these, more clearly than any previous examples, reveal regression, involution, and the ceaseless quest for the mother.

An awareness of passing time, activated by the sight of the shore, becomes associated with the intense sexual significance of the spot. The thrill of a bare foot as it touches the sand, the insistent stroking of the wind, the violent scourging of the water, the dream of vanishing into the waves like an act of slow penetration, or the path through a series of interconnected chambers in a cave all implicitly determined that the beach would become an erotic site, marked by a femininity that was both threatening and redemptive. [51] It is important, however, to beware of anachronism in this respect. The shore had not yet emerged as a scene for the free display of sensual delight. To understand this point, it is important to bear in mind the impact of the code of decency and the taboos that surrounded nudity. [52] Only by using legendary allusions could writers evoke the frolicking, which was usually tragic, of lovers united briefly in 'love chambers' or 'love baths' before being swallowed up by the rising tide. [53]

On the other hand, the Romantics managed to make the sea-shore into the symbolic place of fidelity. This was, paradoxically, the stage on which love adventures unaffected by the passing of time and capable, if necessary, of holding out against death, were played out in an often dramatic fashion. The pathetic figure of the fiancée on the shore, [54] descendant of the Ossianic virgin condemned to die on her lonely island, was a source of deep emotion for Romantic travellers; hardly a one failed to claim to have met her. This touching figure of the young woman facing the death of her lover gradually replaced that of the voluptuously shaped, modest fiancée swept up in the wrath of the neoclassical sea.[55]

THE SENSUAL EXPERIENCE OF THE BEACH

The prodigious enrichment of the emotions provided by the sea-shore, arising a good time after the fashion for seaside holidays, transformed the means of expressing desire. A new intimacy became established between the walker and the elements. The ocean was no longer merely a sublime spectacle to be gazed at from the top of a cliff or a picturesque scene to be framed from the height of a viewpoint offering a clearly separate, dominating vision. The dialogue with the waves or with grottoes suggests that the spectator's position still held by Beattie's minstrel should be abandoned. It ushered in a desire for close contact as a prelude to an imaginary merging with the scene. The Romantic walker avidly appreciated the sensuality of the beach; bathers explored in detail sensations that were previously confined to things not mentioned. All the ways of using the shore were modified as a result.

Painters of the period then developed the practice of coastal journeys, which had been inaugurated by seventeenth-century Dutch artists before being adopted by lovers of the picturesque. In a boat on the open sea, it was possible to appreciate the sight of the sea-shore going by. This fashion corresponded to the vulgarization of a literary device. As early as 1754, Fielding was proclaiming, in passing, the pleasure he found in this gliding journey of the motionless eye. This same vision shapes the descriptions in *Childe Harold*; Byron's ocean is a coastal ocean, and the text offers the reader a kinetic vision, broken up with occasional travel shots, of the coastal spectacle.[56]

This rising mode of delight was reinforced by the growth of yachting. The Royal Yacht Club was founded in 1812. On the Isle of Wight, the first races from Cowes took place in 1826. As early as the following year, they inspired Turner. In both France and England, landscape-painters knowingly increased their production of coastal scenes in order to broaden the circle of 'seascape' lovers.[57] As these artists were often rather poor sailors, they contented themselves with working along the shores.[58]

Excursions on the sea became commonplace, related to the rise of the seaside holiday, forming the prelude to this coastal journey. A play of interactions gradually developed between art history and the history of the seaside holiday, giving rise to a multi-faceted craze for coastal life. Seascape painting shifted its focus to northern shores, and this accompanied the movement of landscape painting towards coastal scenes. These two developments encouraged artists to travel along the Channel and Atlantic coasts in order to set down in their sketch-books the sights they observed during this close-to-shore navigation.[59]

Enriched by the topographers' tradition and the fashion for picturesque travel undertaken according to Hassell, sailing along the coasts was henceforth an event in the careers of the greatest sea-shore painters, including Constable, Turner, and Cotman. This experience culminated in vast systematic explorations. William Daniell began an exhaustive survey of the British coasts in 1813, part of which he accomplished offshore. This long journey was to lead to the publication of his *Voyage around Great Britain* in eight volumes, with 308 water-colour illustrations. In the course of twelve years, Daniell visited the ports and nearly all the beaches of Britain; he crossed lochs and estuaries, gazed at the sea from coastal cliffs, and travelled around a great number of islands. Driven by a need to circumscribe the nation's territory and make its outline tangible to the public, this long process expanded a Crusoe-like adventure to the whole of the British Isles, and took on the form of a hymn to the insularity of a nation that had just overcome terrible trials. More than any travel account, the tremendous labour accomplished by artists along the coasts between 1800 and 1840 is proof of the rising prestige of the sea-shore and the growing interest in all forms of coastal life.[60]

It had long been customary to travel on horseback along the shores of Great Britain, Flanders, and the United Provinces. This practice, sometimes combined with hunting, was part of the range of rural sports and one of the pleasures of the holiday resort.[61] It was also an easy form of travel when roads and paths were in very bad condition. The contributions of Romantic artists to images of the horse[62] and the exaltation of night-time cavalcades along the water contributed to reviving the pleasure of this form of exercise. The joys of horseback riding along the sea-shore became something to describe. Listening to the rhythm of his galloping horse registering a counterpoint to the sound of the breaking waves, the rider was filled with a new emotion as he roamed over the empty strand at the moving frontier between the elements. Byron, who was fond of riding along the beach,[63] displayed this new emotional palette in *The Giaour*. The most prolix of his admirers, Prince Pückler-Muskau, gave a glowingly detailed account of the joys he experienced as he made his horse gallop across sand hardened by the receding waters:

> Then [when the moon rose], I came down from the hills towards the seaside, and I covered the five or six miles that still lay between me and Brighton by following the sandy shore, and making my horse almost dip his feet in the sea. The tide was coming in, and my horse often jumped sideways when a wave crested with white foam came rolling under his feet and then quickly drew back as if it were playing with us. There is nothing I like better than horse riding by moonlight on the vast, lonely shore of the sea with its waves roaring next to me. Only with difficulty can you keep the animal close to the mysterious abyss.[64]

Walking over the beach had long been part of the ritual of 'conversation', and then of the holiday stay. With the Romantics, the heirs to the Ossianic model, it was transformed into a form of *wandering* along the shores. Following a path along the edge of the water, on the aquatic land of the strand, heightened emotions. This explains the new appreciation for, and extolling of, flat coasts across whose surface the tides could unfold broadly.[65] Whereas in the preceding century, the wilderness implied horrible abysses or, at the very least, rocks with a majestic appearance, it now came to include even very ordinary nature, but a nature that refused to record social change, a nature of groves and thin underbrush, of brambles and thorns and vast, flat expanses of sand.[66]

Chateaubriand, accustomed from childhood to walking over the shores of the Mont-Saint-Michel Bay, found his roots in recollections of that unsteady soil:

> Between sea and land, Pelagian plains stretch out like unsettled boundaries between the two elements: there, the skylark flies as well as the sea-lark; the cart and the boat, within a stone's throw of each other, plough the water and the land. The sailor and the shepherd use one another's speech ... Sands of various colours, varied beds of sea shells, kelp, fringes of silvery foam outline the golden or green edge of wheat fields.[67]

On the open strand, the walker, heady as he trod a new shore, could follow that image of desire, the slowly rising tide. He was free to advance against the wind, barefooted at the edge of the water, thus experiencing the threefold caress of the elements.

Between Edwin's wanderings and those of René, the 'Romantic Ossian',[68] whether in Caledonia or along the shores of the sea bordering the monastery in which Amélie, her soul 'stormy like the ocean', has her abode, the direct lineage is obvious. Elsewhere, Chateaubriand imagines the barbarian of olden times: 'wandering over wild shores, listening to the voice that comes out of the ocean, he gradually lapsed into reverie; straying from thought to thought, as do the waves from whisper to whisper, in the vagueness of his desires, he merged with the elements'.[69]

In Germany, too, Ossianic influence could be felt in this respect. Channelled through Klopstock's authority, in conjunction with the influence of ancient models, it affected young Friedrich von Stolberg. The first poems he devoted to extolling the Baltic and North Seas' shores are steeped in the Caledonian model. The poet summons up Fingal, Selma's lyre, and the tears shed by virgins on the graves of heroes.[70] In 1776, Stolberg *says*,[71] he seeks a thrill on the nocturnal shores of the resounding sea; alone, by the light of the moon and the stars, his hero lets himself be rocked by the breezes and by the approaching noise of the purple waves.

The walker enjoys sitting close to the water, a compensation for the impossibility of practising nudity there. He takes pleasure in watching the waves unfurl.[72] As his horizontal glance broadens the field of the sky and the ocean, he soaks up the smell of the seaweed, the taste of sea spray, and the sound of the rollers.

The fourth canto of *Childe Harold*, published in 1818, offers the literary model for this attitude. The ocean that Byron evokes here is indeed that of beaches and bathers.[73] The poem speaks of a sensitivity to the alternation between silence and the loud crash of the breakers, and of an ear for the music of 'the roaring ocean'[74] or the babble of the wavelets on the water's fringe.[75] It also speaks of pleasure in the contact with the salty wind that penetrates one's skin and ruffles one's hair. This awareness of the ocean's harmonic breathing, to which Shelley proves to be particularly sensitive, accords with medical discourse: physicians, more concerned than ever about the working of the lungs, expected that the sea would have a beneficial effect on their operation. The sharper view of the fogs that covered the waters was part of the growing attention to beach phenomena.

All that remains is to provide a better illustration of actual practices that either followed or triggered these literary models. In July 1800, Ann Radcliffe, accompanied by her husband, undertook one of her annual excursions along the southern coast of England. The journey, carefully recorded, regularly included solitary outings along frighteningly lonely shores. The novelist's boldness led her to leave the beaten track and seek out adventure and danger.[76] Custine, a great beach-side traveller in the English manner, fell under the spell of the Celtic twilight as he stood on the edge of the Scottish shore's black water: 'When the darkness of night spreads over these desolate landscapes, the heart of man opens to sadness, and the most melancholy poetry becomes the natural expression of his inner feelings.'[77]

Denise Delouche[78] shows how Chateaubriand's *René*, immediately picked up, commented upon, and illustrated by painters,[79] exerted an enormous influence. Wandering along the shores, advocated by Cambry as early as 1795, very soon became stereotypical behaviour, emerging as a regular feature of the journey to Brittany. At the beginning of the 1830s, this practice suddenly began to spread much more quickly. In this respect, Balzac, Lamartine, Michelet, and then Victor Hugo, as well as members of the Breton Romantic school, share the sensibility of Hippolyte de la Morvonnais, author of *Thébaïde des Grèves* [Thebaïd of the Shores]. It was not until 1848 that Brizeux managed to challenge this model of appreciation and to impose on public opinion the image of the Breton Arcadia.[80]

When Ducos visited England in 1826, he would dream for hours on the beach bordering Dunbar castle. He even returned there at night with his companions.[81]

Long before the escapade by Flaubert and Maxime du Camp, who merely borrowed and redirected an attitude that had already become commonplace, wandering along the shores was built up into veritable walking trips. Young French Restoration artists who read Rousseau's *Confessions* were probably imitating German ramblers as they set out, not afraid of covering dozens of kilometres on foot. The novelty lay in the fact that they were doing so by the seaside.[82] Eugène Isabey, who began to paint Étretat around 1820, went wandering the following year over cliffs and shores. Like his friend Le Poittevin, he planted his easel on the sand.[83] Paul Huet, too, was fond of long walks on the shore.[84] *La Description du Croisic et d'une partie de la côte voisine*, published in 1823 by Edouard Richer, and that of the *Île de Noirmoutier* are nothing but lengthy invitations to travel along the shores and the strand, especially when 'moonbeams bring out the wet muddy bottom that lies before you and give it the colour of water and sand blended together', or else when the walker, lost in the autumn mists in the midst of 'this new ground' strewn with fucus-covered rocks, can hear the crabs 'that glide quickly along ... and their bony limbs that resound in the rocks' cavities'.[85] The hiking trips undertaken by Victor Hugo beginning in 1834 therefore fit into a chain of practices that was already well established and were reminiscent of the coastal journeys of British artists.[86]

The Romantic code also gives new form to the procedures and postures of confrontation. Less appreciative of the excitement provided by the semblance of danger, the Romantic walker likes to stand atop a rock as if rooted there, in a defiant attitude in the face of the elements assailing the headland. This is a heroic posture which at the same time fosters meditation and seems to foreshadow a plan for domination.[87] On a rocky pedestal that jabs into the waves and thrusts into the sky, the solitary, inaccessible walker, like a lighthouse for an instant, feels capable of addressing the ocean that he takes in with his eye. Childe Harold, the poet of *The Revolt of Islam*, and in another register the Victor Hugo of 'Extase' in *Les Orientales* and later in *Les Contemplations* outline models for this heroic posture. The Promethean silhouette of Napoleon as a prisoner within the iron circle of St Helena's rocky coast makes an obscure addition to the fascination of this pretentious figure.

In this respect, it is necessary to distinguish carefully between the emotional strategy of the viewpoint and that of the headland with its massive verticality. Between them stands the same distance as that which separates Mont-Sainte-Catherine from the cliffs of Saint-Valéry-en-Caux, both in Normandy.[88] As the coastline becomes more developed, the use that viewers make of the top of the headland or the lighthouse (a visit that is part of a growing fashion[89]) becomes reversed. Think again of Prince Pückler-Muskau's text. It is no longer the naked immensity that tires his eye as he gazes from the top of the Anglesey cliffs, but the analysis of the picturesque scene that spreads out at his feet. From the sight of infinity in which the elements merge, he expects to relax into reverie. The system of appreciation set forth by Saint-Évremond has been turned inside out.

> The need for rest makes one finally cast one's glance towards the North ... There, there are no further distractions. The vast ocean blends into the sky. For a few more minutes one follows the Anglesey coast, where oaks and walnut trees dip their branches into the sea, and then it recedes, so as *to leave the eye alone*

with the water and the sky, or at most with the indistinct form of some great vessel on the edge of the horizon, or with the strangely shaped cloud that quickly crosses through the air and vanishes. [90]

Breaking with the dry observations of a Fanny Burney, the Romantics dared to analyse and express their impressions of sea bathing. Without a doubt, it is to them that we owe the sexual dichotomy of the figures and attitudes described earlier. They set down in writing two different ranges of emotions. The first accompanies the voluptuous fall. The soothing embrace of the element evokes both sexual union and a mother's rocking. Sea water suggests the aquatic image of beneficent femininity, insidiously accentuated by the proximity of shells. [91] The young man whom Novalis places in his *Disciples at Saïs* releases his 'longing for fluidity'; he boldly praises the sensual delight he equates with liquefaction; he offers to restore the link that had once been established between the quest for happiness and the proximity of the sea. Diving provides an occasion to enjoy the feeling of being attuned to elemental forces and experiencing the coenaesthetic harmony that exists between the movements of the sea and those of the original waters carried within the human body.

The cave bottom in which waves are transformed into lovely maidens gave Heinrich von Ofterdingen an irrepressible longing to swim. [92] Jean-Paul, for whom water suggests bathing women, or Chateaubriand, who hints at the voluptuousness experienced by Mila and Outougamiz as they bathe [93] bear witness to the renewal of the sexual solicitation implied in classical scenes of bathing women. The important point here is no longer the sudden surprise of seduction or the visual breach that leads to the heart of feminine profusion; this is no longer what incites the Peeping Toms at Brighton to bring out their field-glasses. Rather, it is the intense sexualization of the spot and of its substance through the – now symbolic – presence of the beautiful woman who transforms the beach on which she stands, alone and streaming with water, into a magical realm. [94]

In 1834, Balzac contrasted the masculine way of swimming experienced by the narrator of his novel *Un drame au bord de la mer* [Drama at the Seaside] with Pauline's bathing in a 'granite hollow filled with fine sand'. He highlights the underlying eroticism by emphasizing the emotion caused by 'the cry of a woman who emerges from her bathe, revived and joyful' and whose call prevails over the alternating murmur of the water's ebb and flow.[95]

Very important in this respect is the significance of the writings and experiences of young Friedrich von Stolberg in this genealogy of images and practices. He states that he loves to bathe in the straits as the sun is setting and the moon is rising. In his early poems (1776–7), he proclaims the joy and the pleasure of plunging into the sea in the midst of the glowing waves, when water and fire merge.[96] Swimming for Stolberg is like an embrace; it gives rise to the longing to be swept under. 'Get hold of me, thou, Goddess, / Take me into thy mighty bosom!'[97] As early as 1776, and therefore long before Lamartine's confessions, the quest for the mother is clearly designated here as a component of aquatic pleasure.

At the same time, bathing was also perceived as a confrontation, and it borrowed from the aesthetics of the sublime. Stolberg went to the coast in 1777 to face the violent waters amid the rocks. Throughout his life, Byron proclaimed himself a great lover of sea bathing. At the end of his life, he wrote of a lame man's revenge, declaring that he had covered more miles in the water by swimming than had all other living poets by boat.[98] An admirer of the Mediterranean and its shimmering waters, sensitive to the fantasy of thalassic regression, he also applied himself to making bathing theatrical. In it, he saw an opportunity to surpass oneself heroically. In his virile quest, which led him to cross the Dardanelles in 1810 (and which foreshadows the mania for record setting), Byron, mentor of the 1815 generation,[99] shaped the poetic model of masculine bathing as a cavalcade through the waves and a fight against being pulled under that was in keeping with the energetic ways of swimming of the times.

THE PRIVATE PARADISES OF THE SEA-SHORE

With the Romantics, the Crusoe-like adventure, a spreading practice, became more openly than ever before a part of 'the dialectics of back-tracking'.[100] This adventure was henceforth admitted to be a search for privacy. The delimitation of a territory and the reintroduction of boundaries made it possible to mark out a blessed space[101] protected by the abyss. Hence the many feelings that Edouard Richer related in detail as early as 1823 in his description of the isle of Noirmoutier:[102]

One enjoys being closed in within a space whose limits were set by nature herself. Political and moral divisions trouble one's thoughts: they look too much like a form of imprisonment. Those of the ocean, on the other hand, carry within them the ideas of rest and security, and the element that surrounds you seems to have been placed there in order to defend you as much as to isolate you from the rest of the world.

Whether they are protected by a circle of wild reefs or appear to be bathed in a paradisiacal sea, the island shores take on the shape of so many secure shelters, whose child-like innocence abolishes for a time the conflict established between natural desire and moral duty. The quest for seclusion and for a fortunate island becomes transformed by the Romantics into a longing to curl up in the mother's bosom. On the island, that 'mythical image of woman, the virgin, the mother',[103] scenes of regression can be freely played out.

Within the insular space, the creek and, better still, the cave allow the adventurer to elaborate fully the quest for shelter. These inner paradises act as substitutes for the maternal womb; inside them, far from human ears, the stone listens to the water and replies to the sea, increasing the joys of swimming.

The spreading theme of the Crusoe-like adventure and of insular nostalgia was in keeping with the growing quest for privacy within the social body; it was in harmony with the withdrawal into home life taking place among members of the ruling classes. The fascination exerted by the creek and the cave corresponds to the longing for a house, a room, a corner of one's own.[104] The rising prestige of the Crusoe-like adventure is a retort to the toughening of the social game and to the indecisiveness and growing fragility of social ranks.[105]

There are numerous literary models of this renewal in attitudes: Byron indulges in dreaming of Robinson Crusoe-like adventures, and through the account of the love between Don Juan and Haidée on the Cyclades, gives an erotic character to the paradisiacal innocence of island love.[106] Shelley's writings also give a privileged place to the green oasis of the island shelter. Two Romantic artists took particular delight in experiencing island life: Caspar David Friedrich on the island of Rügen and Heinrich Heine in 1826 on Norderney.

In search of the 'spiritual resonances'[107] evoked by the contemplation of the spectacle of nature, Friedrich made frequent visits to the island of Rügen, which his friend Kosegarten considered to be an ideal setting for the saga of the Nordic past. Sometimes accompanied by Philipp Otto Runge, Dahl, or Carus and sometimes alone, Friedrich would take long morning or evening walks, scrambling over the jumbled rocks, following the sheer or vertical white cliffs before returning to his modest inn, under the seamen's astonished gaze. On stormy days, fearing neither rain nor spray, he would rush to be a lonely spectator of the waves' fury as they beat the reefs. On calm days, he sketched the dunes, the beach vegetation, or the rocks. He read Jean-Paul's poems and novels. The island of Rügen was his favourite refuge from anxiety. The range of emotions it offered his nostalgic soul differs from Englefield's picturesque quest on the Isle of Wight, from Townley's therapeutic exercise near Douglas, from the journey of geologists tormented by *libido sciendi* when visiting volcanic islands, as well as from Jean Houel's neoclassical appreciation of Sicilian shores. The diversity of so many contemporaneous longings clearly shows the complex richness of the Robinson Crusoe-like adventure.

Some twenty years later, Heinrich Heine spent a long period of time on Norderney island. The account he published along with the *Nordsee* verses assesses age-old attitudes. Heine recapitulates all recent forms of excursions, reverie, and contemplation in his experience. The influence of Homer can be felt in his text,[108] as well as traces of classical seascapes and Rousseauist reverie. The author sinks into Romantic meditation; he is fond of encountering the people who inhabit the shores. He undertakes an ethnological investigation: he listens to legends, hunts along the coast, walks alone at night by the seaside. The sound of the waves awakens forebodings in him. 'As from the depths of the centuries', he writes, 'all manner of thoughts spring up in my mind, thoughts of primitive, prophetic wisdom.'[109] By 1826, the island had become a magnet for seaside tourism; in order fully to enjoy the pleasures it provided, a visitor had absolutely to avoid the bathing season. This marked the beginning of the withering away of this recent model of Robinson Crusoe-type adventures.

The new emotional strategy governing the Romantic voyage and the modes of recording it inspired also led to a shift in ways of enjoying the sea-shore. Unlike the classical tourist, the Romantic traveller does not intend merely to make a cultural pilgrimage;[110] his aim is not to compare written texts with landscapes, to indulge in the pleasure of recognition, and, if need be, to measure distances. The purpose of his peregrination is to fulfil an individual dream inaugurated by premonition. The trip is filled with ceaseless shifts from the world of reality to the imagined world arising from the confrontation of things seen and things dreamed. In the case of young French artists, whether from the middle class or the aristocracy, their minds are filled with accounts of emigration, and they are tempted to escape from their commonplace post-Napoleonic present: departing is for them the high point of the journey.[111] It becomes important to prepare for the trip, not by a laborious process of documentation, but through preliminary work of the imagination.

The essential work resides perhaps in a long reflection on the longing to travel and the pleasure in and meaning of the journey.[112] The response to this reflection is what conditions the way the journey is recorded. The purpose of the Romantic travel account is not to portray;[113] what matters is to describe not the reality of an object, but its 'effects upon the soul'.[114] Travellers look for the first 'free, natural' impressions, for feelings, not ripe observations.[115] This explains the importance of the loved one whose absence stimulates literary production, makes correspondence necessary, and arouses emotion in the soul. By contrast, the Romantic traveller was revolted by the movement then under way which was turning tourism into a mass industry. Romantics were driven to avoid caravans and even groups, whose chatter hindered the emergence of impressions. 'For four men, a new country, a new nature and new customs are a show; for a single one, they are a conquest.'[116]

Romantic travel accounts produced their own readers. Custine wanted his to be 'interested in recognizing an individual's character through the images he makes of the world'.[117] In the final analysis, the Romantic traveller, who sets out in search of his inner self, speaks only of himself to a reader who is interested in this other individual's dream because an obvious cultural proximity brings it very close to his own. The see-sawing of intentions accounts for the drawing out of the text, which is encumbered with or, rather, enriched by the analysis of aroused emotions. The traveller lingers over the account of time spent in contemplation and daydreaming; he stresses the importance of the fringes, the boundaries, and the shores where the journey begins and of the impressions aroused by the ultimate stopping points leading to its conclusion.

The emotion aroused by standing or walking on the sea-shore is all the greater since, in the course of the Romantic journey, the shore stands out as the boundary where things begin, along which people dream of the journey to be undertaken. Through this initial rite, a break with everyday life is made. 'I was quite sad when I left Paris,' observed Custine on his way to England, 'but when I look at the sea, I feel that the *spirit* of the journey is taking hold of me.'[118] At Dieppe for Nodier and at Boulogne for Custine, the sight of the sea brings with it the hope that the promises of the dream will materialize. This hope is what fills the latter with 'the demon for travel', and starts the metamorphosis of the individual that is indispensable for the success of Romantic peregrination. It is during the inaugural walk along the shore that questions about the meaning of the journey become deeper, and the destructive certainty surges forth that all the expected pleasures will be nothing but illusions.

The Romantic code exerted only a belated influence on the aesthetics and emotional strategy imposed by tourists' guidebooks. Until the mid-nineteenth century, such guides fostered mainly the picturesque approach, combined with a cult of 'the wonders of nature'. This accounts for their relative casualness regarding the sea-shores,[119] unless these were scattered with crevices, caverns, jagged rocks, or desolate dunes.

It is in other writings that the social progress of Romantic sensibility can be traced, especially in books that were intended both as travel accounts and as tourist propaganda whose aim was to attract attention to a region or alter its image. In this connection, Émile Souvestre's accounts are exemplary.[120] The author intends through them to share Romantic rapture and to use it as a touristic asset. Speaking of the Breton shores, his use of the imperative applies not only to things that visitors should do, but also to the 'Romantic' feelings they should experience. The uncertain use of 'you' and 'I' helps the reader to identify with the author and, consequently, to receive his message. Sometimes the reader cannot even be sure whether the author is simply describing his own emotions or calling on the reader to experience them or, at any rate, indicating what he should be experiencing. Souvestre sometimes suggests astonishment, and sometimes contemplation; he makes Romantic attitudes commonplace, and dictates a form of behaviour that is very close to that which Victor Hugo was to develop along the Channel shores.

Part III

The Growing Complexity
of the Social Spectacle

7

The Visit to the Harbour

The practice of walking along wharfs and stone piers, which continued in new forms, expressed the fascination exerted by a stage on which spectators could observe particularly manifest displays of energy, activity, heroism, and misfortune. It fitted logically into the classical journey. Here nature had retreated before the labour of man, who had cut stones and reshaped the boundaries that God had set to the ocean. As the intersection between a fashion popularized by seascape painting and several forms of curiosity that led people to go down to the *quay* in order to watch the veritable plates from an animated encyclopaedia, this didactic place also sought to be a symbol of royal splendour. It was simultaneously a *limes* whose safety must be protected, a haven from which majestic fleets sailed forth, a scene of the pathos of shipwrecks and defeats, and a receptacle for treasures in which a colourful parade of diverse peoples could be seen: this full territory both compensated for, and accentuated, the surrounding void on beaches and rocky coasts.

NUMEROUS WAYS OF SEEING

On this stone bank, vision was managed – not to say enslaved – by a long history of modes of delight and culturalist ambitions. Classical education fostered emotion before the spectacle and the visit to the harbour. Ever since the Hellenistic era, port architecture had called for *monumental beauty*. Alexandria played a crucial role in creating and spreading ancient models of harbour aesthetics.[1] Subsequent representations of the seaside were deeply affected by the impact of models developed in Asia in the wake of the Carian painter Protogenes of Caunos. The lighthouse built by Sostratos of Cnidus, the richly decorated pier mole with an open passage, and the boat that could be seen sailing over a rough sea bordered by a rocky coast comprise the package of stereotypes that was imposed on Roman architects and aesthetes. The Pozzuoli harbour and the Ostia lighthouse commissioned by Claudius deliberately conform to a model successfully imposed on Rome by the painting of Demetrius of Alexandria as early as the era of the Ptolemies and strengthened subsequently by the presence of numerous Egyptian sailors in Italian ports.

In the course of the second century, the Antonines' maritime ambitions stimulated a harbour aesthetics obsessed with symbolism and the need to construct political images. For imperial Rome, imitating the Alexandrians was a way of seizing the prestige associated with the 'Faros' of Ptolemaic Egypt. Moreover, the lighthouse tower was part of the tutelary image of the harbour; it was a sacrifice to the spirits that protected navigation. The Romans had an intense feeling for the significance of this place which was not meant solely for loading cargoes. On the flagstones of the quay, exchanges also occurred between religions, social doctrines, and literary themes.

Ancient images continued to speak to the eighteenth century. Chevalier de Jaucourt, who was responsible for the articles dealing with ports in the *Encyclopédie*, devoted most of his remarks to Greece and Rome.[2] In fact, though, the model had passed through several filters. Claude Lorrain's work, itself partially heir to a Flemish tradition, stood out in the minds of port visitors. At the very moment when seascape painting reached its peak in Holland, Claude was inventing the seaports that he offered to the viewer. He combined the stereotypes of ancient architecture and descriptions from the *Iliad* and the *Aeneid* with the monumentality of Renaissance ports, handled like so many theatre sets. Guided by the wish to give a noble quality to the scene, he focused more on architectural splendour than on representation of the harbour's activity and the jumble of wares with which it was cluttered.

Claude Lorrain's painting invites interpretation; the harbour stands out as a stage on which, more than anywhere else, the world's harmony can be recreated and the unity of its constituent elements re-established. Moreover, he develops religious symbolism there with particular clarity. The evocation of the

Christian's destiny and the allusion to the spread of the divine word underlie the paintings of embarkation scenes and the way ships are represented; they emphasize the journey's expectations, insidiously suggested to the spectator. Aeneas's landing is intended as an illustration of the ancient story, but it benefits from the intense symbolic value of the shelter or haven that welcomes the repentant sinner.

What is most important here, however, is the interpretation made by viewers of the Enlightenment period. They were less aware of the network of Christian symbols in this type of painting than of the evocation of harbour scenes they had read about in novels on ancient sailing, ranging from Fénelon's *Télémaque* to Abbé Barthélemy's *Voyage du jeune Anacharsis* [Young Anacharsis's Journey]. The descriptions of Tyre, Alexandria, Carthage, Syracuse, and Sallentinum were engraved in the memories of cultivated visitors, and these individuals were well aware that the Alexandria Lighthouse was one of the seven wonders of the world.

Vision was not oriented solely by this humanistic culture, though its impact was certainly very great. Classical tradition was combined, especially in British minds, with the harbour scenes in Flemish and then Dutch painting. The latter in particular was a profound source of inspiration for English seascape painting, dominated since 1670 by the presence in London of the two Van de Veldes.[3] Several big-bellied ships from the United Provinces lay at anchor along quays on that side of the Channel, and everyone could go and view them. The port of Amsterdam was a microcosm in which riches arrived from all over the planet,[4] bearing witness to divine blessing and to the heroism that had enabled man to triumph over the raging elements: very early on, it offered a model of enjoyment that spread during the Enlightenment. The attraction that the Dutch had long felt for the sight of great companies' fleets leaving and returning to the harbour paved the way for the pleasure that the 'harbour scene' was to provide for British enthusiasts until the appearance much later of Turner's masterpieces.

During the second half of the eighteenth century, the injunctions of the picturesque code upset previous models of appreciation. The renewal was rooted in the *vedute* of Italian harbour cities, and especially in the representation of Genoa. This art was flourishing by the end of the fifteenth century, and probably inspired Claude Lorrain and Italian seascapes. It was Joseph Vernet, however, who was to establish for quite some time the ambiguity of a harbour scene in which no clear division could be made between the recording of experiences and the mere reproduction of imaginary stereotypes. [5]

Inspired by the topographers' tradition, Vernet made the harbour view into a privileged panorama. In his work, the port is first and foremost a picture that walkers in the hills examined with their spyglasses. [6] The harbour scene was to become the archetype for the viewpoint: the unprecedented vogue for such views from the shores of Naples to the hills of Edinburgh [7] was described earlier. But Vernet, who preferred evening scenes for this kind of picture, wanted to incorporate the picturesque quality of human beings. In the 'Ports of France' series, the variety of costumes, the appearance these give to the bodies, and the gestures of the characters all contribute to animate each picture. [8] In an age of physiocrats and physicians following in Halley's footsteps, the sight of people in motion, like that of commodities, was pleasing to the eye. The code of aesthetics coincided with economic theory and the discoveries made by physiology. Joseph de la Borde, the Court banker, undertook to have his château at La Ferté-Vidame decorated; he commissioned Vernet to paint eight large pictures, one of which was to represent the sun setting on 'an opulent harbour, abounding in towers, lighthouses, and ships, where a crowd of labouring men unload bundles, sacks, and barrels before Levantines who are smoking at leisure'. [9]

For the traveller, especially the English tourist landing at Calais, the harbour was simultaneously a text that abruptly portrayed the picturesqueness of another people and a stage on which the traveller made his entrance into this strange picture. Hence the long, stereotyped pages devoted to an experience that was, after all, commonplace, but that took on the role of an initiatory rite.

In the eyes of someone viewing a picture, as in those of the port visitor, the picturesque quality of the harbour was accentuated by the juxtaposition of characters and by the display of social positions. The quays were part of the itinerary of working-class women; children – and domestic animals – seemed to gambol freely in a space that was an extension and enrichment of the working-class street. The merchant and his assistant came there to keep an eye on their business; the port was their territory. The upper-class visitor indulged in the pleasure of conversation there; if need be, he knew how to show his gallantry. The harbour was a public space in which the theatricality of social positions was played out. The hierarchy of status can be read with particular clarity in Vernet's canvases, in the very way that people and things are jumbled together. The painter manages to suggest obvious divisions, while at the same time applying himself to discovering the particularity of a place[10] and to erasing both its violence and its traditional immodesty.[11]

In the second half of the eighteenth century, the profoundest change in the manner of appreciating the harbour scene nevertheless resulted from the triumph of *didactic ambitions* over aesthetic enjoyment. In the age of the encyclopaedia, the harbour emerged first of all as an instructive spot where one went to study 'the object lessons of the sea'. De Marigny was seeking to satisfy the curiosity of the king, the court, and the city when he commissioned Vernet to paint the 'Ports of France' series. And in fact, above and beyond the wishes of the elite, it was the curiosity of all the country's subjects that was being satisfied.

The spectacle of goods being unloaded, loaded, or transferred, of bundles being carried and barrels being rolled in the perpetual activity of big ports, and the displays that preceded the distribution of goods encouraged the spectator to establish a taxonomy of the globe's treasures. At the same time, the patchwork of ethnic groups,[12] which was especially visible in Amsterdam, London, and Marseilles, allowed observers to travel without leaving home, brought exoticism to their doorstep, and emphasized the civilizing mission of the harbour,[13] which became effectively an ethnological museum. As a topographer, Joseph Vernet devoted himself to highlighting the didactic role of this site of the engineer's triumph. Using images, he explained technology, architecture, and the layout of warehouses and factories. As a painter, he illustrated the specificity of workers' actions. Especially revealing in this respect is his *Port de Toulon*, teeming with working people and officers. In it, the viewer can see butchers at work, livestock being moved around, wine being decanted, and arrangements of vegetables, flour, cheeses, and sugar ready to be loaded. The *Port de Rochefort* explains the process of shipbuilding. His picture of Dieppe harbour, the last in the series, would have fitted well into the *Traité général des pêches* [General Treatise on Fishery] published a few years later by Duhamel du Monceau.

Critics and amateurs were very much aware of the didactic aim of this seascape painting; some sought to learn from painting exhibitions; others, by contrast, discussed the utility of the lesson. In 1755, one visitor took up Vernet's defence: 'From his paintings of the Marseilles harbour and Toulon's naval dockyard, ... one could easily learn things that many people ought to know and about which they would have a hard time finding instruction.'[14] This anonymous amateur expressed the desire that in paintings, viewers might discover 'the history of customs, arts, and nations; these would always be interesting, if true, because they would be useful'. Vernet himself, in the course of his itinerary, stressed his didactic aim and his ethnological interests. The way in which, in agreement with the municipality, he shifted the focus of his plan for a painting of the Bayonne harbour demonstrates this intentional drift. Through a long process that can be analysed in the correspondence between the town's mayor and the deputy, the painter worked to bring out the specificity of the place, thereby contributing to the fashioning of a picturesque image of the site he had been asked to paint. Some time later, the fallacious painting of the quays at La Rochelle, packed with large-tonnage ships, transmitted the erroneous image of an intensely busy port.[15]

Vernet's harbour scenes, widely disseminated as engravings, and then those of his pupil Jean-François Hue between 1781 and 1812,[16] contributed at least as much as travel accounts to satisfying readers' encyclopaedic curiosity. Louis-Philippe Crépin and Pierre and Nicolas Ozanne, who taught the Dauphin's son, played their part in this deliberate production of harbour images. The vast output of the last of these artists,[17] commissioned in turn by Louis XVI in 1775 to sketch the views and draw scrupulously accurate maps of the kingdom's ports, was to serve as a model for French artistic production in the first third of the nineteenth century. The series painted by Garneray between 1821 and 1832 represents a late example of this didactic art.

An understanding of the harbour scene must take into account one final purpose: along the quays, people came both to measure and to contemplate the greatness of the kingdom.[18] This situation calls once again for a classical interpretation, in keeping with the political culture inherited from Rome and the Stoic tradition of heroic landscapes. Eighteenth-century travellers associated the port with the citadel. One of the tourist's first concerns, we recall, was to assess the quality of the defences of the narrows and the harbour. The frequently taken coastal itinerary offered an opportunity to identify the most successful achievements. Travellers were keenly aware of the tragic character of the port; they could imagine the sudden appearance of an enemy fleet or bombing by enemy ships.[19] In the 'Ports' entry in the *Encyclopédie*, Jaucourt stresses the security of various sites more than their activity or the prosperity of their trade. The departure of squadrons, the tragic spectacle of battle-worn ships returning or a vessel in peril as it approaches the fair-way were all subjects greatly appreciated by seascape-painters. Remember that in the Dunkirk region, to cite but one example, the majority of all shipwrecks and most drowning accidents occurred inside the port or the roadway. At a time when various hierarchies were being insidiously, then bitterly, contested within the ailing city, the port was an ambivalent place, both disturbing and comforting. As a space open to the wealth and the threats of the world, it evoked both shelter and refuge, and vulnerability. It combined images of invasion and escapism.[20]

THE PATHS OF DELIGHT

The numerous interpretations of the harbour scene led to an abundance of practices. In the second half of the century of Enlightenment, tourists took up the habit of climbing the hills overlooking a harbour to enjoy the panorama. On the heights that frame the site, a picnic meal among family and friends, such as the one that appears in the *Port de Marseille* [plate 21] painted by Vernet in 1753, might complete the 'pleasure outing'.[21] The languid positions of the bodies here indicates that these people are accustomed to rustic visits. French elites of the time chose the harbour as the place where they went to 'see the sea' and eat fish; this combination seemed so obvious that on this point a close link developed between the pleasures of taste and those of the eye. The merry escapade to Dieppe related by the Duke of Croÿ in 1754, just at the time when bathers were beginning to rush to Brighton, is highly informative in this respect; it provides a good measure of the distance sepa-rating the English aristocracy, already fond of seaside holidays, from the French nobility, who found only brief moments of delight in the view of the sea, moments integrated into a codified series of amusements that make up a pleasure outing. In fact, on 23 May, the Duke and the young Prince and Princess of Condé, with whom he was staying at Chantilly, decided on a whim to go for just a few hours to 'see the sea'. For this sole purpose they covered hundreds of kilometres.[22]

It would be difficult as well as tedious to draw up a complete list of travellers who, driven by the wish to satisfy their many-faceted curiosity, made for the harbour as soon as they arrived in a town. Montesquieu, who visited Italy in 1726, is the finest example of a traveller whose perception of a coastline is obviously shaped by his concern for assessing its defensive capacity. His account lingers over the description of the harbour installations (depth, exposure, access, protection) with a view to a possible conflict. His descriptions of the ports of Genoa, La Spezia, and above all Livorno are nothing more than analyses of their systems of defence. The few lines devoted to the sea reflect this preoccupation.[23] In 1739, President de Brosses, who was passing through Marseilles, made only a few remarks about this city;[24] in 1758, by contrast, Grosley declared that he was keenly interested in the activity of the port of Ancona, its prosperous shops, its warehouses, and the circulation of goods in it. The prosperity of the place alone extolled the glory of Pope Clement XII. Two years later, Marmontel asserted: 'Two things held my lively interest and my eager attention, namely those two famous ports, Marseilles, famous for its trade, and Toulon, famous for war.' Concerning the former, he added: 'The short time we spent there was wholly taken up by visiting the harbour, its defences, its warehouses, and the great objects of this trade that war was causing to flag ... At Toulon, the harbour was actually the sole subject of our thoughts.'[25]

At the end of the century, tourists enjoyed setting about classifying the various goods and attempting to distinguish groups in the crowd that thronged along the quays. In October 1775, Johann-Georg Sulzer stayed in Marseilles. 'The quay', he wrote, 'becomes the meeting-point of the crews from hundreds of ships, and among this multitude, there are people from all kinds of European and Asiatic nations'; traders rushed there because 'it is the site of the Bourse, and curiosity also attracts the city's idlers, churchmen as well as laymen. In spite of this abundance of people, everything happens in an orderly manner and without disputes.'[26] In 1785, Dupaty declared himself dazzled by the goods from Asia and the northern countries assembled at the port of Genoa; 'there is a movement, an activity, a crush that one cannot imagine'.[27] As he had done at Toulon, he insisted on visiting the galleys. The following year, Bérenger stated his wonder at the spectacle of Marseilles' harbour.[28] In 1791, Grimod de La Reynière was in turn enraptured by its 'splendid warehouses'.[29] Travellers who crossed England's southern shores spent at least as much time, if not more, on the defence systems as on the resorts' frivolous attractions. In this respect, the attention paid by Thomas Pennant to Portsmouth's fortifications during his 1793 journey from London to the Isle of Wight is revealing. Among the most methodical visits is the 'crossing' made by André Thouin in 1795 along the quays of Amsterdam. He clearly distinguished three 'classes' or 'nations' of people: the workers, who were then unemployed; the merchants, more civilized; and the 'fishermen and sellers of fresh fish', who were 'less sensitive and less polite', but whose wives none the less appeared indifferent to attempts at seduction.[30]

At the height of the Empire, Millin in turn visited the Marseilles harbour, which still struck him as 'the meeting-point of all the nations of Europe'. He was thrilled by the diversity: 'the different languages, the variety of costumes and of faces make the walk along the harbour delightful for a man who is eager to observe'.[31] Admittedly, this kind of excitement was no longer remarkable, given that Vernet had presented this intense activity to visitors at the Salon art exhibition more than half a century earlier.

Even outside tourist practices, the morning or evening visit to the harbour, combined with conversation, became part of the classical seaside holiday. It provided an opportunity to question merchants and workmen and to meet the common people. Private tutors brought their pupils there in order to study the scene;[32] the members of the upper aristocracy sent their children to the ports in order to complete their education. This was the tradition in the La Rochefoucauld family and even among the royal family. Leading merchants placed the young men in their families as probationers with their colleagues doing business in the ports. This was where their chosen successors received their apprenticeship.

Boat excursions to the headlands that closed the harbour were an extension of visits to the quays. In Mediterranean ports, boats protected by awnings were set aside for this kind of group entertainment. The launching of a ship or the manoeuvring of fleets brought all social classes together on these headlands. On 18 October 1778, Carlo Pilati went to Texel island, and joined the crowd that had gathered in the hopes of watching the departure of the East India Company ships, then awaiting favourable winds.[33]

It is easy to understand the extent to which this set of attitudes stood in opposition to the model of enjoyment of the sea-shores then being timidly developed – before becoming juxtaposed with it. Visiting the harbour made it possible to appreciate the union between town and sea, between cut stone and wave. The wedding of the Doge with the Adriatic, the high-point of the Venetian journey, clearly symbolizes this integration of the harbour visit into the code governing the classical journey.

During the wars of the Revolution and the Empire, the English blockade deprived visitors of this object lesson; the port usually offered merely a spectacle of desolation. On the Continent, stagnant business, defeat, and a hatred of the enemy highlighted the unpleasant features of the place. Barbault-Royer, in an account of a journey made in years VII and VIII,[34] paints a gloomy picture of Dunkirk: 'Everything here reeks of tar, old rope, the nauseating vapour of the sea, and tobacco smoke ... Staying in a port is not very appealing. Avoid resembling the English.' And about Ostend: 'The spectacle of a trading port without trade and without activity, is the most painful thing one can imagine. Ships that have run aground, a heap of ropes, sails thrown down here and there, mutilated masts, and idle, silent sailors.'[35] It is important to bear in mind this traumatic experience in order to understand the evolution of the harbour's image.

With the French Restoration, the ritual walk along the quays continued, but became transformed. Quite naturally, authors of tourist guidebooks incorporated it into their recommended itineraries. This stage coincided with the ravenous appetite which travellers of the time showed for technical achievements and municipal accomplishments. The picturesque character of a place and the harbour's monumental defences and installations attracted people's interest more than ever. In England, this curiosity was stimulated by the boost of recent victories, and in France, by nostalgia for the country's ephemeral maritime greatness in the last years of the *ancien régime*. Little by little, however, change was creeping in. Between 1830 and 1840, the harbour continued to attract Salon painters, but it was no longer the centre of attention for seascape enthusiasts. This was when a 'deepening of harbour themes',[36] as Denise Delouche calls it, took place: an imperceptible shift of attention towards the town's monuments, the maritime approaches to the harbour, and historical evocation. The classical code of the harbour visit ceased to govern its enjoyment by tourists and art lovers. This way of appreciating and visiting the sea-shore that was fading away was one that had in many ways long contradicted the rising desire for natural shores. In 1836, Victor Hugo proclaimed his aversion:

'Decidedly, I set little store by great sea harbours. I hate all this stonework with which the sea is trimmed. In this maze of piers, moles, dikes, jetty heads, the ocean disappears like a cart-horse under the harness ... The smaller the port, the greater the sea.'[37] After this brief but necessary incursion, let us leave the harbour town and its bustling social spectacle and return to a territory neglected for too long. There we shall discover the various peoples who move upon it.

8

The Encyclopaedia of the Strands

The scholar, the 'sensitive soul', and the invalid walk along the strands, far away from civilization and urban society, seeking contact with the elements and indulging in the many-layered quests that reflect the complexity of their desire. But on this unstable territory, they discover a population that presents a strange spectacle. A complex interplay of interpretations and attitudes develops, and becomes increasingly intricate. Initially, travellers have a vertical perception of these labourers on the strand, who are seen through the filter of anxieties that generate an easily identifiable code. Travel accounts and investigations steeped in the pretension of their objectivity endow previously neglected populations and their activities with a textual existence. The picture of the shore is shaped by a desire to give a specific character to their image; the aspects that are outlined and empha- sized are those that the observer expects from this exceptional place. As the decades go by, the way of taking in the scene shifts radically. The claim to objectivity fades away at the same time as the vertical perception of the scene; another awareness, of temporal depth, guides the interpretation of the spectacle. The figure of the strand-dweller loses something of its solidity; it becomes transformed into something transparent, through which observers perceive ancient shadows brought back to life by historical dreams. At the same time, the seaside holiday that domesticates this area both shortens and highlights the distance separating tourists from those who work on the sands and in the seaweed. Soon, the ruling classes would come deliberately, to be seen, while the people of the shores were compelled to yield the beach to a new social scene.[1] But throughout this long period that separates the initial pictures from the elites' ostentatious settings, the eye generally refuses to distinguish individuals from among these communities settled on the border between land and water, who were associated more closely than others with Nature's plan.[2]

THE COMPOSITION OF THE PICTURE

Already in the seventeenth century, in an attempt to obey Louis XIV and his desire to know the true face of his kingdom, special attention had been paid to outlining maritime boundaries. Systematic observation based on scientific measuring instruments sought primarily to establish accurate maps. This exploration of what were then lively coastal regions was the job of astronomers assisted by hydrographers and surveyors, and it fostered progress in the field of oceanography.[3]

Later on, shrill cries of alarm in the same area hastened the composition of this picture. The manpower crisis that obsessed navy officials,[4] the anxiety aroused by the supposed decline of the sea's fertility, and the health threat posed by disgusting emanations from the shore made it necessary to particularize and deepen the investigation.

In fact, since the fifteenth century, the certainty had been gradually spreading along the shores of the Continent that sea life was vanishing. The 'ageing of the sea' suggested that its vitality would progressively wane. 'It is as though', wrote the physician Tiphaigne de La Roche in 1760, 'the exhausted oceans only supply enough fish to make us remember their former fertility with regret.'[5] The alarm became more strident as the extent of the situation was seen to worsen. Fishing statistics and the testimony of fishermen attested to the sudden rapidity of the decline. Between 1716 and 1720, for example, the problem was recorded in Brest, Oléron, along the coasts of Poitou, and in Royan.[6]

On the other side of the Channel, closer to the fertile northern seas, there was less concern. At the end of the century, though, John Knox undertook a detailed investigation along the Scottish coasts that led him in turn to denounce the diminishing vitality of the sea.[7] Johann Christian Fabricius noted the same perception among Norwegian fishermen.[8] This scourge led to studies, and guided the pens of specialists, generating an abundant amount of documentation. Geneviève Delbos even attempted to demonstrate that this renowned disappearance of sea life was more a matter of the history of discourse than actual observed fact. She also detected a correlation between the intensity of conflict over this 'suspended territory' and the amplitude of 'depopulation rhetoric'.[9] In any case, the studies undertaken in France between 1717 and 1720 and the meticulous survey by Le Masson du Parc led by 1769 to the publication of the monumental work by Duhamel du Monceau. In 1760, worry drove Dr Tiphaigne to speak of the need for conservation of species and to decry the adaptation to shortages: 'We have reached the point', he wrote, 'of forgetting the former fertility of our seas.'[10] Nine years later, Father Menc investigated the coasts of Provence, and raged against the fishermen's negligence.[11] On the eve of the French Revolution, between 1787 and 1789, Noël de la Morinière collected numerous observations; he actually began to write a treatise on fishing in Europe, which unfortunately was never to be completed.[12]

The attempt to assign responsibility also aroused debate. Driven by moral economists, opinion gradually shifted from the cosmological explanation to a denunciation of human irresponsibility. Initially, the withering fertility seemed to support the old concept of a declining world, or at least the certitude that its creative forces had dissipated as the time from the world's origins had increased. Nature was 'becoming exhausted' or was 'degenerating', said still other voices.[13] Yet others referred to the cyclical aspects of the sea's fertility. In the eyes of some fishermen, the sea had lost none of its vitality, but it now refused to reveal its riches at its limits. This theory of 'spontaneous sterility' tended to focus attention once again on the divine character of the waters' edge. Along the shores of Norway, people considered the falling yield to be a sign of God's anger at vaccination, which upset his providential plans.[14] Most people, however, attributed the drop in fertility to unpredictable effects of the weather.[15] They blamed the limited number of fine days and the length of the winter, well aware as they were of its increased harshness.[16] 'We scarcely have spring any more,' declared certain fishermen, 'the heat of the summers is short-lived, and the autumns are nothing but tolerable winters.'[17]

Over the years, the diatribe rose against the negative effects of overfishing. There was a growing awareness that fishing techniques were artificial: even at the sea's edge, men and civilization threatened the natural order. A growing nostalgia led to the exaltation of the *age of abundance*, which was situated ever further back in history. Only 'the blind licence of fishermen'[18] was to be blamed, according to Father Menc. The imbalance caused by the quest for profit imperilled the harmonious distribution of Providence's gifts; it 'offends Nature by upsetting her functioning and depleting her generosity'. The use of drag-nets and the increasing numbers of shellfish farms and fisheries, banned in England, were enough to bring barrenness to the French coasts; there was no need to imagine any reduction in the waters' vitality. This diatribe emphasized the urgency of developing a policy of species conservation; it was important to protect not the sea monsters whose reproduction was of no concern, but the schools of small fish that fortunately continued to bear witness to the age-old profusion.

This disaster affected both the rich,[19] deprived of the pleasure they obtained from fresh catches, and the poor, whose very livelihood was threatened by the scarcity of the sea's salt. The remedies seemed obvious. Men must fish, requested Father Menc, 'without offending Providence'.[20] It was the duty of the royal authority, from the perspective of a moral economic order, to ensure the just regulation of activities. Above all, this authority must attentively pursue the application of old laws that had until then proved to be insufficient to restrict night-time fishing and, more generally, poaching on the seas. Between 1584 and 1744, a long series of decrees, edicts, and declarations had actually provided increasingly specific regulations on the shape and type of nets, and had notably banned 'catch-all' systems.[21]

It seems difficult to understand why the government showed so much solicitude without recalling that the coasts, like the moors and the forests, represented a potentially conflicting set of interests. These spaces were not clearly defined; numerous claims to users' rights (including the peasants' right to 'vain poissonnage', or gleaning after the fishing fleets) had become accepted; privileges had been granted; and insidious attempts at monopolization had been made: all this transformed the shores into highly conflict-prone areas. Overlapping exemptions, reviews, and interpretations of the law made any small-time fisherman into a poacher almost by definition.[22]

The second cry of alarm that stimulated investigation reveals the heightened levels of sensibility. It represents the anxiety aroused by contamination of the water, the air, and the earth among physicians inspired by neo-Hippocratic ideas. The fear that the *shores were becoming contaminated* went hand in hand with the dread caused at the time by slaughterhouses, tanneries, and cemeteries, which were thought capable of undermining the health of entire cities. Two agents of infection were accused of endangering the salubrity of the coast, and the debates generated around them trace the circulation of anxiety in various forms within the social body.

The dumping of refuse into the sea had been a traditional and fairly well-accepted practice until then. Suddenly, however, the growth of industry increased the dangerousness of this waste, even as it expanded the quantities produced. It was then remarked that this threatened to 'denature the waters'. Dr Tiphaigne launched an appeal, albeit a nonchalant one, not to soil the sea's purity. Along the coasts of Provence, noted Father Menc,

> there is not a single person living in the city or by the sea who has not taken advantage of this labour-saving situation and assumed the right since time immemorial to dump filthy or useless materials on the neighbouring beaches when it would be less convenient and more expensive to remove them anywhere else. As a result, a part of the coast adjoining Marseilles and its territory from Estaque to Moredon has gradually, to a greater or lesser extent, depending upon the site, become covered in the depths either by factory refuse or by household waste, and its waters have become infected by the mixture of foreign matter with this element.[23]

Within the Marseilles harbour, the water and salt had become *denatured*. Along the coast, poisoned fish had acquired a bad taste. The tuna had deserted the region, and the shellfish had disappeared. Article 33 of the 13 April 1726 declaration, which sought to prevent the *poisoning of the coasts*, like the 18 December 1718 declaration which banned *dumping* on the beaches in order to protect mussel fishing, were dead letters in this case. Dr Tiphaigne regretted that the fishermen themselves were in the habit of tossing waste and filth into the oyster-beds, contaminating them in this way. This type of diatribe, enlivened by the fear of denaturing the elements, took place against the backdrop of pre-Lavoisier chemistry. It coincided with a pious desire to respect the purity of the substances created by God and to preserve his gifts. As such, care should be taken not to interpret it anachronistically.

In France at least, this voice remained relatively minor by comparison with the vibrant denunciation of the *new insalubrity of the sea-shores*. During the last quarter of the century, increasing demand by the glass-making industry and detergent chemistry fostered tremendous growth of soda ash manufacture. As a result, it encouraged the cutting, harvesting, and burning of sea-weed.[24] Furnaces were set up directly on the beaches, and the smoke they released aroused furious debate. Many feared the stink and the emanations, and came to consider the shores on which marine vegetation was burned this way to be unhealthy.[25] They blamed the smoke for the epidemics that raged among coastal communities; they also accused the vapours of causing young plants, trees, and fruits to wither. Gentlemen and shore-dwellers of all social stations issued numerous petitions.[26]

The master glass-makers, the furnace-workers, and certain citizens moved by a sense of humanity refuted these accusations, in their turn signing memoranda against the ban, and certifying that the emanations were innocuous. They warned against confusing inconvenience with insalubrity. In 1769, the affair came before the Rouen Parlement; the attorney-general determined that seaweed smoke was indeed a pestilential vapour, and a decision of 10 March banned the manufacture of soda ash along the Normandy coasts, except within the admiralty of Cherbourg. The glass factories ceased operations, but the Académie royale des Sciences, when consulted, sent Guettard on a tour of inspection along the Mediterranean coasts, and delegated Fougeroux and Tillet to the Normandy beaches.[27]

The documents produced by these investigators are part of the abundant literature of inspections that mounted up between 1770 and 1810 in France during the first golden age of statistics.[28] The practices they reveal sometimes appear unusual. Tillet devoted several months to an extremely detailed study of the beaches. He followed the retreating tides in order to examine the state of the algae, which he studied with a magnifying glass; he gathered quantities of oral testimony; he questioned lords, priests, and farmers along the coast. 'Under the appearances of simple conversation', he wrote, 'we were able to obtain the testimony we desired without using false pretexts.'[29] Like his colleagues during the French Restoration, in particular Parent-Duchâtelet,[30] Tillet was not afraid to undertake dangerous experiments on his own person. 'We stood sometimes for four or five hours at the mouth of the furnaces … we purposely exposed ourselves to the draught of smoke that the wind drew over us … we repeated these tests a hundred times … we breathed this smoke either on empty stomachs or after our meal.'[31] He never suffered any nausea, 'not the slightest discomfort'; furthermore, the seaweed-workers could be characterized by the 'gaiety of their language'. In the admiralties of Barfleur and Cherbourg, it never occurred to the coast-dwellers to complain. The regions surrounding The Hague, formerly inhabited by populations with a 'ferocious nature' and savage character – which explained, for Tillet, the use of the term 'haggard' for them – had been civilized by the work of producing soda ash; this successful custom had banished crime, toned down their manners, and meta-morphosed shipwreckers into rescuers.

In short, Tillet's optimism, based on his belief in the benefits of labour and industry, led him to advise tolerance. The distinc-tion made between inconvenience and insalubrity and the concern for eliminating anything that could hinder the growth of indus-trial activity foreshadowed what would be the attitude of experts under the Empire and the Restoration. In any case, the numerous, meticulous investigations generated by the issue of the sterility and insalubrity of the coasts drove experts[32] to travel tirelessly up and down these shores, encountering official fishery guards, coastguards, and pilots of small boats.[33] In so doing, they map out a path across these shores that corresponds to a novel pur-pose. This series of investigations, the accounts written of them, the questionnaires and writings they generated (especially the treatise by Duhamel du Monceau) all contribute to create a veri-table encyclopaedia of the shore[34] which has been neglected by historians more obsessed with seafaring. Two dominant images shape the description of work along the sea-shore: that of the fields of the sea and that of labour on the sands, ambiguous references to both the hard labour of agriculture and the easy motions of harvesting.

The work of cutting 'seaweed from the depths' and 'shore seaweed', collecting it, and transporting it on dangerous rafts called *dromes*, which sometimes shattered on the reefs or dis-integrated, drowning their pilots,[35] was governed in general terms by the order of 8 March 1720.[36] These various har-vesting activities meant that the shores were considered as part of the land of seaside communities, and that they belonged to the agricultural system of peasants in the area. The peasants came to the shore to find the additives or fertilizers necessary for successful harvests and sometimes, as was the case at Molène or Ouessant,[37] to gather their stocks of fuel. Many of them, as was explained above, also found a source of supplementary income in the production of soda ash in the early industrial age.

Despite the fact that these spaces theoretically belonged to the royal domain, the practices concerning the produce from the shores and the strand were similar to those governing the use of uncultivated pasture land, moors, and parish woods, a matter of custom. Here again, in these unclearly defined spaces, the same types of regulations and the same models of tension and solidarity occur.

In fact, seaweed was not collected everywhere. In Normandy, for example, it was abundant along the shores of the Caux and Cotentin regions, but was rare on the series of beaches between Honfleur and Arromanches or Mezy and La Hougue.[38] While it was worked intensively in western Brittany, seaweed was collected only sporadically and in a disorganized way around La Rochelle and Marennes.[39] By contrast, it was cut and harvested along the coasts of Roussillon. Despite the presence of the local police, the peasants around Marignane piled it up before their doorsteps, 'against the walls of their homes, in the streets, and in public squares to accelerate its putrefaction'.[40] In so doing, they created veritable cesspools. Darluc was outraged by the practice in 1782, and managed to have several of these heaps of compost, which some people went so far as to hide under their beds, removed by force. In Scotland, seaweed was cut every two or three years, dried, and then burned.[41]

The status of the rocks and beaches along which kelp grew, attached itself, or was washed ashore varied from one region to another. In 1771 in the Cherbourg region, each village actually had the use of a stretch of shore, 'but none of the inhabitants possesses anything there that belongs specifically to him. The kelp that washes ashore belongs to the first person who collects it; and when, at the appointed time for cutting kelp, the inhabitants of a village come to the sea-shore, they spread themselves without distinction on the area to which they have a right ... they cut the kelp there where they find it, dry it, pile it in heaps, and reduce it to soda ash.' The scattered seaweed, gathered in heaps or transformed into cakes of soda ash, could remain safely on the shore. There, order reigned without 'the rigour of the law'.[42]

In the admiralty of Fécamp, by contrast, the tradition 'is for each individual to have, during his lifetime, the usufruct of a part of this canton which is once granted to him: a rock that juts out or something distinctive on the cliff serves as a boundary for each part; one of the inhabitants who is recognized for his fairness oversees this simple distribution; the boundaries set in the canton are unalterable, and disputes are rare'.[43] 'Kelp places' could not be transmitted by inheritance. Occasionally, for humanitarian reasons, a beneficiary who had priority would renounce his share in favour of a poor man with a family. Business questions were settled by the village assembly in the presence of a lieutenant of the admiralty. In this way, from Cherbourg to Fécamp could be seen the slow progress of individual appropriation within the framework of still solid community structures.

'Labouring the sands' was the task of 'small-time fishermen'.[44] Barefooted and bare-legged, standing in the sea up to their waists, they would push their shrimp nets in order to catch '*chevrettes*'. Otherwise, on the open strand or among the rocks, wielding hoes, pitchforks, rakes, hooks, or sticks, they gathered mussels, cockles, sea slugs, ammodytes, and 'shelled fish': crabs, crayfish, and lobsters.[45] Some hoed the sand of the strand, others stretched their nets between posts: and still others dragged skeins called '*va-t-en, viens-t-en*' or 'come-and-go'. Some worked continuously on the shore, others came only at the period of the strongest tides, when the strand was covered with thousands of workers. Like the haymaking and the harvesting, the great harvests on the sands involved everyone; on those days, the peasants became 'fishermen on foot'. Gathering pieces of wrecks or wood on small islands without forests or along the Scottish coasts, hunting the birds that frequented the beaches,[46] and in some areas[47] extracting salt by washing the sand completed the wide range of activities of the shore.

As early as the beginning of the eighteenth century, these activities were disturbed and threatened by the increasing numbers of beds, fisheries, reclaimed lands, locks, and other '*escluzeaux*' and '*gorres*' of all sorts. These horseshoe-shaped constructions, some in stone, others in earthworks, whether fitted with fishing nets or not, sprang up all along the strand, despite endlessly reiterated regulations.[48] The purpose of this shoreline architecture was to retain the waters when the tides went out and to capture the fish caught in these traps. The growing number of fisheries, often controlled by rich individuals who hired share-croppers for them, was one example among many other procedures of individual attempts to appropriate or acquire the use of collective spaces that were considered available.

In some places, the fish caught in this way and other fruits of coastal fishing were prepared directly on the beach. Along some beaches, presses were set up for salting sardines, herring, and mackerel. There were also huts in which rays and lamprey eels were dried.[49] In the region around Brest and on Belle-Île,[50] the fishermen set up great buildings by the seaside which housed both the sardine presses and the kegs of wine delivered to the sailors in exchange for the fish. Seven or eight women worked in each of these buildings: 'They skewer the sardines on sticks which they wash in the sea, then return to arrange them in casks and put them under press.'[51] In Scotland, too, fish was salted in huts built on the shore.[52]

The shore was the meeting-place between work on land and work on the sea. This drifting border allowed activities to overlap or be juxtaposed. Because the seasons for fishing, working the sands, and carrying out agricultural labour were complementary, a subtle rhythm for working the sea-shores could develop. Consequently, the distinction made by experts between 'small-time fishermen' on foot and 'coastal fishermen' seems somewhat artificial;[53] when the days of great tides came, the strand provided work for everyone; fishermen laboured side by side with peasants with their heavy carts full of kelp. Even clothing was not always a reliable key to their identification. At Oléron, the 'small-time fishermen', shod in wooden shoes, were dressed like peasants; at Blaye, they also wore tarpaulin aprons.[54] By contrast, in the Nantes region, they wore a smock of heavy canvas, like the sailors, and put on boots to test the waters.[55] In the Royan region, they dressed in shirts and trousers of sailcloth and could be confused with 'coastal fishermen'.

EXPANDING THE ANTHROPOLOGICAL PLAN

Half a century after this encyclopaedia of the shores was pioneered by Le Masson du Parc, and then by Duhamel du Monceau, plans for a more specific portrait began to emerge that went beyond the mere description of techniques and clothing. The territory uncovered by the ebb-tide and those who dwelled in it were caught in the gaze directed at the fishing population. 'Fishermen on foot' and 'small-time fishermen' attained, as a group, the status of a social spectacle. Observers finally began to write, 'we can see them ...'.[56] 'Their customs', noted Darluc about the populations along the coasts of Provence, 'their lifespan, and their diseases, dependent on a vigorous, masculine constitution, *are worthy of our research*.' To tell the truth, the procedures that shaped these descriptions of shore-dwellers could hardly be expected to be original; their particularity lies in the specificity of their imagery. In addition, the logic underlying this portrait and the deliberate sequencing of its components guided what was said, and made any grasp of reality somewhat random. The many aims that drove observers to the beach encouraged them to make their perception of the individuals found there coincide with the image of their own desires. Consequently, the wish to praise what was natural helped to preserve certain prejudices; it inspired reports that exaggerated the strength, fertility, and longevity of shore-dwellers, as well as the vitality of their children.

From the outset, this vision was subjected to the anthropological canons inspired by neo-Hippocratic ideas. These individuals so closely exposed to the elements, living in contact and confrontation with them, were the finest illustration of the logic underlying medical topography. Consequently, there is nothing astonishing about the precociousness of the works devoted along these lines to coastal regions.[57] A few years before the work published by Lépecq de la Cloture[58] and the great project designed by the Société royale de médecine, Desmars wrote a long treatise devoted to the constitution of the inhabitants of the region of Boulogne,[59] and H. Tully penned an *Essai sur les maladies à Dunkerque* [Essay on Diseases in Dunkirk].

As applied to the populations of the coast, this medical discourse was not in complete contradiction to the approach simultaneously used to vaunt the salubrity of the shore to invalids. Often, the two attitudes overlapped and reinforced one another. Townley explained his presence at the seaside by the studies he was performing on the longevity of local populations. But the symmetry was sometimes upset. French scientists, in particular, were not as optimistic as their British colleagues. Desmars, for one, could not help joining in the interminable debate over the salubrity or harmfulness of sea water, though he hesitated to judge.[60] This uncertain approach positioned the French beaches on the North Sea and the Channel at a mid-point between the beneficial northern shores and the unhealthy Mediterranean beaches, between the image of an absorbing and invigorating sea and that of an 'exhaling' sea.

Unlike tourists, 'small-time fishermen' were exposed continuously, from their earliest childhood, to the air, winds, water, and sand of the place. Therefore, the elements could not be expected to produce the same effects on these rustic, uncivilized beings as on the melancholy individual who came to seek vigour from the waters of the sea. In short, Boulogne's bathers and its fishermen ate the same food, but their habitus required scholars to carry out separate analyses of the two groups. This remark applies to all the social portraits inspired by neo-Hippocratic concepts; members of the ruling classes were more easily protected from the impact of the elements. The 'influence of the earth', wrote Dulaure in 1788, 'only acts powerfully on the behaviour of members of the common folk; in general, no matter what land he lives in, *a man becomes stripped of local character as he becomes educated or capable of reason*'.[61]

The portrait outlined by Desmars began to be filled in with detail: the same specific characteristics became highlighted as the description moved from the coastal populations to fishermen, then to sailors. These features became more intense the more the figures were exposed to the elements. By the same token, the sailor came to designate the ultimate stage, representing the archetype of these populations. Among those who comprised these three groups, an analogy developed between the organs and the local characteristics of the water, air, and soil.

The powerful winds that blew along the shore put human fibre to the test. Consequently, men's skin became thicker, darker, and drier; it could then no longer allow subtle exhalations to escape. Hippocrates had once noted the harmful way in which bodies exposed to the setting sun were both 'burned and discoloured'.[62] Muscles, too, tended to become thicker. This hardening was the cause of longevity. 'The thickness of the cuticle indicates solid organs', and 'bodies capable of offering a long resistance to the action of the elements.'[63] Darluc in 1782 saw the inhabitants of the Provence coast as a race toughened by the wind and the waves;[64] later on, in 1794, Cambry was convinced that the wild inhabitants of Léon, 'men of the coast', suffered no infirmities, and enjoyed long lives because they were constantly attacked by the air and the water.[65] The typical image here is that of the coastal tree, bent, hunched, and forced constantly to resist the wind. Desmars, for his part, noted that the inhabitants of the Boulogne shore were small but solid.

The weakened[66] and humid atmosphere, the proximity of sea water and abundance of volatile and septic salt, and a putrid diet based on fish[67] all ensured that these people had a specific nosology. Rancid sweats and stubborn skin diseases were frequent among them, as was blindness.[68]

According to the Hippocratic tradition, it was easy to deduce the temperament of these peoples subjected to a watery atmosphere and uneven seasons on an open landscape. Classical medicine was believed to have explained the features later scrutinized by medieval observers:[69] the sailor, the fisherman, and the shore-dweller were characterized, in order of decreasing intensity, by a delicate balance of ferociousness, courage, and piety. Their whole being was caught up in the sea's temperament. Medical discourse wove endlessly back and forth between images of human beings and of the elements. These people were characterized by their vivacity and love of dancing, boundless passion, a veritable raging storm within, a taste for drunkenness, a thirst for freedom, scorn for moderation, piety, and repentance, just as the ocean was marked by the alternation between the hurricane and the calms.[70] Like a rebellious sea controlled by man, these populations knew how to resist civilization; they had kept the savour of human origins: candour, honesty, moral purity, and faithfulness to their memories. They also tended, however, to be superstitious.[71] In short, this portrait simply reproduced characteristics that were shared by the common folk as a whole, while accenting them. The paradox was that the proximity of the shore kept civilization at a distance; it hindered the flexibility of the imagination, and did 'not facilitate the acquisition of purely social talents'.[72] The view of, and contact with, this primitive environment thrilled sufferers from melancholy who decided to come to stay by the sea-shore and, if necessary, face the unhealthy vapours rising from it.

In this way, a shift occurred to incorporate the description of these labourers of the shore into an idyllic discourse; these were simple folk, but there was something heroic in their piety, and they provided a pretext for exalting a rustic way of life that counterbalanced the horrors of urban blight. On this subject, the fundamental image is that of manna. The strand was like a table filled daily by Providence – or Nature. This natural generosity made industriousness unnecessary. On the damp sands, the only work required was simply collecting and gathering Nature's fruits. Like the Hebrews on their way to the Promised Land, these dwellers at the sea's edge had only to reach out and take this daily manna. The shore uncovered by the tide appeared to be a place of abundance[73] – and this made the decline in the waters' fertility all the more disturbing. The strand was also seen as an ephemeral land of equality: poor people seeking nourishment rubbed elbows with the rich who came in search of distraction to participate in the in-gathering.[74] By the sea, it was often repeated (though not without betraying reality), there is no property. During the strong tides at the equinox, fishing on foot harked back to the early days of human existence.[75] Small-time fishermen were ignorant of speculation. Nature set an example here of just regulation, by avoiding two dangers: she made impossible either total misery or excessive abundance such as would encourage humble workers to leave their station. Along the coasts of the Channel and the North Sea, there was no true poverty, and no beggars.[76] The strand played the role of a charitable organization governed by Nature. In Marennes, the poor lived on oysters; in Oléron, island-dwellers and soldiers benefited from the shore's abundance.[77] The sea in its attentiveness even varied its gifts; the shellfish season prevented diets from becoming monotonous.[78]

By this means, Nature ensured that order – the *status quo* in terms of social position and the distribution of roles by age and sex – was maintained. Most descriptions of the sea-shore served to recapitulate a classification whose obviousness was a pleasure to the viewer. In so doing, they echo the tradition of Dutch painting. Remember that, after spelling out the code of the philosophical journey, Diderot fell into this idyllic portrait when he began to describe the coastal fishermen of Scheveningen: their simplicity, their honesty, their piety, and the serenity with which they carried out their roles according to their age or sex called forth images in his soul of the earliest ages of the world's existence.[79]

In order to maintain the just equilibrium of Nature's gifts, mankind must first stop applying himself to upsetting this balance. Fisheries that made the fortune of their owners created shortages among the 'fishermen on foot'; in so doing, they disturbed God's plan. For this reason, Tiphaigne called for the elimination of 'all individual rights'. 'The sea is the heritage of fishermen, and it should be free from any taxes of this type.'[80] This diatribe both repeated and exceeded the one attacking the usurping of collective resources.

The idyllic representation of the shore-dwelling peoples played a function in the social portrait drawn by those who transformed them into a spectacle. The crowd of fishermen on foot and small-time fishermen – less frightening than the sailors' world – offered an image of good folk. Gatherers on the shores were not affected by social mobility, which had not yet made its way into their midst. These workers of the shore, toughened by contact with the elements, reasonably hopeful of enjoying long lives, made fertile by their diet of fish,[81] attached to their religion by regular recourse to its consolation, were reassuring to city-dwellers who felt the contestation of social hierarchy rising in their midst. The sailors' piety and the solid faith found in these coastal parishes, in which seafarers' families were united with the 'shore community' through empathy, were in keeping with this serene interpretation.[82] Age-old stereotypes reinforced this idyllic perception which stimulated both the tourists' longing for the shore and the pleasure they found there: they observed this reassuring population and felt little inclination, for the moment, to carry their ethnological investigations any further. The delight procured by this spectacle was all the more freely enjoyed since there was no need to fear in this setting that the social distance would be reduced.[83]

Observers who thrilled to the scene of a people who mirrored nature so spontaneously also rehashed another, symbolic stereotype: few among them neglected to point out both the naïveté of the women and their bare legs.[84] The feminine quality of the shores would be deliciously provocative were it not for the simplicity and social distance that defused any eroticism. Tanned, prematurely wrinkled, reeking from the fish they handled,[85] these women showed their legs not to excite desire, but because they remained close to their wild state. Their piety, the marital fidelity they practised, and the absence of any jewelry or accessories demonstrated that their intentions were pure. Before this naïve, hard-working nudity, the antithesis of the courtesan's calculated scanty dress, no traveller would ever admit – or should even feel – that his sensuality had been awakened.

Until the end of the 1780s, neo-Hippocratic theory and the idyllic code prevailed almost completely in this social portrait of the northern shores of France.[86] Tourists were not yet pausing to examine the picturesque scenes on the strand; in this area, the port continued to be the primary focus. To my knowledge, it was Cambry who was first charmed by the diversity of scenes from popular life on uncovered beaches. This is how he closes his account of his tiring travels, with happy pages that correspond to the serene relaxation he was finally enjoying, in the company of his cultivated hosts.[87]

After 1780, by contrast, bits and pieces of ethnological enquiry are to be noted here and there that are little influenced by notions of medical constitution. Some travellers saw this as a specific way of intensifying their Robinson Crusoe-like experience. They attempted to inventory the customs and practices that highlight the peculiarities of island peoples whom civilization has not yet detached from their local roots. The Atlantic islands were the first to be studied in this light 'because – closed and isolated – they received the projection of utopias about good and bad savages',[88] before the arrival of Republican-minded travellers seeking the virtuous island.

In Year VIII [1800] Thévenard, anxious to document the decline of custom,[89] traced an ethnological portrait of the isle of Ouessant. Adopting the then traditional approach,[90] he compares the inhabitants to the Hurons and the Algonquins. Cambry's account of his adventures along the Breton coasts combines the range of observation techniques used by travellers during the Enlightenment. His book contains a medical topography; the author makes a careful distinction between 'coastal man' and 'inland man'; he enquires about superstitions; he sees the incarnation of a 'sea-shore savage' and a typical inhabitant of the shores of the Orinoco in the poor shipwrecker Thomas Yvin, known as Philopen, an outcast living with a concubine in the midst of the Penmarch rocks. Cambry weaves a network of comparisons extending to include the Kamchadals, the Lapps, the wild peoples of the far North, the California savages, and those of Tierra del Fuego. Cambry met with the coast-dwellers; he questioned the elders about their customs and their former practices; in the speech of the shore-dwellers, he hoped to find the trace of an original form of language.[91] Already, the figure of the Druid was mingling with that of the savage; Celtic culture was beginning to compete with the fascination exercised by exotic cultures. Insidiously, a new sensitivity worked its way into the text of this child of the Enlightenment. His portrait of a young girl pleading with the sea-gulls to bring her fiancé back home, or even better, his fantastic rendering of the processing of seaweed, leads his readers away from the idyll and drops them down among the people living on the romantic shore, whose description was to be influenced by the horror novel.

'Imagine ... a woman obliged to work in the winter night, in the midst of storms and the ocean's fury, in deepest darkness, on a slippery rock, sometimes standing in water up to her waist, sometimes hanging over the abyss, as she collects the seaweed brought in by the sea with a rake.'[92] Cambry claimed to be objective, but he chose to ignore that this marine vegetation was collected by day, that the calendar for its harvesting was carefully determined, that the gatherers certainly never hung over the abyss, and that they laboured when the weather was fine. But reality matters little here; the historical interest lies in the shifting images and in the savour of new emotions.

9

A World of Transparent Characters

WAYS OF LOOKING FOR MEANING

As the way of viewing shore-dwelling populations evolved,
so did the traveller's vision of the landscape. The images,
schemes, and practices they inspired overlapped, alternated, or
became combined. It is important, however, not to overestimate
the degree of innovation and, before discussing changes of
interpretation, to emphasize the weight of continuity. Medical
topography pursued its career into the nineteenth century, and
guaranteed the survival of neo-Hippocratic thought for several
decades.[1] Villeneuve-Bargemon's description of the 'Marseillaise
race' living on the coast between Fos and La Ciotat is a clear
example of this rigidity.[2] The pages Dr Bertrand[3] devotes to
the people of the shore of the Boulogne region are patterned on
the model developed a century earlier by Dr Desmars; they refine
the catalogue of stereotypes that determine the portrait of the
sailor as Dr Forget[4] in turn decrees it must be seen. As late as
1844, Armand de Quatrefages de Bréau did not hesitate to contrast the northern inhabitants of the miniscule isle of Bréhat with
those from the south. The former, he writes, 'have throughout
their being something rough, almost fierce, which contrasts with
the politeness of southern people'. They do not speak the same
language, or have the same accent or the same customs.[5] In the
middle of the century, the quest was still on for the connection
believed to develop between 'peculiarity' and 'locality'.

Within this analytical framework originally based on climate, however, and rooted in the specific features of the air, water, and soil, social distinctions were confirmed. This explains why Dr Bertrand dwells at length on the influence of education, the type of work, and the particular habits of each 'class' in the Boulogne region; he modulates the effects of the climate in the light of this sociological data. Furthermore, the doctrine of temperaments, updated once more, added a new theme. The predominance accorded the blood system contributed henceforth to explaining the irrepressibly passionate nature of these coastal populations, given to drinking and fighting.[6]

Although scarcely anyone spoke any more of the sea's losing its vitality, the fear remained that its depths were becoming depopulated. In France, this alarm fitted with Restoration ideology; it was fashionable then to criticize the negligence of Revolutionary authorities and to emphasize the legislative void created by the elimination of the admiralties in 1791 and by the declaration that fish were *res nullius*, something which had no owner.[7] This led to a call for renewed implementation of the former regulations.[8] The debate was especially sharp between fishery-owners, whose numbers had grown since 1789, and coastal communities. From 1816, in this area as elsewhere, an attempt was made to return to the old order; however, unlike in other disputed zones, notably forests and rivers, the tensions here did not give way to the publication of a code. In this indecisive period, profound changes were under way. Along the shores, many of the coastal fishermen who until then had followed the schools of fish, were beginning to settle near the canneries. In the same period, the fishing trades were being organized as part of the battle against poverty.[9]

At the same time, scholars and government agents were detailing the descriptions of labour on the sands. Milne-Edwards and Audouin devoted voluminous pages to the study of fishermen on foot in the Mont-Saint-Michel bay, a region that was becoming the focus of attention among enthusiasts for the shore at low tide. In Granville, with its 7,212 inhabitants, more than 1,000 people were involved in fishing on the strand.

What now became most important, however, was the necessity that travellers felt to deepen their understanding of these shore-workers. Despite the numerous obstacles, there was a growing desire to reduce the distance and to plunge into the midst of this primordial humanity that was both barbarous and innocent.

The quest undertaken by travellers as a result generated profoundly different procedures along the northern coasts and the shores of the Mediterranean. Near the beaches of antiquity, a cognitive system that had been less completely overhauled operated between the two competing (and sometimes blurred) figures of the heir and the outsider. The most common interpretation, except within territories still under Ottoman domination, nevertheless remained faithful to the classical image. The populations of antique shores continued to engrave the mark of their customs and practices on the seaside, where travellers lingered who, like Swinburne, hungered for a continuity of traits. As seen by Villeneuve-Bargemon, who undertook to establish an exhaustive portrait of the Bouches-du-Rhône district, the fishermen of the 'Marseillaise race'[10] were none other than Phocaeans who had managed to perpetuate their race. Their language, their costumes, their 'inclinations', and the physiognomy of their women all provided ample evidence. Only the common people knew how to preserve this mark and escape the destructive effects of time. Therefore, they were the only documentary evidence available about the Greeks. Things became more confused at the next step up the social pyramid. At the same time, anthropology and archaeology became intertwined, or at least supported one another. For this reason, Villeneuve-Bargemon undertook a joint investigation that combined digging, the study of archives, oral history, and the recording of rituals, as a result of which, he handed down a notably beautiful description of the festivities of Saintes-Maries-de-la-Mer.[11]

As for the figure of the outsider who settled, or rather stumbled, into the realm of antique ruins, it was totally different. This image aroused reticence and sometimes disgust among viewers, and even hostility which was soon sharpened by combat with the philhellenists. Peoples whose origins have been lost were opposed to the heirs, to a clear and prestigious filiation. It was important for travellers to identify the two parties clearly. Lavallée, who in 1801 presented Cassas's journey along the coasts of Dalmatia, asserted that there, philosophers could study 'peoples who breathe in the proximity of ruins' and analyse 'the impression made on their souls by the pompous debris which they trod upon every day'.[12] The issue was no longer to distinguish an identity that had been maintained, or at least a close correlation between a population and prestigious remains; rather, it was to measure a distance, to analyse an echo. By staging a confrontation between palaces and humble cabins, between Roman baths and the rotten straw on which Dalmatian women slept,[13] the shores of Istria and Dalmatia made the penetration of time and the intensity of the sacrilege painfully manifest.

Travellers were to avoid becoming lost in the medley of 'heirs' and 'outsiders' settled on these coasts. At Sebenico, for instance, they would encounter the most 'idiotic' and 'the most imbecile' race. These coastal inhabitants 'feed solely on insects, fish, and shellfish which the sea deposits on its shores'. They remained seated all day on the rocks or before the doors of their miserable cabins. 'Their faces are sunken, browned by the sun, blackened by poverty; their eyes are full of fear, their hair is black and unkempt, ... they are more brutal than ferocious: the simplest ideas never germinate in their minds; they are equally incapable of understanding, remembering, and imitating.'[14] These were not savages; they had neither the pride, nor the touching candour, nor the independence. They were degenerates: 'they have lost their origins.' 'The savage is the first link in the species; these people seem to be the last.'

The radical difference between the historical paths followed caused a similar distinction in the way of viewing 'heirs' and 'outsiders'. Between two worlds of poverty, and indeed misery, the traveller's system of appreciation created an unbridgeable gap: the distance separating admiration from spite and disgust. The Dalmatians of Sebenico were ruins, too, in the eyes of Lavallée, but ruins without meaning. Their erratic presence seemed an aberration among peoples who continued to incarnate classical beauty and simplicity and who were the only ones capable of rendering texts, monuments, and landscapes comprehensible. Consequently, any ethnological investigation had to begin with a triage, a radical division that would determine two emotional directions. These were certainly antithetical, yet they were both governed by the code of the classical journey.

The northern coasts, whose innocence had not yet been undermined by holiday making or tourism,[15] provided travellers with an accessible stage on which the imagination's games could be played out. With growing intensity, a quest was under way along the Baltic, Scottish, and Breton coasts for anything that could bestow meaning on the worn rocks, the mysterious ruins, and the curious beings who dwelled among them. Initially, the resonance of Celtism[16] and the Ossianic mode swelled. Originating in the work of Scottish scientific societies and popularized by Dr Johnson and Boswell's travel account, this practice of recording ruins became codified only belatedly in France, when the

Académie celtique[17] established its programme. Scholars set to work, searched for traces, noted customs, described rituals, sketched monuments, and attempted to rediscover the link that connected the various objects of their curiosity. Through a population which they perceived as a transparent window, they watched for, imagined, and sometimes believed they could see bards and Druids. Satisfied at having found riches to rival those of fervent amateurs of classical antiquity, these Celtic enthusiasts were not yet haunted by nostalgia, but were rather spurred on by a (partially fanciful) awareness that customs were vanishing.[18] They let their joyful vision leap back across the centuries that had rolled by, convinced as they were of being able to read a direct, living image of these origins in the face of a Caledonian island-dweller or an Armorican peasant.

The people of the shores found themselves implicated in this quest for primordial life. The quest in turn led to the creation of stereotypes that were to spread far and wide across the decades, until they finally reached the popular reviews whose number multiplied in England at the beginning of the century and in France following the Revolution of 1830. Bretons, for instance, the true Celts, seemed then to be anthropological fossils that had emerged intact from the beginning of time. In 1832, Habasque introduced his readers to the coastal moors of Plouha, Keroisel, and Morgat: 'If a Druid', he wrote,

> managed to shake off the dust that has covered him for twelve, thirteen, or fourteen hundred years, and came to visit some of these out-of-the-way cantons, he would still understand the language, he would not be too unfamiliar with the traditions, and he would recognize the dress. True, the ancient forests have disappeared, but he would find the moors, the gorse, the rocks, and the heather; he would see the dolmen again, the cromlech, and the menhir, though he would be astonished that we had forgotten everything, even the purpose and use of these sacred monuments. Yet he would note with joy that the Breton peasant still reveres verbena and mistletoe ... In Lower Brittany he would find the people just as credulous as in his own time.[19]

Along these beaches, the primordial heartland, the inalterable standards were different from both Swinburne's 'heirs' and the degenerate 'outsiders' observed by Cassas on the Dalmatian coasts. 'They are almost like a tribe of Celtic fishermen,'[20] noted Fulgence Girard in an article for *La France maritime* on the subject of the coastal inhabitants of the Mont-Saint-Michel Bay.

After Morven and Staffa, two sites emerged most intensely as
high points in the search for the Druids:[21] the island of Tom-
belaine located a stone's throw from Mont-Saint-Michel[22] and,
better yet, the isle of Ouessant, on which the few remains of
the temple assigned to Belenus summoned up the image of guardian
virgins. Stendhal borrowed from the same stereotypes when he
noted, on 6 July 1837: 'This morning, at five o'clock as we left
Vannes for Auray, the weather was truly Druidic.'[23] At the
same time, however, the initial forms of this fascination with the
'beauty of the dead'[24] and the 'silent page of the dolmens'[25]
were already being enriched by travellers seeking more subtle
procedures for travelling through time.[26]

The way that the Romantics interpreted the populations of
the shores was based on nostalgia. More strongly than ever
before, this interpretation was stimulated and guided by an
awareness of how fragile speech, rituals, customs, fairy-tales, and
legends were. This discovery of their precariousness and of the
meaning that was being lost engendered the suffering necessary
to enjoy the place. This new aim sharpened the procedures for
listening. Travellers dreamed of entering into the daily life of
small-time fishermen; they attempted to hear – or contented
themselves with dreaming of hearing – these folk in the inns,
in their huts, or along the shore. As this desire became more
complex, it gave new meaning to oral investigation. The purpose
of this research was no longer merely to collect fragments and
to save ruins and remains, which had been the ambition of the
Celtic archaeologists portrayed by Samuel Johnson. This time,
the quest was a prelude to an immersion in the freshness, innocence, and energy, in the powerful sap of a humanity that had,
through fairy-tales and legends, preserved the language, myths,
and rituals entrusted to its care, and maintained both the trace
of its origins and the landmarks of its path through the bygone
centuries.

When he watched the shoreline populations at Norderney, then
a rapidly growing bathing resort, Heine was pained at the
thought that these people would not retain their anthropological
status much longer. The little fishermen watched the spectacle
of the wealthy lodging at the resort's hotel, and enjoyed observing
their gestures, expressions, and even their facial contortions; but
this aroused new aspirations in them and new 'desires' in their
wives. Already, some of the children in this threatened population
were being born with the faces of bathers. The foreshadowed
withering away of this link to the past engendered a desire in

the poet to participate in the 'communal directness' of the fishermen, to experience the promiscuity of bodies close around a hearth, and to grasp the words that were understood 'even before they could be spoken', lying in wait for gestures, facial expressions, or 'intonations of their voices'.[27] Heine dreamed of joining the 'community of thought and feeling' which was the only real means of communication and the indispensable condition for happiness and magnificence in art. Yet the poet suffered from the tension that developed between his desire to bridge the gap separating him from this people standing 'on the threshold of a new age' and an irrepressible feeling of disgust, born of habit, that he felt before their ugliness and above all the repulsive odour of the women.

In 1820, Nodier set forth a clear recording programme dictated by the urgency of the situation. The sentiment that all this would soon disappear heightened his emotion. 'The generation now reaching its end', connected to the vestiges by a 'feeling of indescribable community in decadence and misfortune', wanted to 'enjoy the fleeting vision of a picture that time is going to erase.' The priority in this conservation effort went to popular poetry; it called for listening attentively to the 'narrations of the rustic guide'[28] and scrupulously examining *ex votos*. The fisherman's hut or boat and the seaside community's chapel seemed to Nodier to be the most favourable sites for collecting traditions. Wakes among coastal communities, intensified by the pathos of the sea and the grave bond that united the fisherman's fiancée so quickly with death, enlivened by accounts of far-away lands and by savage exoticism, filled the listener with a 'moving and living history'[29] that only oral recounting could render.

Romantic travellers delighted in imagining all the intermediate scenes that stood like a series of screens between present times and long ago. Their dreams centred on the ages revived by historical anecdotes. In this way their pleasure mounted before evocations that emphasized the extent of time and peopled the solitary shores with a succession of dreamy characters who gave an intermittent meaning to the ruins, the mossy rocks, or the sea-gull's cry. This delight in the multiple layers of historical time coincided with the new understanding of coastal geology, as it was then recognized that the shores had been worn down over the course of millennia. The heroism of this cast of touching imaginary figures distracted the travellers' attention briefly from the poor labourers on the shores, who became a part of the scenery.

Then the theory of *Littus Saxonicum*, already announced by Lépecq de la Cloture and Noël de la Morinière in the previous century, was revived again.[30] In 1823, Edouard Richer mentioned the debates aroused by this hypothesis; two years later, Morlent declared it to be scientific truth.[31] From Calais to southern Brittany, the coastline had been dominated by wild Saxons; indeed, 'the Saxon blood and Saxon character, which a long series of centuries could not alter', could still be recognized in the workers on the salt-marshes around Le Croisic. History therefore explained the contrast that was growing up between the coastal regions and the inland areas. Travellers, reassured by the relative backwardness of the French coast,[32] thought that they could see Basques, Celts, Saxons, and Normans here and there.[33]

The pleasure of a journey vascillated between visiting the port and seeing the historic scene on the beach. It grew with this alternation between the spectacle of current activity and the dreamy recapitulation of former landscapes.[34] The growing prestige of Mont-Saint-Michel was due to the ease with which travellers there could imagine the many filters standing between the Druids and the industrial age. Fulgence Girard set them up, and explicitly made them shine in the imaginations of his readers. 'Druidic rocks' had known the 'sombre passions of the cloister', the alternating or simultaneous splendour of 'mail coats and monks' robes',[35] and the successive horrors of oubliettes and central prisons, before tourists joined the ranks of pilgrims to the site.

In travellers' eyes, the populations of the coasts became the guardians of coastal legends. The stakes in this process were high, and it occurred later than was long realized. By establishing a topography of legends anchored in accounts of this space, this process reinforced the indeterminate, vacuous nature of the territory. It imposed an image of a cultureless place over which an unchanging nature reigned. It would be interesting to perform a closer analysis of the correspondence that developed between the coastal territory and the body of legends that was being collected. Clearly this work of the imagination required a denial of the appropriation or the working of these places, whose access was ensured by fairy-tales and legends. From this point of view, the shores, like the moors and marshes, created an opportunity for a discourse rejecting modernity.[36]

Along the Armorican coast, the menacing rocky skeletons were transformed into an enchanted sculpture. Morgan Le Fay[37] drew her name, so it was said, from her sea birth.[38] On the pathos-ridden, mournful coast bordering the Baie des Trépassés [Bay of the Dead], the Kernewotes shared the complaints of the ship-wrecked souls who held a lugubrious reunion every year near the beach.[39] While Nodier was planning to investigate Normandy, Heine set to work collecting the legends that were traded back and forth along the coasts of the North Sea. The legends of the island shores of Denmark inspired Nyerup. In 1837, Xavier Marmier introduced the French to this population of fairies, water sprites, elfs, giants, and warlocks[40] and this realm of damned hunters and licentious old men of the sea, these tales of treasure guarded by dragons, of phantom ships and profane churches swallowed up by the waters.

There can be no question here of drawing up a list of these legends and beliefs,[41] and even less of undertaking their analysis. It is enough to realize how much the attention that was paid to them revived the possibilities for coastal reverie. 'I often walk here by the edge of the sea, and I think of those marvellous tales that the sailors hand down,'[42] reported Heine in 1826. For the readers of popular magazines such as *La France maritime*, the beaches from the 1820s to the 1840s were peopled by le-gendary characters borrowed from folk stories. The marine depths occupied an important place in these stories. From Diaper to Southey, underwater voyages had generally enjoyed a very limited audience; but now, sunken cities endowed with a legendary his-toricity found their way into every imagination. Many of these tales and legends made of the sea a place of punishment, thereby reinforcing an ancient stereotype. The old image of hellish storms survived in the accounts of Ashverus the sailor, of phantom vessels, and of the laments of souls and the lugubrious songs that rose from watery Sodoms and Gomorrahs. Some of these tales endowed the shores with a sexual charge that had been forgotten since the disappearance of marine images of Venus, Triton, and the Naiads. Sea men with green beards called in the evening from the waves' edge to lonely young girls in order to lead them into the depths of their crystal grottoes.

Listening to the populations of the seaside stimulated reverie, and made it possible through imagination to reconnect with humanity's and each individual's buried past. It gave a foundation to the newly confirmed homology between the depths of the sea and the depths of the psyche. The beach, once deserted, became dense with fantastic beings who invited those who so desired to rediscover the openness of childhood and to share the primitive beliefs of this child-like population. Here, in a single process of social regression and psychological involution, popular legend and children's stories came together, and merged into one.

The revelation of this imagery brought about a new way of delighting in the shore, and heightened sensitivity to the contact of the elements. The way of appreciating the solitary sound of evening waves or the shadows cast by rocks across the shore became loaded with images produced, or so it was thought, by the minds of simple fishermen. Readers, and especially cultivated travellers, found a way to simulate the sharing of beliefs and, in so doing, to satisfy their desire for encounter and their thirst to communicate. Sea bathing took on an added dimension: an imaginary immersion, which was not so much an escape from society as a refreshing self-renewal in a humanity postulated to be intact and primordial – and which, its observers foresaw, would inevitably be perverted, and all the more quickly through the interest being paid to it. This was apparently when the fiction took root that shore-dwelling peoples are lovers of the sea.

As early as 1830[43] some individuals attempted to unveil the farce and reveal the artifices that underlay these touristic practices. Voices began to rail against city-dwellers in search of Druids;[44] at that time, however, the search for traces, the prestige of travelling through time, and the ground covered in moving closer to the common people had sharpened and redirected the longing for the shore.

THE ADVENTURES OF THE ANTHROPOID

Near the strand, that indeterminate place of biological transitions, the links connecting mankind with the mineral, vegetable, and animal kingdoms can be seen with exceptional clarity. Observers there can easily deduce the metamorphoses[45] and recapitulate the different facets of mankind, from his lowliest ontological aspects to his purest heroics. A new reading of the 'enigmatic grip of place'[46] enriches the simple stereotypes of neo-Hippocratic medicine, which were gradually rendered obsolete. The harmony detected between the texture of the prevailing rock and the structure of personality quickly becomes generally recognized.

After having described 'the always desolate edges of the ocean', during the excursion to Le Croisic in *Béatrix*, Balzac created the image of the individual of granite in *Un drame au bord de la mer*.

Cambremer, 'the man with a vow', laid low by remorse because he had killed his own son, sits all day long on a chunk of stone; 'his stoic immobility could be compared only to the inalterable attitude of the piles of granite that surrounded him ... Why this man in the granite? Why this granite in this man? Where was the man, where was the granite?' wonders the narrator, unaware of the tragedy.[47] Dauvin orients his description of the inhabitants of the Île des Saints around this homology, which is not merely a metaphor. To an observer, the island-dwellers appeared 'sombre as the sky that hangs over their heads; like the waves that rumble around them, they are defiant, scarcely communicative, credulous ... never a smile on their lips, never a light in their eyes! You would think they were men of stone'[48] like the Celts who built the megaliths. The inhabitants of Ouessant live in 'this final state in which man establishes a difference between his existence and that of the rock where he lives only by material movement'. This passive life, resulting from incomplete humanization, is defined by a combined absence of reflection, sensitivity, imagination, and hygiene. The existence of these island people is not so much beast-like as mineral; men there are identical to the menhirs on which history has left no trace. This explains their lack of vice or chronic disease, and as a result, their great longevity. A sort of vegetative or mineral happiness reigns on these shores, and the scene there provides a solid test of civilization, whose presence is hollow.

The obsession that emerges from the sight of slime, mud, and the shifting soil of the strand casts a shadow over graphic and pictorial representations of the small-time fishermen along the Normandy coast.[49] Their ramshackle huts are delicately piled on the soft marl hills or in unstable cracks in crumbling cliffs; they are scattered in no particular order near the shifting beach, as if the artist wanted to emphasize the understanding that exists, not between man and granite in this case, but between the precariousness of the soft rock and the frail condition of these prolific beings so dependent on the gifts of Providence. Writers, too, from Crabbe to Balzac,[50] highlight this homology. Sometimes the prevailing vegetable nature is what shapes the description. On the English sea-shores where Philarète Chasles claimed to have spent part of his childhood, the sea moss had covered rocks, the old church, and the homes of the poor with 'a garment of age-old lichens, grey and purple, green and bluish, an imperceptible but eternal vegetation'[51] that observers enjoy describing with a brand new precision and which suggests mildew attacking souls.[52]

In this description, the image of the wild animal's den, the presence of the anthropoid, and the threat of the horde are already perceptible. By the sea, the animal nature hidden in man erupts with particular ferocity. Ever since Cambry made the eyes of the 'tiger'[53] gleam before his readers, the character of the ship-wrecker had haunted sea-shore literature. This fitted in with the fashion for horror stories and, at least in France, with the fear of a return to animality that gnawed at people's minds during the French Restoration, a period obsessed with the memory of regicide.[54] The looter on the shore was one of the most terrible and fascinating figures of the threatening common folk; he was a prelude to the developing portrait of the dangerous classes lurking in the shadows of the city.[55] From Cambry, Corbière, Bonnelier, and Sue to Souvestre,[56] soon to be followed by Michelet, and from Tillet to Dauvin and Philarète Chasles, the stereotypes were trotted out with monotonous regularity. The animal wiliness of the community fostered the image of a group of cats slinking behind the rocks. Consider Philarète Chasles, inspired by Richard Southey, on the subject of English ship-wreckers: 'When the weather is bad, you see men climb down along the whitened reefs and lurk for hours at a time, all covered with foam, hidden in some tall grass, who wait for the ocean to toss them some wreckage of men or riches.'[57] When these remains arrive, the horde rushes in, and picks apart the ship and the bodies of the sailors. It haggles over them viciously; it tears away at the wrecks.

Revealing of the accentuation of this animal image is the evolution of Philopen between the publication of Cambry's account (1795), the appearance of *Vieilles femmes de l'île de Sein* [Old Women of the Isle of Sein] by Hippolyte Bonnelier,[58] and 'La Cornouaille' [Cornwall] by Souvestre. The shore-dwelling savage with his touching and primitive innocence metamorphoses into a shipwrecking animal. Already in 1826, Hippolyte Bonnelier made him into 'an unimaginable man', a 'wolf with flexible haunches, strong paws, and a well-armed jaw'. At the same time, the cut-throat old women of the isle of Sein, knives in hand, formed a 'bloody horde'[59] on the beaches of Plouvan. For Souvestre, Philopen is a 'lynx' who, once his brutality is satisfied, 'slinks' across the rocks, drops with his animal mate on to a bed of dried algae, or goes fishing, armed with his 'shipwrecking hook', 'with that uneasy swaying of a seal from the polar seas'.[60]

The looters of wrecks make repeated appearances in Turner's work,[61] where their image is in conformity with that sketched by writers. His *Wreckers – Coast of Northumberland, with a Steam Boat Assisting a Ship offshore 1834* [plate 16] – offers a gripping evocation of the animal ferocity of the shipwreckers who burst on to the strand.

In this area as well, this territory of the void, where property was abolished and objects recovered their primordial accessibility, appeared to be a legitimate gathering ground. 'The sea', said the peasant Kernewote, '... is like a cow that gives birth for us; what she deposits on her shore belongs to us.'[62] In this register, the energy of these barbarous shore-dwellers is released on this stage before the eyes of observers or in the readers' imaginations; their 'powerful sap'[63] compensates for their lack of civilization.

Observers were forced to admit, however, that after the first third of the century, the coastal populations ceased to cause shipwrecks and contented themselves with pillaging them. Actors and observers then seemed to hesitate between the two antithetical roles of the shipwrecker and the rescuer. On this subject, the juxtaposition that Turner presents in his 1834 painting is revealing. One year earlier, Souvestre showed the same uncertainty, which suggests that a metamorphosis was under way.

> At the cannon's first blast of distress, men, women, and chidren rushed toward the sea with lanterns and burning faggots; they could be seen running across the shores, climbing down along the headlands, a thousand lights accompanied by bizarre and terrible cries; soon the guns of the customs agents shone, and the voices of fishermen and boats' pilots rose above the storm, sending warnings or signals back and forth ... while on the cape, in the light of the fires, a thousand ardent faces watched the ship, and a priest who had come running to stop the plundering repeated the prayer for the dying under his breath![64]

This metamorphosis, which is ever possible, facilitates full ascension to humanity. The transformation of shipwreckers into rescuers is accompanied by a process of individuation. The scene of the rescue always ends with the acknowledgement of a man of the people. His heroic act draws him out of an anonymous community whose brutal cohesiveness was emphasized by the pillage scene. In this way, to the observers' eyes, the individual is promoted to the status of a person, but without having been tamed or in some way put into the service of the traveller/author. The animal-like behaviour of the group and the actions of the good shore-dwelling folk were dramatized and made aesthetic in an antithetical way. The account of the rescue also drew growing relevance from the correspondence that was developing between the sublime nature of the shore scene and the equally sublime heroism of the people. Clearly, the rise of this theme and the progressive refinement of the portrait of the rescuer on the shores also fulfilled the function of demonstrating, even more bitterly than the idyll of the waning *ancien régime*, the barbarian's potential for metamorphosis. His heroic action bore witness to the existence of a good folk that could be contrasted with the dangerous and vicious working classes that social investigations were then revealing in the city's underworld.

Already in 1777, Tillet had praised the metamorphosis of the terrible inhabitants of the shores of The Hague, but he attributed it solely to the growth of industry. During the Restoration, the theme developed of the coast that could be made safe. It coincided with the philanthropy of the age. A series of technical advances ensured greater safety along the western coasts. The shores of the Atlantic and the Channel were dotted with modern lighthouses; rapid progress was made in mapping the coasts and the depths, as was demonstrated in France by the work of Beautemps-Beaupré; navigators perfected their skills; stations along the shore were equipped with unsinkable boats. On the other side of the Channel, and later on the Continent, the number of sea-rescue organizations multiplied. In 1826, a group of rich Englishmen founded *La Société humaine* in Boulogne, whose purpose was to rescue people in danger of drowning. Life-guards were organized to watch swimmers, and resuscitation techniques were improved. [65]

The fragile life of the imperilled sailor, anonymous and pious, aroused compassion. The wave of Romanticism and, in France, the atmosphere of the Restoration, combined to reactivate the Christian symbolism of the sea at the same time as the animal or barbarous face of the seaside communities. The sailor, a living image of sacrifice, was perceived as a lightning rod; for the happiness of all, he drew to himself the thunder of nature or God. [66] 'In the help that men give to their brothers', affirmed Ducos in 1826 on the subject of sea rescue, 'there is something of Providence, something that *tightens social bonds*.' [67] This distant, fertile people, subjected to the whims of the sea, had everything necessary to arouse compassion without inciting too much fear.

The tragic isle of Ouessant, located at the tip of the most dangerous entry into the Channel, had once paradoxically been represented by Billardon de Sauvigny as a social paradise, and it now became a symbolic place of sea rescue.[68] According to Rouget de Kerguen, the inhabitants of Noirmoutier, former ship-wreckers, had been transformed into heroic rescuers.[69] Those of Cayeux, a tragic site along the shores of the Channel, were instinctively good.[70] In England, the figure of Grace Darling, a young girl whose heroism had saved dozens of passengers on the *Forfarshire* in 1838, took on symbolic dimensions.[71] The range of accounts of such devotion expanded; people began to speak of the hard life of lighthouse-keepers[72] and to exalt the self-denial of ship's captains. In 1823, Jouy vaunted the intrepid character of navigators from Quilleboeuf,[73] and set the stage for the hymn that Michelet was to dedicate to the navigators of the Gironde region.

The Romantic literary figures of the corsair and the pirate[74] also play a role in their own way in this extolling of shoreline humanity. In paintings that include them, there is a juxtaposition of group ferocity with the seductive mystery of sublime individu-ality that stands out from it. Monique Brosse has thoughtfully analysed the slow refinement of the corsair's image: 'a man of desire and eagerness', 'condemned to fulfil himself through imme-diate action',[75] sombre, taciturn, and enigmatic. She shows the firm grip of this mysterious figure on the imagination of the times. In this area too, Byron proves to be an effective 'reservoir of images'. Balzac's pirates (*Annette et le criminel*, *La Femme de trente ans*), those evoked by George Sand (*L'Uscoque*), Eugène Sue, and all the star writers of the Western maritime novel, not to mention Fenimore Cooper, follow in this tradition.

The will to live for the moment finds its logical stage on the open sand with its ephemeral forms. The complicity established between instantaneous desire and the emptiness of the shore contributes to promote the fascination exerted by the place.

In actual fact, this harmony developed only gradually. In the eighteenth century, the dominant image was a ferocious one of groups of bandits and pirates, in both Defoe's *Robinson Crusoe* and Baculard d'Arnaud's *Makin*. The crimes of barbarous pirates much feared by all those who travelled along the Mediterranean coasts[76] and the extortion practised by corsairs during the Re-volutionary and Imperial wars reinforced this disparaging image. For some time, heroes could find their place only surrounded by frugal humanity slouching in the caves of contraband runners drunk on brandy and weary from carnage.[77] The scene of pirates on the shore extended the sublime picture of supposed brigands by Salvator Rosa and accorded with the brutality of Schedoni in Ann Radcliffe.[78] Furthermore, the corsair is not really a character belonging to the shore. His desire for freedom forces him to take the sea as his territory; the pauses on the beach merely mark the intervals between sailings.

His powerful figure nevertheless made a profound and rich contribution to the imagery of coastal shores and caves, and nourished the reverie of walkers fascinated by the mystery of these exceptional beings, who were both terrible and liberated from the weight of the ambient animality. It is also worth noting that sometimes the Romantic character of the honourable bandit, a close relative of the pirate, was also fond of periods ashore. The bonds that connect Nodier's Jean Sbogar with the beaches of the Adriatic bear witness to this proximity.

THE GOLDEN AGE OF THE SHORES

During the first third of the nineteenth century, the literature of shores, cliffs, and caves emerged hesitantly to begin with, then became generously prolix. Novels, and short stories even more so, thrilled to these places that opera and light opera would also soon contribute to popularizing. In England first of all, then in France, the number of maritime reviews multiplied. Their content did not deal solely with navigation – far from it. It was a combination of historical anecdotes, laying the foundations for discovering the shore's prime sites; accounts whose purpose was to illustrate the feats or crimes of shipwreckers, contraband runners, ship's captains, and pirates; and toned-down genre scenes, among which the fisherman's return was the endlessly rehashed model; in addition, they offered melodramatic short stories that testified to the renewed pathos of the shores and were sometimes spiced with tragic eroticism played out in grottoes. The end result outlines an immense field of study whose surface historians have barely scratched.

Describing the motions of everyday work provides an introduction to popular sociability. There was not yet a strict dichotomy between the mingling of the circle of men and women busy on the shore,[79] hunched over their nets, or grouped around a fire of wood from a wreck or a small boat to be tarred, and the strictly masculine world of the island inn.[80] Shore women enjoyed a certain status, and had powers and knowledge that enabled them to resist the masculinization of forms of sociability better than the wives of peasants. In fact, on all these subjects, most of the period descriptions provide information primarily about the sensibilities of travellers and about social imagery; one fine example is a knowledgeable account by Charles Rouget de Kerguen. With his head full of literary and pictorial references, the author dared to brave the disgust inspired in him by the odour of drinkers gathered together at a fishermen's inn in Noirmoutier, 'a shack built of pebbles cemented together with sand and kelp and covered with seaweed' that resembled a cowshed in Lower Brittany. Inside he discovered a 'fantastic picture in the style of Hoffmann or Callot', featuring 'purple-faced drinkers wearing large jackets and tar-covered trousers, with rough and wild physiognomy'. In the hearth burned 'a fire of peat and seaweed'; men, women, and children 'refreshed themselves with brandy from a large tin pitcher'.[81]

Along the beaches of Brittany, Cornwall, Caledonia, and Norway, the cutting and gathering of seaweed turned into another tirelessly reused stereotype.[82] Souvestre, the first to make a mockery of Druid hunters, found in this image an opportunity for a delirious text that mixed the centuries together.[83] The cockle-gatherers in Mont-Saint-Michel Bay inspired Nodier with some of the most gripping pages of his *Fée aux Miettes* [Crumb Fairy].

Yet the idyll survived, and made a contrast with the harsh portrait of this work. The genre scene broadened to include the evocation of popular rituals. In the city, the middle class came to make merry at the cabarets just outside the turnpikes; and observers came to the seaside to discover the *joie de vivre* of a once terrible people. They paused in wonder before the 'pleasure parties' whose primitive innocence or latent eroticism reduced their animality. Hunts, the bivouacs of young nest-robbers, village wedding dances on the hardened sand of the strand, and group bathing by the populations of the Basque mountains who came down to Biarritz once a year[84] were all components of this mellow picture.

Already the pleasures of the ruling classes were making inroads into this collective jubilation. As yachting was developing, at the same time equestrian competitions were being organized that transformed the beach into a racecourse. This practice grew up early in England, and spread later to the French coasts. The shores of Cesson and Langueux, where pigs normally roamed free among the sea-birds, were covered with tents three days a year. 'Temporary open-air cafés filled with drinkers. The prefect, the gendarmerie, the troops and a bright procession'[85] came to watch the races, surrounded by thousands of spectators seated on the

rocks. This event, inaugurated on 31 August 1805 and held regularly since 1807,[86] attracted 'elegant jockeys' and peasants from
the neighbouring communities. It offered an occasion for unusual
confrontations.[87]

Between 1810 and 1840, as this literature fascinated with
beaches, grottoes, and wet sand was developing, sea-shore painting
went through its brief golden age. Turner, the heir to the Dutch
tradition handed down by seascape-painters, stands out as an
initiator in this area. Between *Calais Sands at Low Water: Poissards
Collecting Bait* inspired by his 1802 journey and *The City and
Castle of Scarborough: Morning, Young People Catching Crabs*,
which dates from 1811, many of his canvases, water-colours, and
sketches encourage the viewer to look at the spectacle offered by
workers on the strand. In France, Eugène Isabey took an interest
in the small-time fishermen comparable to that which his friend
Géricault showed for grooms. His regular stays on the Normandy
coasts gave him the opportunity to paint them as much as he
could wish. His portrayal culminates in his *Plage à marée basse*
[plate 24] of 1833.[88] His main point is no longer the confrontation
with the anger of the elements, or the merging of sea and sky
in the style of Huet or Bonington, or the strand's inconsistency,
as in certain paintings by Turner or Constable. Isabey stops
at the land's end, and paints the people who cling to this
crumbling edge, who work and live in an environment made up
of mud, cob, and wood. Between the rock slides, the collapsed
embankments, and the scarce fish flesh whose consistency was
reminiscent of that of the pasty soil, the wooden posts, the frames
of the houses, and the vaults of boats turned upside down on
the beach seemed desperately to check the earth being devoured.[89]

The Swiss Johann Jakob Ulrich, equally charmed by the shore-
dwellers of the Trouville coast during a journey made in 1824
and 1825, returned ten years later to study the activity of fishermen on foot in Brittany.[90] Already, though, the omnipresent
domination of holiday resorts was contaminating the scenes of
shore labour. Revealing on this point is the confrontation and
tension to be seen in Turner's *Margate*, dating from 1822. After
1845, paintings of labourers on the strand had become an anachronism.[91] The leisure class, in its quest for ever new ways of
being seen, became the subject to be represented; paintings ceased
to provide a direct expression of their longing for the shore.
The social scene and the pleasure it procured became the focus,
to the great regret of solitary Romantics. This gap continued to
widen and soon became a painful issue for Michelet.[92]

Henceforth, the people of the shores were scarcely visible except through a process of domestication that brought them into the service of members of the leisure class. This metamorphosis is too complex to analyse here. The model for it came from the cicerone on the classical Neapolitan journey, who also placed his children at the disposal of tourists. Very quickly, these practices made their way into travel conventions for the Caledonian journey. When peace returned, their insistent soliciting already appeared to be just as overwhelming along the shores of Scotland and Wales as it was on the beaches of the Mediterranean.[93]

As early as the end of the 1750s in English resorts, male and female bathing guides, fishmongers, and boarding-house keepers had offered their services to cure-takers, an early sign of this metamorphosis. This process cannot be fully explained by the need for the services of these workers. The pleasure of this 'taming' process also comes from the opportunity it creates for social contact. Consequently, the benefits attending the proximity of the ordinary people's vigour increases the desirability of the relationship that develops with the bathing guide. This relationship also offers the pleasure that comes from dominating these formerly terrible beings.

In his novel *Un drame au bord de la mer*, Balzac imagined how, fortuitously, a little fisherman could be led to place himself in the service of a couple of walkers wanting to follow the beach to Guérande.

In his short story 'Le Traîneur de grèves' [Hauler on the Shores] from *Scènes et moeurs des rives et des côtes*, Émile Souvestre attempted in 1852 to retrace the origins of this process as he imagined it on the tiny beach of Piriac at the end of the Restoration. Louis Marzou, his hero, had lived until then on the sea's offerings, and was the first among the poor of the village to learn this new trade. 'Whenever someone needed a messenger for Guérande, a bathing guide whose experience would prevent any danger, a guide who knew every little curiosity around the bay, Marzou was always ready.'[94] And this, according to the novelist, did not fail to arouse the hostility of the local population, shocked to see this beachcomber lose his innocence.

A look at the popular journals devoted to the world of the sea and published during the last forty years of the century shows the fascination exerted by the shore and the powerful evocativeness of this theatre of the void criss-cross with shadows: it was a kaleidoscope that juggled composite mosaics and created fleeting series of characters, stony men who had not changed since the times of the bards and the Druids, virgin priestesses of Belenus, strong-blooded barbarians whose ferocity satisfied readers' sadistic sides, heroic rescuers, anxious women gathering the sea's manna on the strand and the rocks and awaiting the fishermen's return, and the moving spectacle of young girls questioning the sea-gulls about the destiny of their beloved. All these polysemous figures nourished the contemplation and reverie that filled tourists' souls and which had already been refreshed by the legends collected from the shore-hauler's mouth or from the old fisherman. In short, the quality of the place, at the meeting-point of sea, sky, and land, facilitated the mingling of images and an additional feeling of travelling through time and space. It was a springboard for the imagination, producing a body of literature and painting whose richness between 1810 and 1840 exceeded that of rustic imagery. This was all the more true since this territory so open to evocation was also a stage for pathos between 1750 and 1840, a theatre whose functions and methods were extensively renewed in that period.

10

The Pathos of the Shores and their Metamorphoses

THE DRAMATURGY OF FEELINGS

Towards the middle of the eighteenth century, the shipwreck became the second most evocative figure of catastrophe, after the earthquake, and sensitive souls were moved by any mention of it. Of course the theme had already been worked over during the sixteenth and seventeenth centuries in Spanish and Portuguese literature. It spread all the more effectively in France and England because it accorded with political symbolism[1] and found its place in the protocol of the classical storm. In the 1740s, however, its grip became stronger, and the emotional strategy which it made possible became more complex. It also gained the rhetoric of pity.

Lafont de Saint-Yenne reported the intense emotion aroused in visitors at the 1746 Salon by the sight of Vernet's shipwrecks.[2] In 1762, Falconer's *Shipwreck* received exceptional acclaim. The author, who was inspired by the sinking of the *Britannia* off the shores of Candia, was lost at sea seven years later. His tragic destiny furthered the success of the poem.

Shipwrecks were the most common form of accident then. Quantitative studies show how frequent they were. Between 1779 and 1791, thirty-five ships sank within the roadstead or harbour of Dunkirk, only nineteen off the city's shore; thus 64.8 per cent of these catastrophes took place within the roadstead or the harbour.[3] As a result, most drownings took place near a coast, within a bay or an estuary. In calm weather, the strong currents that hindered access to a channel made boats drift slowly toward coastal reefs. On his return voyage from Sicily, Goethe detailed his anguish and that of the other passengers on board a felucca that seemed to be carried inexorably by the waters toward the rocks of Capri island.[4] In heavy weather, a captain would sometimes decide as a last resort to turn the boat sideways and cast it on to the coast; in so doing, he hoped it would settle on its side and protect the shipwreck victims from the violence of the waves. The spectators could then watch the tragedy unfold from the shore and hear the cries and prayers of the survivors climbing or clinging to the masts. Between those who were perishing and those who were watching, torturous dialogues could sometimes develop. On 15 February 1739, the *Hareng couronné* sank at the mouth of Dunkirk harbour; the sailors clutching 'the ropes raised their hands to heaven and begged for help from the people they saw on land'.[5] The frequency of this spectacle of pathetic gestures engendered familiarity with, or at least stimulated the public's curiosity about these tragic scenes. The sinking of the *Britannia* and that of the *Saint-Géran* on 17 August 1744 off the coast of Mauritius merely crystallized the horror of a visual experience that was in no way unusual.

This makes it easier to understand how this image of catastrophe made its way as a stereotype into the emotional palette of painters, of the engravers who distributed their works, and even of story-tellers, who were thrilled to have an occasion to show off their own heroism as travellers.[6] Between the plastic arts and literature, so much reciprocal borrowing went on that it would be vain to look for antecedents in this area. In 1777, the account of a shipwreck that opens Baculard d'Arnaud's *Makin* seems to be a monotone rendering of a work by Vernet; but twelve years later, the same artist painted *La Mort de Virginie* [Virginia's Death]; engravings of this painting, which enjoyed a broad distribution, helped to maintain the success of Bernardin de Saint-Pierre's novel for more than half a century. Falconer's *Shipwreck* went through more than twenty-four editions in England before 1820, and continued to inspire artists until about 1832.

Popular art also began to express the pathos of the sea in a more lively manner. Shipwrecks do not appear before 1740 in the series of *Provençal ex votos* developed by Bernard Cousin;[7] but later on, the theme was expanded upon in coastal regions. Bit by bit, the handling of the disaster changed. The votive composition focused on the ship in trouble; it placed the spectators at the centre of the event. In this type of *ex voto*, the heavenly space disappears especially quickly, while the area devoted to the moment of danger expands. It is also interesting to note that shipwrecks were one of the themes of the literature distributed by peddlers, and it nourished a prolix oral tradition in and around the major ports.

Readers of travel accounts and Salon enthusiasts did not consider the shore as merely a place from which to view the sublime anger of the elements; they also experienced it as a vast stage, surrounded by headlands, with the infinite expanse of water as a backdrop. Contemplating nature's excesses created the dramaturgy of feelings. There, gestures of farewell, nostalgic posturing,[8] the collective vigil for the return of ships,[9] and especially the horrors of shipwrecks could be easily staged.

The extent of this flat stage made numerous dramatic scenes possible, all juxtaposed with fallacious simultaneity. In this setting, struggle and mourning could be placed side by side. The increasing lure of the shore would remain incomprehensible without a brief look at the conventions and roles played by pathos on this territory exposed to the intrusion and unfolding of disaster.

In order to understand clearly the interpretation that an amateur of the Age of Enlightenment would make, this rhetoric must be placed in the context of a dense network of practices. It is important to take into account the attitudes of the theatre-going public, the success of sentimental comedy, the emphasis on posturing, and the conventions and gestural language of pity when confronted with the declamation of pain, paradoxically accompanied by restraint. Even more important are the significance, the function, and the historical usage of tears.[10] In order to analyse a shipwreck scene by Vernet or Loutherbourg, it is necessary to recall that tears could then be a sign of joy or pain; that the sexual dichotomy that would soon dictate ways of crying did not exist; and that demonstrations of chagrin were not yet strictly reserved for the intimacy of private space.

The history of self-presentation in the Enlightenment generated a theatricality that belonged to the street, as well as specific expressions and social forms of effusiveness. Understanding them makes it easier to appreciate the representations of dramatic events with the eyes of spectators from this century who were so easily moved; it helps to track and understand the evolution of the rhetoric of postures and gestures revealed in the coastal scene.

Remember, there were many lovers of painting who had never seen the sea, or in any case, had not experienced its storms and the emotions that such weather might have aroused in their souls.

A lack of visual experience of reality, however, did not prevent them from enjoying art. Diderot trembled at Vernet's shipwrecks long before his trip to Holland. The Mediterranean shown by painters, the Indian Ocean in *Paul et Virginie*, and the neo-classical sea of *Jeune Tarentine* spoke to the soul through the heart. The representation of nature shaped by art could be even more moving, as was the case earlier with Marmontel, than the spectacle of tangible reality. Vernet, as Diderot noted, managed to move art lovers who had been left cold until then by the view of the raging sea.

The influence of the critics increased as the public for the Salons developed; with enlightened amateurs, they established a clever dialogue above the artist's head which refined the analysis of emotional tactics, and demonstrated the virtuosity of ingenuous sadism among seascape-painters. The shore facilitated an easy construction of dramatic scenes that affected the characters in them. These figures were no longer there to mime a symbolic scene, as in Lorrain's work, or simply to enliven a picturesque painting, as Gilpin recommended; they had a role to play, one which fitted into the list of theatrical tasks. In a painting, spectators could read terror, prayer, effort, fatigue, devotion, and the unhappiness of individuals struggling against the excesses of nature as it sought to crush them. They could also analyse the subtle interplay of emotions among the survivors, the wounded, and the dead. Finally, their glance could rest on the figure of the simple observer, who provided a model for attitudes and feelings. Generally, the artist used this angle to foster pity and provoke an outbreak of tears. The enthusiast's eye would slide quickly over the spectacle of the raging elements toward the scene of the disaster; he would savour the game of identification that took place on multiple levels. The viewer of a picture was first encouraged to identify with the spectator of the shipwreck so that he could eliminate the distance separating him from the event and suffer with the victims. In this way an insidious propaedeutic of tears operated, whose efficacy was enhanced by the artist's power of illusion.[11]

Lafont de Saint-Yenne was not alone in describing the shivers that passed through the audience of art lovers upon seeing Vernet's shipwrecks. Nineteen years later, the critic for the *Journal en-cyclopédique* remarked: 'We noted two shipwrecks in particular that inspired terror.'[12] In 1759, Diderot, fascinated by the power of illusion wielded by the painter, imagined he could hear the cries of those who were perishing.

The multiple appeals to the senses gave force to the painting's rhetoric. A coastal shipwreck was an occasion for playing with a variety of taxonomies through which the spectator could grasp his own destiny more intensely. The subject lent itself easily to the temporal divisions then being rehashed by painting. It thus found its place in the series of the four parts of the day, which had to be shown together in the same room.[13] Often, it was midday that proved to offer the most pathos; this was the case in the commonplace series composed of a calm scene, a shipwreck, a port, and a ship beneath the moon. The motives behind this stereotypical tactic, as explained by Valenciennes, were discussed earlier. A catastrophe was an opportunity for injecting movement into this monotonous hour and for wrenching 'involuntary tears' from the viewer. The shipwreck is actually only one solution among many. From the neoclassical perspective embodied by Valenciennes, other possibilities include the burning of a village, the sufferings of a young girl bitten by a snake, those of a child attacked by a dog, or even the anxious wait for a vessel ripped by cannon fire. These various types of disaster in painting were adapted to 'the diverse feelings of the viewers; pain, pity, fear, curiosity, and even nonchalance'.[14]

The range of dramatic scenes allowed spectators to review the faces of misery. At a glance, they could go through the ages of man (or woman) and could sample the illustrations of their respective qualities. As individualism was working its way deeper into the social body, seascape-painting, like accounts of catastrophes, became attuned to the growing need to pass in review the trajectory of human existence.[15]

The representation of shipwrecks also provided an opportunity both to enumerate human feelings and to praise them. Maternal, marital, and filial love, as well as friendship, were declared here with special emphasis. For those who stopped to analyse it, painting evolved into a hymn to the family. On a more general level, *it exalted the ties that bind*; by the same token, it invited the sensitive soul to take part in this ostentatious dramaturgy of unhappiness which erased animality, ferocity, and egotism, and was content to display the strength of its attachments.

Here, the artist could easily offer the viewer examples of a series of postures and gestures, combined with indications regarding the feelings they expressed. Based on a catalogue of emotions recognized by all art lovers, the painter distributed the set of roles in this space. Paintings of coastal shipwrecks, like theatre shows, served to stage the prevailing scientific convictions of psychology. The images of terror, torment, and gratitude, exacerbated by the suddenness of the catastrophe, provided the spectator with the sublime pleasures of the fallacious perils defined by Burke. Contemplating the shore soothed the passionate desire for self-preservation, and each viewer found an easy means of satisfying his death-wish.

Paintings of the shores did not yet offer openly erotic scenes; this would be the task of artists of the holiday resorts and the civilized beach. This kind of sexual appeal, however, was already slipping insidiously into the shipwreck scene. The sea's action was used as a pretext for stripping away clothes. It allowed the artist to display partial nudity: a foot, an ankle, or even a leg. It was an excuse for pressing allusions to threatened modesty. It was her refusal of the rescuer's naïve embrace that caused Virginia's death. Wet clothing that clung to the body revealed the beauty and sensuality of its curves, and aroused sadistic desires. The shipwrecked woman, complacently collapsing in the arms of her spouse or lover, seemed to imitate a scene of ecstasy. Virginia's fatal modesty pointed readers to the sensual homology implied by the fiancée's body being washed out to sea on the eve of her wedding or being torn apart on the jagged rocks. The pleasure procured by this spectacle of female immolation culminated in the troubling portrayal of the green paleness of the corpse. The beauty of death without bloodshed, especially when the death scene was bathed in bright moonlight, accorded with the unmentionable fantasies that made Romantic-minded observers dream before the *Enterrement d'Atala* [Atala's Burial].

Baculard d'Arnaud did not bother with the usual precautions. He traced an initiatory journey through the fainting of Makin's fiancée, trapped with her lover in a narrow row-boat, the slow drift, the shipwreck, and the passage into a dark cave. This is a prelude to a metamorphosis, laying the groundwork for an Edenic desert-island adventure for the couple, built around passion and unimagined pleasures, spiced with the foreseeable incest of their descendants. The account closes with a second shipwreck that brings in a priest, re-establishes the social bond, and restores marital and family standards – and reassures the sensitive woman who had been endlessly tormented by this deviation, driven to sin as she had been by the wrath of the elements. In 1812 *L'Heureux naufrage ou le cri de la nature* [The Fortunate Shipwreck or Nature's Cry] by Madame Ménage hinted at the eroticism induced by a maritime catastrophe which successively exposes the abandoned wife to the horrors of seduction and makes possible the foreseeable reunion of the married couple.

The bond that develops between the tragic evocation of the shore and the history of dream-forms would be worth analysing. Accounts of shipwrecks provide a model for nightmares which, from the eighteenth century, opened up the depths of an unconscious world thought to be related to the rumbling of the viscera. The tale which Diderot spins out in detail in the guise of a criticism of the 1767 Salon includes an account of a shipwreck in the form of a nightmare which bore witness to the 'watchfulness of the intestines' and drew tears from the narrator.[16] In 1769 and again in 1771, the mention of the trembling[17] caused by Loutherbourg's seascapes gives Diderot an opportunity to stress once more the dream-like content of this dramatic form of painting.

A quarter of a century later, in the manual he prepared for young artists, Valenciennes ingenuously presented the sadism of such emotional tactics. To his eyes, the proximity of the shore defuses the pathos of the shipwreck; it weakens the cruelty of the scene. The glowing lighthouses and the coastal captains in their lifeboats check the emotion: as the author notes regretfully, one can sense that the victims will be saved; whereas on the open sea, the viewer knows that the ship is sure to sink. He can measure just how inane the victims' struggle is, and he understands better the justification for their terror before the inevitable.

> The spectator sees the shipwreck victims in this cruel position; he follows them with his eyes; his heart softens and is distressed. Tears come to his eyes; the chagrin he feels does not even allow him to imagine the hope that these unfortunate people might reach a shore from which Providence would seem to reach out in order to pull them from the abyss and save lives that might be precious to their children and useful to the country.[18]

Valenciennes, who seems to forget that most shipwrecks occurred near the coast and who neglects the emotion procured by an identification with the witnesses to the catastrophe in the painting, outlines the limitations of sea-shore pathos, just as he introduces the pleasures of imaginary deviation.

THE BLOOD OF THE SEA

At the time when Valenciennes wrote his treatise, the cruelty of these evocations was heightened. A shift took place that drove seascape-painters to focus on the horrors of naval battles rather than theatrical mourning over the misfortunes of shipwrecks. As a result, both the means and the functions of pathos were modified. Once again, maritime history sheds light on this turn-about in representations. Between the beginning of the Seven Years' War (1756) and the battle of Trafalgar (1805), there was little respite from sea warfare (1763–78 and 1783–92). During this period, it tended to become fragmented. The massive confrontations between powerful squadrons that were characteristic of seventeenth-century conflicts continued; examples include the battle of Saintes or of Ouessant. But progress in artillery complicated the task of attacking enemy ships, and increased mobility made it more difficult to destroy the enemy's fleet, thereby leading to a rise in the number of combats between two ships.[19] These battles allowed artists to present the suffering in greater detail. Above all, the battles depicted by seascape-painters moved close to coastlines trimmed with jagged reefs, and they soon provided specialists with opportunities for complicating shipwreck scenes.

From then on, representations of battles accentuated their horror. This loss of the reserve which artists had traditionally shown by preferring to emphasize attitudes and to stage mourning and lamentation gave free rein to the painting of physical suffering. The frightening glow of fires revealed streams of blood spilled by bullets, swords, and explosions. The sea, already red, prepared to swallow up the shattered victims. Not enough emphasis has been given to the unbearable character of this painting which abandons the declamatory theatricality of shipwreck scenes to play up the horror of human ferociousness.

In the same period, many shores were transformed into iron ramparts as a protection against invasion. Along the northern coasts of Brittany following the English attack and the battle of Saint-Cast (1758), and even more visibly along the British coasts facing the Boulogne camp, in 1804, numerous redoubts sprang up, and the garrisons swelled.[20] The tragic nature of the coast was reinforced by the terrible pontoons[21] on which, a stone's throw from the finest bathing resorts, the bodies of prisoners rotted slowly. The strategy of naval warfare was to reduce the adversary's manpower by holding captives, and this increased the sailors' mortality rate, and spread a new mortifying stench along the shore.[22] At the same time, members of the French nobility forced to emigrate discovered a new range of experiences of the ocean. Social classes that had been unaware of this type of peril found themselves abruptly confronted by pirates, storms, and fires on the open sea.[23]

Understandably, this rise in the collective experience of the ocean's tragic character stimulated artists, and shaped spectators' tastes. From 1781, crowds in London rushed to see the mirages of Loutherbourg's *Eidophusikon*. Among the scenes that visitors could see, the storm, the shipwreck, and the naval battle were well placed. Catastrophes were also among the favourite themes of the 'panoramas' that began to appear in 1787.[24] Between 1792 and 1815, naval warfare haunted the minds of the British. Battle painting enjoyed an unprecedented fashion among them; it repeatedly served up the successes of Jervis, Hood, Rodney, and especially Nelson. It helped to whip up nationalist sentiment, which was based more than ever on mastery of the seas. Naval victories were commemorated by porcelain and pottery pieces, and battle-painters received commissions and official acknowledgement. Loutherbourg adapted his talent to the new sensibility, which he in turn stimulated. The 'glorious first of June (1795)' [plate 27] and the battle of Camperdown (1799) gave him opportunities to paint the suffering of soldiers with new precision and to combine the wrath of the elements with human ferocity, reaching new heights of horror. France, too, had its technicians in this field: Louis-Nicolas Van Blarenberghe, the protégé of Choiseul, and the brothers Pierre and Nicolas Ozanne, who knew how to represent battles with scrupulous exactitude and who contributed to codifying the materials of this pathos in the form of nightmares.

Curls of smoke, fire-breathing cannon mouths which spat out their balls, spurts of liquid released by the impact of projectiles, gigantic explosions that lit up the space which they scattered with their debris, the glow of hellish fires, and floods unleashed beneath a sky streaked with lightning all became part of the systematic composition of horror, sometimes highlighted by the livid light of the moon.[25] Naval battle paintings no longer hesitated to display bleeding and already rotting flesh; it prepared the delight that the Romantics would experience upon rolling out the battery of tortures. Though its theatre was often the open sea, this art stressed the tragic character of the shore.

SHIPWRECKS ON THE SAND

The Romantic generation was obsessed by shipwrecks.[26] These were one of the nightmares of the age. Even after peace returned, there was no drop in the number of accidents. Every year, some 5,000 Englishmen perished at sea. Many families lost a dear one, and there was a long list of catastrophes that cast a shadow on the collective imagination: the *Saint Géran* and the *Britannia*, of course; but also the *Aurora* (1769), which took Falconer down with it; the *Nancy Packet* (1784); the *Halsewell* (1786); the *Lady Hobart*; the *Dutton* (1796), which ran aground on rocks near Plymouth; the *Abergavenny* (1805), which sank off the shores of Weymouth despite the efforts of its captain, Wordsworth's brother; the *Peggy*; and the *Minotaure*, which sank in the night of 22 December 1810, causing the death of 570 passengers out of a total of 680, and which inspired Turner, the *Méduse*, and the *Forfarshire* (1838), some of whose passengers were saved thanks to the heroic Grace Darling.[27]

The favour enjoyed by accounts of shipwrecks was strengthened by the late vogue for horror novels. The terrifying nocturnal catastrophes described by Maturin[28] were one example, as was the frantic sea of *Han d'Islande*, though it is true that its author had not yet seen the ocean.

Between 1815 and 1840, accounts of shipwrecks were fashionable. While English historians attempted to popularize and celebrate the accomplishments of the Royal Navy,[29] the influence of catastrophe literature from the Age of the Enlightenment extended its influence through numerous reprintings of *Shipwreck* and *Paul et Virginie*, the publication of several tales by Crabbe, William Cowper's *The Castaway* in 1799, and finally *Naufrage* [Shipwreck] by Sarah Burney, published simultaneously in 1816 in Paris and London.

Beginning in 1829, however, innovation took over; this is when the maritime novel, whose development has been minutely traced by Monique Brosse, from Fenimore Cooper to Eugène Sue and Melville, started to come into its own. This literary genre revived the depiction of perils and battles at sea. Never in France had this so inspired the imagination as it did in the early 1830s. An 'infra literature' drew heavily on travel accounts from the eighteenth century and from the *Histoire des naufrages* [History of Shipwrecks] published by Deperthes in 1781; and it reworked accounts of catastrophes, dwelling with predilection on those that occurred near the coast.[30] Collections of official statistics were started. In 1832, Audouin and Milne-Edwards drew up a first analysis: they distributed the 1,508 recent accidents of the past twelve years according to place and year.[31] Specialized journals such as *Shipwrecked Mariner*, *Nautical Magazine* (since 1816),

Le Navigateur, Journal des naufrages (1829), and *La France maritime* fed their readers with accounts offering great technical detail, often including a list of the victims' names. Between 1835 and 1841, *La Société générale des naufrages* published its own magazine. In the year 1829 alone, out of the 113 articles presented by *Le Navigateur*, twenty-seven were devoted to accounts of these catastrophes. [32]

Naval battle painting maintained its following, even as the means of portraying shipwrecks evolved. In England, a clientele of shipowners, officers who had made their fortunes in combat, and sailors of all stripes, including many retired from the sea, commissioned artists. [33] Thomas Luny, Thomas Whitcombe, [34] and many other painters of lesser talent continued to exalt the Navy's achievements and to stage the horrors of the sea. Turner attempted in his seascapes (which were widely distributed as engravings) to combine the wrath of the elements with the rage of the fighters. [35] In France, too, painters sought to outdo each other dramatically, and worked to highlight even further the demonstrations of naval violence. The Brest artist Gilbert, Jean-Antoine Gudin, who rose rapidly to fame for the excessiveness of his shipwreck scenes, and then the lithographer Ferdinand Perrot all knew how to satisfy the demand of a clientele eager for the sight of suffering and horror. Attention in France focused on the danger of the Breton coast. Under the influence of artists, Saint-Mathieu Point consequently became the epicentre of maritime catastrophe; it came to symbolize a new tragic vision of the coast which Michelet would contribute to popularizing in 1833. [36]

Votive art shared this taste for paintings of accidents; *ex votos* of shipwrecks proliferated all along the shores of the western Mediterranean. [37] These works of specialized painters who, like the Roux family in Marseilles, did not hesitate to sign their compositions, achieved a new level of detail. The date, hour, and circumstances in which the ship was lost, the position of the vessel, its name, that of the captain, and sometimes that of its owner were indicated on these little paintings, which took the form of chronicles evoking the tragedies they appeared to illustrate. The spread across social lines of this taste for the horrors of the sea coincided with a growing fascination among the public for all forms of catastrophe. This was demonstrated by the brilliant success of 'yellow' journalism and the overflowing columns devoted to accidents by the local press.

Thus was the shipwreck promoted immodestly to the rank of a spectacle. It was indicated as one of the tourist attractions of

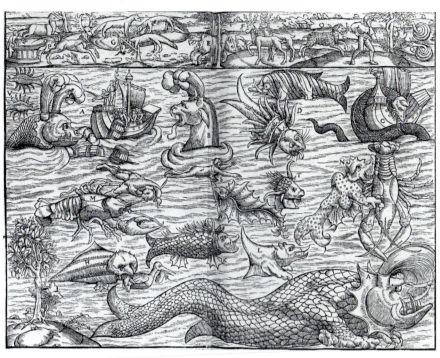

1

2

4

5

9

10

11

12

13

14

16

18

19

22

23

24

25

a coast. It ceased to be the monopoly of travellers eager to strike the hero's noble pose. Tourists, and even simple bathers who enjoyed watching a storm unleash its fury, could hope to take in the efforts of the crew of the sinking ship.[38] 'What a curious and gripping spectacle it is', declared Émile Souvestre in 1833, 'to see a shipwreck at night in our bays.'[39] Watching the accident was part of the voyeurism that flourished openly in bathing resorts; it was one of a series of distractions made possible by the construction of marinas, dikes, and jetty promenades. In short, the tourist infrastructure popularized and added variety to an experience previously reserved for coastal populations. In 1836, Félix Pyat, aware of how commonplace it had become, recounted humorously how, during a visit to Ostend, he and the customers of a café set up on the dike had used their field-glasses to watch a probable catastrophe unfold.[40] Dr Hartwig mentioned the pleasure which bathers at that resort took in 1845 as they watched two shipwrecks taking place on the beach while a dance was being held at the bathing establishment's clubhouse. The whole crowd, noted the author, followed 'this moving drama', and listened to the cries of distress.[41]

For chroniclers and journalists, the loss of a vessel was the occasion for an ordinary report. The suffering of the poor sailor drew pity; reporters no longer hesitated to seek exclusive accounts from those who escaped or to question the family of the unfortunate captain or fisherman.[42]

The experience and the accounts of shipwrecks became increasingly commonplace, and this event was no longer the only tragedy that captured the imagination of readers or travellers eager for maritime disasters. The pathos of the shore was becoming more complex, spurred on by the demand for new thrills. The fate of the *Méduse* and the tragic experience of the passengers on its raft left a deep mark on the collective imagination; this was drama that suited Romantic sensibilities. The tale evolved into a horror saga. Here, the shipwreck was merely the prelude to a diversion that would allow a sadistic recapitulation of tortures. On the raft, as well as in the lifeboat, the new refuge for unspeakable things, the terror aroused by the fear of drowning and of the shark's jaw was joined by the horrors of hunger and the slow drift towards cannibalism.[43] The twofold menace of being devoured, combined with that of being swallowed up by the sea, aroused a fear in Restoration France of a revival of animality; people realized that this tendency lurked at the bottom of the human heart ever since the execution of the King, the massacres of the Terror, and other abominable crimes had revealed its persisting existence. It may also be that the lowered threshold of tolerance for torture and any form of imposed physical suffering made any depiction of the tortures inflicted by nature all the more exciting.

Events accompanied and stimulated the course of imagination; the resounding echo of the adventure of the *Méduse* was based on the accord that developed between the unfolding of this tragedy and the new protocol of horror. The diversion of the lovers in *Makin* preceded the vessel's accident off the coast of Senegal, but this event in turn inspired Byron's *Don Juan*, whose influence is well known, as it was reflected in Delacroix's painting *La Barque de Don Juan* [Don Juan's Boat].

But for the purposes of this work, the fundamental point lies elsewhere. In this same period, the shore, that theatre in which helpless mourning was played out, also became a locus of horror. It was no longer navigation, or even drifting on the waves, that triggered the dramatic event, but the very strand itself. Its uncertain, confused topography made it easy for a bather to be cut down by the undertow, and allowed the rising tide to surprise its victim. Tragedy resulted from a betrayal by the earth, which simply abandoned the swimmer. The sense of this shiftiness grew at the same time as the belief in the unalterable purpose of the sea's boundaries was fading away, and the new time-scale, proposed by actualist geologists, was becoming accepted. The sands, constantly worked over by restless underground waters, became treacherous through their commerce with this wily element,[44] and they acquired a consistency that corresponded to the new uncertainty in values and in the social order.

In this way a protocol of accidents on the sands was outlined and gradually detailed. Such accidents were often caused by the blind progress of a 'shroud of fog'[45] that accompanied the tide. It was no longer navigation, but the route along the dangerous and shifting strand, that represented an adventure heavy with symbolism.

This transfer of the scene of the tragedy required a new familiarity on the part of its readers. To be understood, accounts of such disappearances into the sands necessitated a knowledge of the consistency of sand on the beach, of its treacherous nature, and of the back-and-forth motion of the waters that crossed it. This phenomenon would have been poorly understood by the French public at the eighteenth-century salons, who were familiar with the pictorial pathos inspired by the Mediterranean shores.

Unfamiliar with the sickeningly soft feel of wet sand under bare feet, they would not have been able to grasp all the horror of quicksands. [46]

This new tragic image came with new forms for nightmare literature, which abandoned the age-old figures of infernal tortures. It was attuned to the implications of dream-like fluidity. [47] The range of such nightmarish impressions might include being pursued by the sea with one's feet riveted to the ground, feeling the terror aroused by the conviction that a monster is catching up with one, or having the intuition that a reptile-like network of canals [48] has cut off all possible lines of escape.

This new scenario heightened the anguish generated by the idea of disappearing without a trace. Sinking into quicksand, even more than drowning [49] or being cremated, was a radical form of disappearance. It ran counter to the rising desire for individual tombs which would ensure the perpetuation of memory and promote family cults of the dead. Habasque trembles when he speaks of the terrible robbers along the desolate northern shores of Brittany, who bury their unfortunate victims in the sands of the sea, thereby erasing any trace of their misdeeds. [50] Among the Freemasons, being buried beneath the surface of the strand was a punishment; it was a destiny reserved mentally for one who had betrayed his brothers. [51]

The couple sinking into the sands, [52] like the image of lovers being swallowed up by the waves, also harmonized with the representation of Romantic love, which death could not sunder. The image of being ingested by the sand, even more than that of being absorbed by the waters, satisfied a desire for regression; at the same time, the young girl or fiancée being devoured was a new way of fulfilling old sadistic dreams. The image of being sucked down or inhaled by quicksands, like that of watching the irresistible and cruel tides rise around the victim, created an opportunity to rehash the fantasy of the good creature being inevitably swallowed up.

Writers were tempted by such scenes. When the narrator discovers Virginia's remains, they 'were half covered with sand'. [53] Forty-four years later, Nodier forcefully depicts the mud of the Mont-Saint-Michel Bay in his *Fée aux Miettes*. The page that Victor Hugo devoted to an individual sinking away and disappearing remains one of the best-known images of *Les Misérables*. It was, apparently Walter Scott's *The Antiquary*, however, and more specifically, the rescue of Sir Arthur Wardour and his daughter, surprised by the tide in the company of an old beggar, that served as a model for writers of lesser talent. This subject haunted the abundant sea-shore literature that made of maritime magazines such a success. Flood-tides and sand-bars covered by the galloping waters of a 'cold and vicious'[54] sea contributed to the popularity of Mont-Saint-Michel with tourists. Fulgence Girard told the tale of the tragic fate of a 'child of the shores', a beautiful young girl devoured by the tide one day in 1816 because she did not wish to yield to the advances of a scoundrel.[55] Most of the time, such accounts claimed to be authentic, though they were difficult to verify. President Habasque reported an accident that occurred in 1828: 'People were dancing in a circle on this treacherous shore (between Ploulech and Lannion). Suddenly, the mobile ground on which they stood disappeared from beneath the feet of a young maiden. A captain of the merchant marine to whom she was engaged rushed over and saved her, but he himself died, a victim of his devotion'; 'a young man aspiring to become a notary also perished on this occasion'.[56]

The Goa passage, used at low tide to reach the isle of Noir-moutier, could be employed to play on a wide range of emotions in the minds of readers. Rouget de Kerguen applied himself to this end when he told the story of an excursion in the company of a few friends. A 'sand guide' offers to bring the little group back to the mainland. Half-way there, the cicerone loses his way; like an Indian, he

> threw himself down on his stomach, stuck his ear against the sand, then stood up coldly: 'the tide, my boys, the tide is coming in! Run! Do you hear me? Run!'
> The sand was already moving, the mud was already shaking under our heavy steps. A profound terror grabbed hold of us. We heard the sea roaring in the background: it was perhaps a league away, and we perhaps half an hour from death ... I could see the Ocean opening its gaping jaw to devour us.[57]

Night falls. The despairing guide gets down on his knees, and prays out loud in the centre of a circle formed by the young men. Happily, the travellers manage to climb on to a buoy; the next day, a 'sardine boat' comes and collects them. One companion who had lagged behind is swallowed up by the sands and perishes: a tragic conclusion to what started as an ordinary outing on the shore. The stories of forests and towns overrun by sand or the sight of villages being devoured, such as the little town of Saint-Michel, which President Habasque mentioned long before Zola described Bonneville,[58] rooted this image of sinking into the sands all the more deeply in the imagination.

Painting of the time called forth the same emotions. At the 1837 Salon, a little-known painter exhibited *Une famille occupée à la pêche, la domestique s'est laissée surprendre par la marée* [While a Family is Busy Fishing, the Maidservant is Surprised by the Tide]. Two years later, art lovers could contemplate *Paysans Bas-Bretons surpris par la marée montante* [Peasants of Lower Brittany Surprised by the Rising Tide].[59]

The attraction of vast iridescent shores fostered tourist activities monitored by 'sand guides'. As early as 1775, William Wraxall had used the services of a specialist to cross the Mont-Saint-Michel Bay.[60] In the middle of the following century, a whole network of cicerones offered their skills to tourists who wished to attempt to cross the sand-bars at little risk.[61]

The discovery of the dangerous beauty of the shores refreshed the pleasure that individuals could experience in simulating their own destruction. Sinking into the sands outlined the protocol of a new nightmare which was easily soothed by the mere modicum of risk. This original and tragic interpretation of the strand enriched to good account the changing forms of people's desire for the shore.

11

Inventing the Beach

The means of appreciating the sea and the vision of the populations that dwelled on its shores were not solely a matter of individual habitus, cultural level, or sensibilities. The manner of being together, the complicity among tourists, the signs of recognition, and the procedures for making distinctions also determined the means of enjoying this place. The ways in which time was used and space constructed were shaped by the forms of sociability that were organized and then spread across the seaboards. The range of distractions, pleasures, and obligations that resulted dictated the form of the seaside holiday, then still nascent. It is interesting to analyse how this new social scene developed, spurred on by the desire to enjoy the shore, and how former practices were reshaped in accordance with this new purpose.

THE GENEALOGY OF PRACTICES

The genealogy of practices proves to be highly complex, in so far as the initial models have undergone numerous successive reinterpretations. At the outset, however, it is important to mention the ancient *otium*, without which the chain of influences would remain incomprehensible. It is of course difficult to provide quantitative proof of a direct relation; but the weight of Latin literature from the end of the Republic and the first two centuries of the Empire is well known. For his part, Daniel Roche[1] notes that 'Life under the Enlightenment was the daughter of the *otium*.' Cultivated men were not unaware that the sea-shores, abandoned and thought repulsive for so long, had once been places of meditation, rest, collective pleasure, and unbridled voluptuousness. The image of Cicero in his retreat at Tusculum or in his Cumanum, Pliny the Younger at Laurentis near Ostia, Pollius Felix's villa at Sorrento described by Statius, and Seneca's *Suasoriae* all suggest the character of an age of cultivated leisure.

On this point, it is important to guard against anachronism. Antique *otium*[2] as it appears in the minds of men of the Enlightenment was not synonymous with idleness; it differs profoundly from the rest, necessitated by the later rationalization of time, which we call holidays. The primacy of an ethical purpose creates an *otium, cum dignitate*, seen as a means of self-development. In Cicero's work, *otium* refers to a selected leisure, reserved for *optimates* who briefly forget their quest for positions of authority. It is a corner of private life that the individual organizes as he sees fit, seeking to avoid the two pitfalls of laziness and boredom: a place for relaxation which allows for the play of intelligence and, if necessary, prepares for future action. It is a time for renewal which paradoxically[3] fits in with the prevailing ethics in Whig England at the time of the Glorious Revolution of 1688. Somewhat later, Seneca's *Suasoriae* encouraged the identification of *otium* with the contemplative life, in the Stoic manner.

Otium implies the *amoenitas* of the leisure villa whose image it helps to shape. Pliny's home overlooking the shore and the sea at Laurentis[4] offered a clear view which permitted one to enjoy the intermingling of earth and water. The Roman elite liked – or so it was said, then – to have its ears caressed by the pleasant sounds of nature, the murmur of a fountain, the breath of the wind in the trees, the rhythmic breaking of the waves on the shore. When necessary, they sampled the pleasure of treading the sand beneath their feet, there where the last ripples of water petered out.[5] *Otium* implied variety: reading, the pleasures of collecting and correspondence, time set aside for contemplation, philosophical conversation, and walking were savoured alternatively. Relaxation in the open air was sometimes accompanied by child-like games that were possible on the beach: fishing or collecting stones and shells. These various activities were held together by the bonds of friendship and hospitality.

Otium was often a way of living by the sea-shore. Rich Romans, who owned several villas, would ensure that at least one of them looked out over the coast. From the end of the Republic until the middle of the second century of the Empire,[6] seaside resorts flourished along the shores of Latium and Campania, which rode the crest of a veritable fashion wave. At the time of Pliny the Younger, an almost unbroken line of villas stretched across the coasts near Ostia, between Terracina and Naples, and along the shores of the bay. The rich favoured Antium, Astura, Caieta, and especially Misenum, Pozzuoli, Pompei, or Baiae, which Statius and Martial considered the jewel of resorts and which Seneca and Propertius saw as a den of vice. Caesar, Pompey, Antony, and many others, like Cicero, owned villas near Pozzuoli. They came there during the season[7] to enjoy cooler weather and take sulphur baths. Pleasure-boating, water sports, open-air banquets and music, contributed to making this into a life of delight, that distanced some of these resorts from the model of *otium*.[8] In the imagination of travellers, the grand tour and the stay in Naples enlivened or updated a model of leisure that was already helping to fashion social rituals in the English countryside and in inland spa resorts.

In France, where the play of influences appears to have been less complex and the holiday-making tradition was more recent, a relationship can be traced more clearly. Daniel Roche has clearly shown the weight of the *otium* model, and more specifically, that of the academy at Tusculum in this academic social practice, and he has analysed the spread of the practice to the French provinces at the end of the *ancien régime*.[9] The networks of friendship and the ritual of hospitality that developed within the cultivated, well-travelled elite that was engaged in self-discovery in Enlightenment Europe is not completely comprehensible without referring to the model from antiquity. After visiting Edinburgh, the trip through the Highlands in the manner of Samuel Johnson involved the attentive, enlightened hospitality of the lairds. As had once been the case during the classical voyage to Italy, it was vital to visit the celebrities.

It was in Revolutionary France, however, that I have chosen the most instructive case. In 1795, after his journey to Brittany, Jacques Cambry[10] paused at Kerjégu, the home of his friend Mauduit. It overlooked the sea; from its windows one could see Groix, the Glénans, and the jagged coast. A path led to the woods, to the orchards, and to a flower garden bordering the shore. Mauduit, 'a noble sage', had Tibullus, Juvenal, Martial, and Virgil in his library; he read Tasso, Ariosto, and Petrarch with ease. His hospitality was boundless; he considered it a duty to receive any traveller 'who is visiting the coasts'. His wife, his daughter Roxane, a young engineer, and a few other young ladies made up his circle of friends. They walked on the beach, and bathed as a group in a cove of rocks they called 'Diana's baths'; they visited the ruins of castles built along the coast. In the shade of a beech tree, guests at Kerjégu read Rabelais, Bayle, and Molière; they tasted oysters washed down with 'wines from Ségur and Ay, so preferable to those of Falerne'. At the same time, Cambry worked. He organized his notes, and prepared his travel account. In short, through the very writing of the pages devoted to Kerjégu, this civil servant of the French Directorate imposed on his readers the reference to *otium, cum dignitate*.

Until the middle of the nineteenth century, the sands and rocks of the beach were an occasion to recreate the primordial circle of family or friends; as the circle grew, it gave shape to the sociability of the bathing holiday, then in its infancy. This concentric growth was very clearly suggested by Thomas Mann later on in his descriptions of Travemünde in *Buddenbrooks*. Nevertheless, between the ancient model of *otium* and this practice of friendship and hospitality among a cultivated elite, stand a series of filters that complicate the genealogy of these practices.

Without forgetting the phase of the Italian Renaissance villa and the well-known influence of Palladio on English aristocratic architecture, consider another metamorphosis that held a bright future: La Reynière defined it in 1791 as 'the pleasure that follows work',[11] and people sought it in the *bastides* (small country houses) in the Marseilles region that were so numerous as to surprise travellers. In fact, the social membership and the pace of life it implied kept *otium* at a distance from this custom which foreshadowed the modern working world's weekend as it would soon be practised in certain bathing resorts close to the major English conurbations.[12]

'No Marseillais the least bit well off', noted Millin in 1808,[13] 'would know how to do without a *bastide* ... Even craftsmen have huts that they call country houses ... You go there on Saturday evening; you spend Sunday there with the friends you invite, and you return on Monday morning.' In the merchant circles of Marseilles, there was a very strong desire 'to live in a place other than that in which one had one's occupation, to feel distant from business matters, from people who might speak of them, and from everything that could be reminiscent of them'. Millin did not hesitate to assert that the inhabitants of Marseilles worked during the week only in order to be able to go to their *bastides* on Sunday. 'More than 50,000 souls leave the city that day', and scatter among 5,000 country houses, a number of which were built on the side of the *Vista* and had terraces with a view of the sea.

The fashion for country houses was spreading throughout northern Europe in various different ways. The British and inhabitants of Provence had no monopoly on the idea. Travellers noted the same thing in Holland and in Denmark. But the distance separating Kerjégu from the 'merchant *otium*' of Marseilles *bastides* is apparent. Seaside holiday-making regrouped these multi-faceted practices into a whole.

In this area, English inventiveness nevertheless remains fundamental. The model of bathing holidays at inland spas weighed heavily on the invention of the beach. In many respects, Brighton appears to be an offshoot of Bath. In both cases, the primacy of therapeutic purposes is evident. The same effect of fashion worked successively in favour of the two resorts, which were initially frequented from June to September by members of the aristocracy and the gentry. For them, a stay in Nice, Bath, or Brighton was generally less expensive than the high society life they were expected to lead in English country houses. It was to save money that the Elliots, the heroes of Jane Austen's *Persuasion*, decided to spend the season at the most famous spa. Bath, and later Brighton, with their strictly codified and ritualized social life, created a more contained and less expensive space than London, in which one could verify one's social position and attempt to 'catch a husband'. The narrow circle of visits, walks, and outings, the small number of meeting rooms and theatres facilitated encounters that everyone hoped would appear to be fortuitous. At the same time, the abundance of bookshops and boutiques, the quality of the visitors, and the concentration of beaux added flavour to a stay which physicians insisted was healthy. Last but not least, it was easy to escape the supervision of mature adults, who sometimes hesitated to make the trip, and the control of the rural clergy, often considered overly concerned with the virtue of quality members of their flock.

The distractions and the daily rhythm at Brighton were familiar to regulars at Bath and Tunbridge Wells. Seaside resorts, like inland ones, had bathing establishments and bookshops with reading-rooms; the simplest ones offered mobile libraries. Each spa offered a network of walks and a range of excursions. Bathers could visit Celtic ruins and enjoy the views. In addition, at the sea's edge there were outings in boats and, above all, yachting, which grew fashionable at the same time as the seaside resorts.[14] Balls, conversation rooms, and gaming rooms made for pleasant evenings. At Brighton, the *Castle Hotel* and the *Old Ship* competed to attract clients. In 1766, the former built a ballroom. The next year, the latter[15] offered a set of meeting rooms that included a dance-hall, a card-game room, and a concert gallery. In 1783, the first horse-races took place on the Downs. In 1793, the promenade at Grove, the first public garden in the resort, was opened. Between 1770 and 1807 William Wade, master of ceremonies and imitator of Beau Nash of Bath, reigned over Brighton's social life. It was he who made introductions, settled disputes over precedence, and determined etiquette and manners within the salons of the *Old Ship* and the *Castle Hotel*. He was the one who kept registers in the bookshops in which newcomers could sign their names so that they would be published in the society column of the local paper.

The visits of invalids to their doctors and the meeting of writers, artists, and celebrities gradually enriched the ritual of holiday resort life over the years. The diary of Frances[16] – known as Fanny – the second daughter of Dr Burney and author of a well-received travel account, offers a good example. When she stayed in Brighton in May 1779, she was a single young woman of 27; the following year, her *Evelina* enjoyed a certain success. Frances Burney frequented the spas; she moved between Bath, Tunbridge Wells, and Brighton, where she stayed again in October 1779 and October 1782, this time accompanied by Dr Johnson. As seen earlier, this young girl admitted how much pleasure she found in bathing almost every day, beginning at sunrise. By contrast, she says nothing about the picturesque seascapes; at most, she notes in passing that the sea is located a few metres from the Thrales's house in West Street, where she was living. On the other hand, she declares herself to be very sensitive to the cool breezes during walks along the Steyne, and these give her the sense of coming to life again.

The oppressive social ritual played out in Brighton makes the young girl a focus of attention. Frances, always accompanied by others, never finds a minute for herself. On 1 November 1782 she decides that she will no longer leave the house, as she is weary of the monotonous succession of gestures and rituals. Her days are punctuated by conversations, of which she records some snatches: conversations over tea or whist, exchanges in bookshops that often concern literary works, about Pope, Gray, or Dryden. Sometimes during these meetings, a public reading is given of a work in progress. From her diary emerges the importance of the glances exchanged: young people and the members of their families stare at Fanny. People find that she has a certain charm and a French look, which can be explained by her maternal heritage. She is so bothered by this inquisitiveness that she considers no longer going out. For her part, she does not fail to notice the good-looking young men and to apply great psychological insight in her judgements of the young beaux of the resort. The fact is that social control, efforts to charm, and the task of catching a husband were openly admitted, with no hesitation whatsoever. Mrs Thrale confesses that she virtually lives on the Steyne, in order to observe the strolling couples and the purpose of their visits, all day long. The bookshops that do business there contribute to making this promenade on the rock the centre of the social stage. Individuals there are assessed and designated by their yearly revenues. There, as at balls or in the gaming rooms at the *Old Ship* during the evenings, young naval officers enjoyed great prestige in that period of war with America. The theatre proves a detestable experience, and Frances is pained to see Dryden's *Tempest* there, which, according to her, is a massacre of Shakespeare's play. In short, anyone who didn't know that this young girl liked to swim in the sea would find very little in this life-style that would give her any pleasure in being near the Channel.

This impression of life in Bath transferred to the seaside engenders Torrington's disgust with Weymouth, although the beach there was the focus of resort ritual more than at Brighton. The Baronet came to visit during the 1782 season with his wife and two of her friends, who settled into a lodging-house. He savoured the pleasures of exercise along the shore with every muscle in his body; on the other hand, he detested the style of social life.[17] It must be said that, unlike Fanny Burney, Torrington did not rub elbows with writers. He abhorred the excessive femininity of the place, the imaginary invalids, the premature weariness of the women, the presence of the beaux, the dowry-hunters, and the lack of virile youth, who were off fighting in America. He was revolted by the conversations and the teas, where no one spoke of anything but the 'season'. Torrington detested excursions and the obligation to make them. When he did yield to fashion and watch the regattas or agree to organize a tour of the isle of Portland, he soon regretted it, criticizing a fad that had already perverted the island-dwellers, who were formerly so virtuous. Our Alceste found excessive the prices charged at resorts, and wished that such places be reserved for the infirm and the convalescent. What he missed were hunting, gaming, and drinking, activities so dear to the gentry. This diatribe of a grouchy traveller of virile temperament, who had participated in the American War until the battle of Yorktown, is interesting in that it highlights how formalized the social life had become at Weymouth in 1782.

Between 1755, when Dr Russell settled in Brighton, and the end of the 1780s, a pattern of seaside holiday-making developed in England, and made its way, after a delay which cannot be explained by the naval war alone, along the coasts of the Baltic, the North Sea, and the Channel. The Continental resorts, unlike those in Britain, which were constructed gradually in response to needs and desires,[18] were generally designed as complete projects, according to programmes sometimes sponsored by the authorities. They were focused around a single bathing establishment intended to be the centre of all therapeutic activities, entertainment, and festivities in the place. The process took place in three stages. Between the Peace of 1783 and the renewal of naval hostilities in 1792, a few miserable establishments, at Ostend and Boulogne for instance, set out to satisfy the English clientele who had settled near ports that served Dover. Between 1792 and 1815, during the hostilities, numerous resorts were created at the demand of German rulers whose states lay on the Baltic and the North Sea. Where an English clientele had developed, however, a decline set in. Ostend, Scheveningen, and especially Boulogne were much too closely involved in the conflict to allow a true holiday bathing industry to develop. After 1815, once peace had returned, the rush of English visitors once again stimulated the growth of resorts, which Continental dwellers had also grown accustomed to visiting. From this point of view, the 1820s marked a turning-point. This was when the first major bathing establishments were constructed, and a specific social life developed, based on the English model but reinterpreted on the basis of national temperaments and customs. This outline needs to be filled in, however, in greater detail.

The seaside holiday gained its popularity earlier in Germany than in France.[19] German scientists were already aware of writings about the benefits of sea water when Dr Georg-Christoph Lichtenberg introduced this novelty in 1793.[20] In 1774 and 1775, this physician lived in Margate, and was very impressed by Deal, which he took as a model. Satisfied with a stay that he considered very beneficial, Lichtenberg raised the question in a resounding article in the Göttingen almanac:[21] 'Why doesn't Germany have sea-bathing establishments?' even though Central Europe had a prestigious network of hot spring spas.

In 1794, a decisive year, a debate arose between partisans of
the Baltic and those of the North Sea. The arguments of the
latter, strengthened by Lichtenberg's support, included the amplitude of the tides, the powerful waves, the water's saltiness, and
the fine sand. Defenders of the Baltic vaunted the merits of a
more accessible sea that was calmer, with water made warmer
by the slighter variation in tides. They were the clear winners, due
in part to the support of a celebrity: Dr Samuel Gottlieb Vogel.

Under his direction, construction of the first great German
resort, at Doberan, began in 1794, on the lands of the Grand
Duchy of Mecklenburg-Schwerin. Each year, Dr Vogel published
the results obtained in the resort. In 1797, and again in 1809,
the great physician and hygienist Christoph-Wilhelm Hufeland
sang the praises of Doberan. In 1822, an anonymous traveller
wrote a detailed description of the already intense social life
that enlivened the resort. [22] It then included a club for the bathers,
a theatre, and a landscaped promenade near the castle. The
Grand Duke's orchestra gave a concert every day from noon
to one o'clock at the music pavilion. A library supplied the
daily newspapers. Tea-parties and grand balls were organized
for the 240 bathers who came in July of that year, and whose
number included the Grand Duke's son. In addition, the inhabitants of Rostock came to spend their Sundays in Doberan.
On the beach stood the colonnade of the bathing establishment,
built in the middle of a garden. Visitors could stroll over the water
on a pier.

Since 1794, a number of resorts had been founded along the
Baltic: Travemünde, founded in 1800–2, near the free town of
Lübeck, was already very lively when Edouard de Montulé visited
there in 1822.

> ... they have established sea baths there; the building constructed
> for this purpose is considerable and is better maintained than
> those at Bath in England. On a beach jutting into the sea, where
> the waves come and expire on the sand, there stands a lovely
> monument with columns, in which the warm baths are housed.
> Two hundred feet behind it sits a great hotel, with an immense
> café decorated with an airy gallery. The whole place is surrounded
> by an English garden, in which there also stands a house divided
> into innumerable, lovely apartments. It is covered with thatch,
> which does not detract from the picturesque image of the setting.
> These baths are frequented by all the wealthy inhabitants of the
> north. [23]

In 1802, the king of Prussia encouraged the construction of sea baths at Colberg. Rügenvald (1815), on the coast of Prussian Pomerania; Putbus (1816), on the island of Rügen where Friedrich took refuge; Zoppot (1821), near Gdansk, developed under the impetus of Dr Haffner; and Swinemünde were the other great resorts on the Baltic, not to mention Warnemünde (1805–21),[24] near Doberan, and two beaches of the Grand Duchy of Holstein, then in Danish possession: Apenräde (1813–15) and Kiel (1822).

Sea-bathing enthusiasts were already numerous in 1819 at Swinemünde[25] even before a modern resort was built between 1822 and 1826, for which the King of Prussia showed his solicitude. In 1827, 2,200 bathers frequented its club and bathing establishments. Most of them found accommodation in lodgings; others lived at inns. Everyone rented for the season. The club offered a buffet and refreshments. Guests could play billiards and read magazines. A music room and a great concert hall with 130 seats completed the facilities. Excursions were organized to the islands or to the mouth of the Oder.

Along the coasts of the North Sea, many fewer resorts were created. None the less, in 1797, Dr Van Halem, drawing inspiration from the Doberan model, founded a resort on the island of Norderney, despite the initial resistance of its inhabitants. In 1800, during the first season, 250 bathers stayed there; in 1820, one year after a state-owned 'sea bathing establishment' was built, there were 832. The island of Wangerooge, which belonged to the Duchy of Oldenburg, had received bathers since 1801; it developed its facilities three years later. In addition to these resorts, Cuxhaven (1816) on the territory of the Republic of Hamburg, Wyk (1819) on the island of Föhr[26] off the western coast of Holstein, and an establishment on the island of Helgoland (1826), then under English domination, were also developed.

The late development of the Dutch coasts is astonishing, given the long history of baths at Scheveningen. The first bathing establishment there was not built until 1818, and it was just a tiny structure erected by a fisherman. Zandvort, near Haarlem, was what Stierling[27] considered the country's brightest resort. Almost as late was the movement to develop resorts along the coasts of the Austrian Low Countries, which were to become territories of the Empire, of Holland, and then in 1830, of Belgium. Yet bathing in the sea was an age-old practice at Ostend.[28] During the reign of Joseph II, many Englishmen had settled there, attracted by its status as a free port. As early as 1784, in what constituted a novelty on the Continent, William Herket obtained authorization from the Emperor to build a hut from which he sold refreshments to bathers. In 1787, a literary circle opened in Ostend. The war put a stop to the resort's growth. During the Empire, taking the waters became a common but erratic practice. Peace gave them a new impetus. A rivalry developed between Spa and Ostend that was not unlike the rivalry between Bath and Brighton. Yet it was not until Belgium obtained its independence that the resort became properly equipped. In 1837, the first casino opened, shortly before the first bathing establishment. With that, bathers rushed to Ostend. Karl Marx stayed there in 1846, and soon after, in 1854, Dr Hartwig devoted a significant but late work to this resort, which was to inspire Michelet.

When Daniel Lescallier visited the city in 1775, he noted that there was a large English population living in Boulogne. Some came, as Smollett had done earlier and as Townley would do, to take the sea waters and look after their health; others merely made a brief stop at the end of the grand tour; still others contented themselves with sending their daughters there to complete their education in a French convent, or organized exchanges between families of the type still arranged today in the name of language skills.[29] Three years later, Brissot spent some time in Boulogne with Scottish friends[30] before his departure for the British Isles. When Samuel Ireland described it in 1790, the city had just acquired an establishment for both hot and cold baths.[31] Its construction, followed attentively by the royal family (which was prevented by events from coming to Boulogne), proved to be long and costly. The builder, who had just returned from a journey to Italy, had been inspired by a Mediterranean model, rather than the open sea baths so dear to the British. This may explain the establishment's small success, despite its beautiful appearance.[32]

It was only once peace – and the English – had returned that Sir Viersal decided to build baths near the entrance to the harbour. This establishment, 'of a Roman Doric order',[33] opened in 1824. It was inspired by constructions at Brighton, Ramsgate, and Dieppe. Female bathers had their own private drawing-room, a resting room, a refreshment room, and a music room. The men enjoyed a meeting room, a billiard hall, and several salons. The two wings led on to a great assembly and ball room decorated with ionic pilasters and columns. Settled comfortably in the midst of elegant furniture, ladies and gentlemen found reading material and games. The building had a peristyle opening on to the sea ·and a porch opening on to the town. A stairway provided access to the platform that covered the building, from which visitors could watch the sea under the shelter of elegant canopies and, when the weather was fine, see the coast of England.

This site, however luxurious it might be, paled by comparison with the veritable hymn to the sea on which the architect Chatelin began construction in Dieppe in 1822 for Count E. W. de Brancas.[34] This was a revolution. Before 1822, tourists who wanted to bathe directly in the sea found only a few bathing cars and a small number of tents scattered randomly along the beach. Now, however, the regular line established with Brighton in 1824 and the proximity and beauty of the picturesque sites in the Caux region and the Seine valley inspired many British tourists to take this route when going to Paris; consequently, the idea of a luxurious open sea establishment, combined with a hot sea bathing facility within the town, became a necessity.

The baths at Dieppe, like those at Boulogne, were a veritable complex intended to order the entire social life of the resort. Before the sea, a gallery measuring more than 300 feet long stretched out, taking the form of a tent held up by lances. 'The vault of the portico is covered with caissons and rosettes along its entire length. Outside, Greek niches set in the pillars enclose four statues representing the major seas. Cabinets built into the corners are used to deposit books and newspapers and to distribute entrance tickets.'[35] The gallery, broken in the middle by a portico in the shape of a triumphal arch, ended on each side with a square pavilion. The ladies' pavilion consisted of a large drawing-room 'used as a meeting room before and after the bath'. It communicated with two rooms for rest and for attending to 'lady bathers whose condition required particular care'.[36] These rooms looked on to the sea and an English garden which was used as a promenade. The men's pavilion was identical, except that 'the main room is used as a billiards hall'.[37] A circular staircase led to the terrace which crowned the portico. Telescopes were available to look at the sea ... and the beach. Across from the pavilions, pontoons edged with railings were installed, along which the bathers walked to the sea, under the supervision of authorized guides. 'At the foot of these pontoons are mobile tents in white canvas and twill', where 'one leaves and picks up one's clothing'.[38] To the west of the garden, a restaurant-owner provided board.

To compete with Folkstone and Brighton, France under the Restoration offered these two prestigious establishments, whose ambitious and orderly organization contrasted with the sometimes chaotic exuberance of English resorts. But aside from these two rival centres, there was virtually nothing,[39] except for Granville[40] where a 'cabin' had been set up in 1827 for bathers, and Royan, which began to attract tourists at the beginning of the 1820s, and Biarritz, whose originality is well known. The first sea bathing establishment at Sète did not open until 1834; and it was a temporary structure of wood, built on stilts, which was replaced in 1839 by another structure, also built of wood, and no less temporary.[41] The emptiness of the beaches around Calvados is clear proof of French backwardness in this area[42]; to realize its extent, one need only read the account left by Dumas of his stay at Trouville in 1832, at the inn of Mother Ozeraie.[43] This modest establishment, which was also frequented by Paul Huet, Eugène Isabey, and especially Charles Mozin,[44] seemed cut off from the world, in the heart of a poor fishing village. The sporadic practice of sea bathing that was noted here and there, in Courseulles, Luc, Carolles, and Pornic, could not disguise the absence of the aristocratic type of social life that then defined the seaside bathing holiday.

The case of Biarritz is complex enough to merit some discussion. As seen earlier, hedonistic aims initially overshadowed therapeutic purposes. The inhabitants of Bayonne, joined on certain days by villagers from the Basque country, had long been in the habit of coming to splash in the waves during the summer, amidst the Biarritz rocks. The importance which Biarritz acquired during the Spanish War heightened the appeal of the resort. The Emperor came twice to bathe there in June 1808, though not before the area had been reconnoitred to prevent his becoming a victim of an English incursion.[45] During the July Monarchy, the complexity of the beach's physiognomy increased. To older practices was added a resolutely English fashion. According to Auguste Bouet, who described Biarritz in 1837,[46] Parisians, as well as guests from Bordeaux and Lyons, came to rent lodgings there for the season. Beginning in 1835, the flow of refugees, victims of the Spanish civil war, and later Carlists swelled the ranks of the clientele. Biarritz became the resort of Castilian dukes, English lords, and French counts. These guests encountered the craftsmen, the working-class girls, and the 'handsome sons' of Bayonne, who commonly organized pleasure outings there, with the exception of those who preferred henceforth to retreat to Guétary or Saint-Jean-de-Luz. On Sunday mornings in July, there were traffic jams around the Spanish Gate; in the evening, an unbroken line of carriages rolled along the road to Biarritz. But here there was no organized social life as there was in Brighton, Dieppe, or Doberan. Already in 1784, the mayor and the municipal judges had refused a carpenter 'the authorization to build sheds or shelters on the Old Port to receive bathers',[47] and in 1837, all that tourists found on site were a dozen little wooden sheds in which to change their clothes.

It is important, then, to make a careful distinction between, on the one hand, these natural sites, where the pleasures were spontaneous and often of a working-class nature (the same situation existed in a slightly different form around Le Havre and Saint-Brieuc[48]); and on the other, the carefully planned installations on beaches intended for a distinguished public whose emotions were carefully guided, controlled, and magnified against the backdrop of a luxurious setting of stone, greenery, and sand. The neo-classical sea palace erected at Dieppe, which imposes a clever identification between female bathers and nymphs, is an ode to beauty, to discreet sensuality, and to feminine fragility. It is just one part of a brilliant orchestration that aims simultaneously to arouse and to channel a longing for the sea, which is assuaged there through glittering luxury and modest physical ostentation.

ATTUNING SPACE WITH DESIRE

In 1822, establishments on the Dieppe model, with their expansive terraces and pontoons, enabled bathers to harmonize the space in which they moved with their longing for the shore. Yet at that time – and indeed already in earlier years – a broader process was under way in England whereby space was becoming attuned to impulse. It is worth noting that the deliberate development of these sites occurred some time after the proliferation of statements about the picturesque sea, the quest for views, and the admission that contact with sand, water, and rocks was a source of pleasure. It was not actually until the 1820s that seaside architecture began to flourish, moving towards the cumbersome magnificence that is now familiar.[49] Its growth sanctioned the rising desire to breathe sea air, in an age when Romantic consumption was enjoying its greatest prestige.

On this subject, Brighton is the clearest example of the ground in which innovation took root.[50] At the end of the eighteenth century, the famous Steyne on which Fanny Burney and her friends strolled was nothing more than a vacant lot, a communal property. Fishermen came there to dry their reeking nets; little black pigs browsed freely; and a dirty stream ran across it, sometimes swelling with the rains and transforming the area into a swamp. Walkers could barely stray from what was still a simple path. For quite some time in coastal resorts, the network of walks proved to be paradoxically indifferent to the picturesque sea view. At Dieppe, almost all the houses rented by bathers during the Restoration turned their backs to the sea; the first 'villa' at Biarritz was built in 1841. By the beginning of the nineteenth century, however, a model of development emerged, typified by Brighton, which allows us to follow its genesis and form.

The classical visit to the harbour, the already old practice of strolling and conversing on the dunes or the beach, and the habit of exercising after bathing all took place according to each individual's fancy. Beginning in the 1810s, however, the construction or improvement of a pier promenade became a necessity for any resort with the slightest ambitions. For sailors, it gave the illusion of standing on a ship's bridge; for tourists, spectators at regattas, and enthusiasts of yachting and sea excursions, it recalled delicious sensations. Spontaneously, the old port jetties, with their masses of defensive stone, became an opportunity to extend the traditional walk along the beach. This was the case very early on with the Cobb, the jetty at Lyme Regis on which the Elliot family strolled in Jane Austen's *Persuasion*. It was during this walk that the sea-breeze restored the bloom of youth to delicate Ann, who had momentarily lost it. And of course the jetty at Douglas was one of the favourite destinations of Richard Townley on his walks.

In this respect, it seems that Margate, rather than Brighton, took the first initiative. The new pier, built in 1815, was the first example of a jetty designed as a *parade*.[51] For a penny, tourists who so desired could spend the day looking at the sea, and on fine days could sway to the music of the orchestra. Brighton had no true quay. Passengers arriving from, or sailing toward, Dieppe had to ride on launches. But the Regent refused to have a trading port built, because its sooty dust would dirty the waters in which he bathed. In 1821, there was talk of erecting a jetty to be used exclusively for strolling and for docking packet-boats. Captain Samuel Brown proposed a structure suspended from chains hung from a series of towers. The construction, measuring 1,134 feet long and 13 feet wide and ending in a platform, was inaugurated in 1823. Despite the aggressions of storms, which caused considerable damage, especially in 1824, 1833, and 1836, the jetty was not destroyed until 1896.

The Chain Pier constructed below the new Steyne connected with the Marine Promenade via a stairway cut in the rock. At the entrance to the structure, located at the bottom of the cliff, tourists paid two pence if they had not purchased a season pass for the price of one guinea; this gave them access to a sitting-room, a library, and a reading-room. There, the day's entertainments and weather information were posted. On the jetty, set in the towers, walkers passed boutiques selling 'souvenirs'. At the end, on the platform of the Head Pier where the orchestra gave its concerts, a telescope and a camera obscura were available; beyond, tourists could reach the floating baths. In the evening, the jetty was often illuminated by displays of fireworks.

The sick and the weak came to the Chain Pier to inhale the salt air prescribed by their doctors, at no personal risk and, especially, without having to fear seasickness. On stormy days, as Adolphe Blanqui[52] noted in 1824, crowds gathered there to admire the sublime sight. The great figures who did not despise social mingling would stroll along the jetty in the company of their friends. In 1829, the future William IV,[53] then Duke of Clarence, liked to walk on the Chain Pier, as it reminded him of the bridge of his ship. When he became King, he continued to come there when he was residing in Brighton, from mid-November to mid-February. He did not hesitate to speak in a familiar manner with other walkers. La Garde Chambonas, who stayed at the resort in November 1833, reported that his host, Lord Holland, Fox's nephew, went to the Chain Pier every day from one to two o'clock on horseback; his wife accompanied him in a sedan-chair. The lord would sit at the end of the jetty, where, surrounded by his friends, he held an 'open-air club'[54] in which they discussed politics. Numerous in those days were the fashionable walkers on the 'elastic jetty'. La Garde Chambonas noted the presence of Princess Paul Esterhazy, wife of the Austrian ambassador, the Duke of Devonshire, Prince Koslowsky, and Lady Uxbridge. This was where visiting sovereigns sometimes came ashore; Victoria visited the Chain Pier in October 1837 and in 1843, before the exiles Metternich and Louis-Philippe. For the aristocracy, it was a new parade. Here, the social setting that unfolded in the inland spas and in the parks of London encountered the desire to see, feel, and experience the sea.

The Continental resorts, too, made a point of acquiring jetty promenades, which were often old structures quickly made over. During her stay in 1824 at Dieppe, the Duchess of Berry walked up and down the beach and the jetty,[55] and Félix Pyat, as seen earlier, admired a shipwreck at sea from the shelter of Ostend's promenade. At the miniscule resort of Portrieux, reported President Habasque in 1832, the jetty was the focus of the promenade.[56]

The use and arrangement of space dictated by these new ways of experiencing the sea have their own history, which cannot be reduced to the history of the jetty. Plates 8, 9, and 10 provide examples of the complex systems of esplanades, terraces, and marine parades with which British resorts became equipped in order to facilitate the admiration of the natural scene and the ruling classes. Once again, the search for models is a sinuous one. The influence of London and the inland spas alone is not sufficient to explain this innovation.

Certain Mediterranean practices observed and mentioned by travellers obviously weighed heavily on the arrangement of this coastal space, as is suggested by names like 'marina' and 'montpellier'. The image that should come to mind is not so much Venice, despite its famous night-time promenade on the waters of the lagoon. The Neapolitan Chiaia, by contrast, was the site of a parade that deserves further discussion. Once a year, on 8 September, it was the duty of the King and all his court to progress, in carriages only, along the quays of the Chiaia. On other days, the nobility came there to amuse themselves. 'One of the greatest pleasures in the city', wrote the Abbé de Saint-Non in 1781,

> is to form processions or promenades on the quays and along the edge of the sea; the numerous carriages hitched with teams of eight and ten horses that gather there form a very amusing view; in addition to the grand equipages of the nobility, there are innumerable little carriages called *Calesse*, and though these are public wagons rented by the common people at little cost, they are almost always golden and equipped with very good teams. All this movement taken together creates a very lively spectacle, and it is further embellished by the view of the superb gulf of Naples and the shore of the sea, which is almost always covered with innumerable little boats and fishermen.[57]

The marina at Palermo, however, seems to be the most evocative model. Almost all the travellers who visited Sicily declared themselves to be fascinated with it.[58] Paved with great flagstones and bordered with a carefully worked parapet decorated with statues, the marina followed the sea-shore to the right of Porta Felice. It was shady; a series of fountains and comfortable seats made the walk more pleasant. Visitors could refresh themselves inside shops dug into the ramparts. An orchestra played in a marble theatre. Generally, visitors followed the promenade in carriages, because the Palermitans had a fancy for 'being carried'; however, the *banchetta*, a raised path between the sea and the carriage road, was available for promenades on foot.

Along the marina, Palermitans came to seek fresh air and coolness and to satisfy their need for boundless horizons.[59] The parading carriages, both here and in Naples, made the promenade into a stage and a focal point of gallantry. No husband, it was said, would dream of forbidding his wife to walk at night in the shade of the marina. In April and May, and later in October, the nobility resided in the countryside; but in summer, they came to savour the coolness, sheltered by the walls of Palermo. These aristocrats were also heavily influenced by English fashions, and this complicated the play of social interaction; they were accustomed to rising at noon and taking their walk at about three o'clock in the afternoon on the Cassaro.[60] Then came the time for dinner. At six o'clock in the evening, the nobles began arriving on the marina. The orchestra, which until then had played to please popular tastes, tuned up for society's leaders.[61] These individuals trod the promenade even in bad weather, if only for half an hour, wrapped in heavy coats. From nine in the evening until one in the morning was the time for 'conversation' or a performance at the Opera. Then they returned to the marina until about four or five o'clock. 'One never goes to sleep in Palermo', noted Vivant Denon, 'without having taken a turn on the marina ... this promenade is cloaked in the most mysterious and most respected darkness: everyone confuses somebody with someone else, loses someone, looks for him, and finds him again. Suppers are organized there, which are eaten as picnics, and on the hour, at caterers' establishments along the walls of the ramparts.'[62] 'The marina is truly the meeting-place for all of Palermo', 'the interesting refrain for the day'.[63] This nocturnal stroll, taken in the dark, stimulated the fantasies of travellers even more, it would seem, than the Neapolitan night or the gallant life in Venice.

Now back to Brighton. Until around 1805–10, half a century after the beginning of the fashion for salt-water bathing, no one, states Antony Dale,[64] seemed to have imagined that it could be pleasant to watch the view and stroll along the heights overlooking the sea. People were happy merely to breathe the healthy breezes on the Steyne and, when necessary, to confront the waves. Yet in a quarter of a century, primarily between 1822 and 1828, the period in which Wilds and Busby produced their greatest works, a majestic seaside architecture was to blossom, after the area of the Steyne had been reclaimed (1793), landscaped, paved (1806), and built on. The number of residences began to expand, following the example of the half-Indian, half-Chinese marine pavilion built by the Prince of Wales beginning in 1786, with windows facing the sea. Soon it became possible to walk endlessly while overlooking the waves.

The fourteen lodging-houses on the Royal Crescent begun in 1798 and finished in 1807 were the work of Jean-Baptiste Otto, in which the influence of Bath was again visible. Despite their modest proportions, they represented the foundations of seaside architecture. In 1822, King George IV solemnly opened the connector, created on the sea front between the residential zones that had developed to the east and the west of the town. Between 1823 and 1827, two sea terraces, the square, and the two wings of Lewes Crescent, comprising Kemp Town, were constructed,[65] according to the plan drawn up by John Nash, who derived his inspiration from Regent's Park in London. During the same period, a seaside residence was built in the parish of Hove which was to become, as is demonstrated by the harmony of Brunswick Terrace, the purest example of processional architecture realized by the seaside. Finally, in 1829, the Grand Junction Road was completed from East Street to Marine Parade.

In 1833, J. D. Parry remarked, 'The sea line of houses in Brighton now extends from the eastern extremity of Kemp Town, to the Adelaide Terrace, beyond the western esplanade, a good three miles; a line of banked-up buildings, of which only St Petersburg had an equivalent.'[66] This line, the levels of the esplanades and the terraces, and the seaside parades fascinated travellers. On the terrace rising more than 200 feet above sea-level, there was, as Adolphe Blanqui noted in 1824, 'a continuous train of carriages and men and women on foot or on horseback' moving back and forth 'a hundred times on the same site'.[67] And La Garde Chambonas wrote *circa* 1834: 'A crowd of carriages, elegantly mounted horsemen, and people on foot, decked out with care, rushed there as though it were a holiday. Landaus, coupés, and phaetons mingled there with a large number of those small carriages, almost skimming the earth, that are called flies, and whose only merit is their great lightness.'[68]

Of course Brighton was the site of a royal residence at the time, and grand high-society promenades existed in all the capitals; consider the ritual of the Bois in the Paris of the Bourgeois Monarchy. But in Brighton, from the Chain Pier to the Marine Parade, the purpose of the installation was to breathe the sea air and gaze at the Channel waters. Along the esplanades, the sites of pedestrian promenades and the starting-points for bathing,[69] benches were installed facing out to sea. People came there to read, to do embroidery, or to let their children play.

That the intentions of Brighton's architecture are particularly clear should not overshadow the many other examples. In this genealogy of places, northerly Scarborough, the other ancestor among resorts, is also of great interest. The New Spa promenade gallery aroused Granville's admiration.[70] With its neo-Gothic style, decorated with little towers and opened by windows looking on to the sea, it enabled visitors to use the promenade, breathe the breezes, and view the open sea even in rainy weather. Scarborough offered still other interesting constructions: 'Fronting the sea here, some neat houses appear, which are let as lodging, and are called the marine houses. They have a small adjoining building for cold and warm baths – the sea at spring tides reaching nearly to the threshold of its garden front.'[71]

All the resorts rivalled one another in the art of organizing space, including the recent and much appreciated resort at Blackpool, which in 1841 had some 1,500 houses of lovely appearance available,[72] built below the cliff. In addition, there was a terrace and a marine promenade via which pedestrians and riders could reach the vast sandy beach which Granville considered to be unequalled anywhere.

THE INITIAL PRIMACY OF THE ARISTOCRACY

The above discussion provides a glimpse of the significance and the social arrangements behind the spread of the seaside holiday. This represented a combination of practices initially limited to the concentric spheres of the royal family, the high-ranking nobility, talented individuals or celebrities who were 'in', and the gentry. Though some members of the clergy did not disdain to stay by the shore, they nevertheless criticized the forms of sociability practised there. When the industrial and merchant middle class began to frequent these resorts, they brought with them other rhythms and other customs, despite their desire to imitate the noble model. This initial primacy of the aristocracy must be emphasized: it has become very common among Continental sociologists to analyse the ritual of the bathing holiday as a response to the aristocratic model of castle life, a ritual invented by a middle class seeking to reinforce the legitimacy of its power by displaying it in new ways. This erroneous hypothesis is obviously influenced by the belated splendour of Ostend and Deauville, or perhaps by that of Proust's Cabourg. In England, in France, and sometimes on the shores of the Baltic, it was the highest aristocracy who played the pioneering role; very often, the royal families themselves decided on the creation or vogue of a resort. In any case, they were the ones who launched the fashion.[73]

This process actually follows directly from what made the fortune of the spas. Granville emphasizes that in 1733, the list of bathers at Scarborough read like a *Who's Who*,[74] and the transitional role played by this resort has already been discussed. More than a century later (1841), Scarborough was attracting a more democratic range of visitors, but it was still frequented by distinguished guests. During the season, the town still had its gazette which was determined to set the trends, and which published the list of new arrivals at the resort.[75]

Even more revealing is the case of Weymouth, although this beach was located 130 miles from London. In 1780,[76] the Duke of Gloucester, the King's brother, spent the winter there. He built Gloucester Lodge, which was to become a royal residence. For medical reasons, King George III came there to take the waters; in 1789, his first sea bathe was the occasion of a touching aquatic ceremony.[77] From then on, the sovereign stayed regularly at Weymouth. An analysis of the chronicle in the *Gentleman's Magazine* reveals the royal family's daily timetable from 25 July to 14 August 1789, when they left for a fairly long stay in Plymouth in order to visit the city, the port, and the citadel and to review the naval fleet.[78] So that year, the royal couple, the princesses, and the Duke of Gloucester, who arrived somewhat later, stayed at Weymouth. Their actions, and even their menus, were published in the press. Pitt and the ministers of the Privy Council came when affairs of State made their presence necessary, but they did not share the life-style of the royal family.

What is striking first and foremost is the appeal of the seaside for the guests of Gloucester Lodge. Of course the highest-ranking members of the aristocracy had been enjoying the pleasures of yachting for almost half a century. During the twenty days, the royal family once took their tea on the *Magnificent*, and went on three outings aboard the *Southhampton*, not to mention a longer excursion to some neighbouring castles. By contrast, they were present at just one 'select party at home', for the birthday of Princess Amelia. A whole range of other distractions were also available. George III bathed twice in the sea, in the company of the princesses or the queen, and following the first bathe, he savoured the fresh morning air. On two occasions, the royal family took an evening promenade on the Esplanade and the New Terrace that passed Gloucester Lodge; their walk lasted until nine o'clock in the evening. In addition, the King went on outings on horseback or by carriage across the sands or into the surrounding countryside. As for the princesses, they visited the shops in Weymouth.

While Louis XVI was confronting the Parisian Revolution and facing the abolition of privileges in tumultuous Versailles, where there was little time left for playing shepherd and shepherdess during that summer of 1789, the King of England and his family were enjoying natural pleasures, derived for the most part from the proximity of the sea and their subjects.[79] Certainly there was a therapeutic necessity hanging over the King, who was still capable of fulfilling his role despite his depression. But this is not enough to explain the difference between these attitudes. The distance between the political regimes and cultures of the two nations was great, but so was the gap between the two aristocratic manners of appreciating nature and its pleasures.

Even more apparent is the impact of royalty on the destiny of Brighton, a town much closer to London than Weymouth. During both the Regency and the reigns of George IV and William IV, the resort was virtually a second capital. From its beginnings, it had been frequented by the aristocracy.[80] In 1756, the Duke of Gloucester made a short stay there: the next year saw the arrival of the Duke of York, another brother of the King. In 1771, the youngest of these brothers, the Duke of Cumberland, began what proved to be regular visits to a resort that Princess Amelia, the King's aunt, adopted in 1782. It becomes clear, now, why so many aristocrats rushed to take lodgings near the Steyne. In September 1783, the Prince of Wales went there, despite the disapproval of George III, to visit his uncle, the Duke of Cumberland. He went riding on the dunes, and played cricket. He returned in 1787 and bathed in the sea. His physicians believed that this would help to cure his goitre. From then on, the Prince remained faithful to Brighton; he stayed there for forty-four consecutive years, settling into the Marine Pavilion, which is the centrepiece – more than Gloucester Lodge – of the Marine Palace. The attentive presence of the Prince, the Regent, and then the King was to transform a thermal resort into a holiday and pleasure centre[81] whose freedom and gaiety long contrasted with the somewhat stiff style of Windsor. In short, for the first time, a shift took place from therapeutic aims to hedonistic ones, and this was to characterize all the great Continental resorts during the nineteenth century. The transformation was completed, in another place and much later, by the Duke of Morny.[82]

Until the end of his reign, in 1837, William IV in turn wintered at Brighton in the company of Queen Adelaide. Victoria, by contrast, quickly found the resort too noisy, and was wearied by the crowd packed into it. The Queen, who detested the Pavilion's architecture, could not lead the private life to which she aspired in the bosom of a constantly growing family. Prince Albert could barely garden in Brighton, and it had become difficult to bathe in the sea there. After 1845, although she continued to send some of her children there, the Queen preferred Osborne on the Isle of Wight. In this she maintained the contrast with the new French royal family, which continued to disdain to visit the sea and its shores. As a side-note, Metternich spent the better part of his period of English exile on Brunswick Terrace after 1848.

Initially, every resort needed the presence of a member of the royal family in order to attract distinguished guests. Worthing gained this advantage from Princess Amelia's stay in 1798; Southend developed after Princess Charlotte of Wales was sent there by her physicians. The visit from George III in 1791 gave Sidmouth its start. The Isle of Wight benefited from the arrival of Charles X in exile, from his sad walks along the beach at Cowes,[83] and even more from the presence of Queen Victoria.

During the 1830s, the social life of a great resort like Brighton remained focused on the aristocracy. La Garde Chambonas gave a precise description of the *otium dulce*[84] which made it possible in this setting to overcome spleen while fulfilling the trying rituals of high society. The fashionable gentleman there was interested primarily in horse riding, hunting, yachting, and cricket. He danced at the *Old Ship*; he took note of the time at which he could cross paths with the greats on the Chain Pier. He eagerly collected the anecdotes and news titbits that filled out his conversation. Ladies devoted their mornings to reading, music, and drawing. The afternoons were for visits, shopping, charity work, and the marine promenade; the evenings were for 'conversation' and dancing. [85]

The same was true in France, where the seaside holiday was taking root. During the Restoration, the resort at Dieppe, closely connected with Brighton, offered the fantastic bathing establishment described earlier. This was no longer merely a place where, according to tradition, Parisians came to see the sea and eat fish. During the season, it became an aristocratic residence. On the last day of July 1824, the Duchess of Berry came to visit for the first time. Until the monarchy fell, she returned every year, except in 1828. The Duchess sought to establish a vogue for sea bathing and to put the royal stamp on this resort, which she wanted to make into a symbol of the Bourbon dynasty. Louis XVI had shown great solicitude towards the maritime world. His visits to Le Havre and the victories won at sea against the English fleet (which contrasted with the resounding defeats suffered during the wars of the Revolution and the Empire) provided reasons to praise both the sea and the dynasty. The château at Arques, scene of the most famous victory of the first Bourbon king, was nearby, and the Duchess of Berry had a pavilion built close to it, thus reinforcing this desire to promote Dieppe through symbols. The aristocracy of the waning Restoration gathered there in the summers, and indulged with delight in what was referred to as 'Anglomania'. [86]

The Duchess paid careful attention to setting the scene for the part she played. [87] Dressed like an Amazon in a white blouse gathered tightly at the neck with a large black silk tie, wearing a hat trimmed with a floating veil, she cut a figure that burned itself into the memories of those who saw her. Each 'season', the Duchess made a regal entrance into the town. Light shows within the walls, cannon salutes, popular acclaims, and verses celebrating her arrival were followed by a distribution of gifts and then a comedy and a ball both in her honour. On 3 August 1824, Marie-Caroline, who allowed the bathing establishment to be placed under her patronage, was driven to the sea, and 'exposed to the waves' by the inspector of the baths, Dr Mourgué. [88] After that, the Duchess took the waves regularly. During a stay in Boulogne, she bathed one morning at eight o'clock, surrounded by naiads. [89] Otherwise, she walked around the harbour, enjoyed the sea view from the end of the jetty, took part in openings, and made excursions, in the English manner. Marie-Caroline intended to strengthen the bond between the dynasty and the common people, whose picturesque nature and contact she enjoyed. In the streets of Dieppe, she mingled with the crowd; at the town hall, she received gifts from the *poissardes*, the good womenfolk of Polet. She even took part once in a rescue in the harbour, a heroic action which was celebrated the next year at the Salon exhibition. In short, the Duchess of Berry brought to Dieppe a model of behaviour inspired by the English monarchy that could have altered the image of the French dynasty if events during the Three Glorious Days of July 1830 had not decided otherwise.

As the Count Apponyi, secretary to the Austrian ambassador, noted in his diary, Dieppe lived to a twofold rhythm. Until about mid-July, before the massive influx of aristocrats, it was the realm of doctors and bathing guides. During this period, Dieppe was frequented only by those seeking to rest or to regain their health. At the beginning of July 1828, Count Apponyi bathed, played badminton, sketched the château at Arques, went mackerel fishing with a friend, and took boat rides on the sea. Lady Granville's daughters sketched in their albums; the charitable ladies went calmly about the business of easing the misery of the poor. Social life could be summed up in these few circles of young English ladies and girls; but 'no visits, no dos, no societies. Ah! How deliciously restful for the body and the mind.'[90]

After mid-July, another rhythm took over, and the mass arrivals from Paris and Brighton began.[91] The local press, here as in Boulogne, published the list of new arrivals, which the Count took care to reproduce with great precision.[92] This led to the necessity of paying visits, including those that must be made across the Channel. The pace of festivities and entertainment accelerated. One reception followed another; at the château, the Duchess held masked balls in the manner of Henry IV. For one of them, Rossini, who spent two summers in Dieppe, composed cantatas. Marie-Caroline's 'club' met at the sub-prefecture; meetings took place throughout the town. Mediocre amateur concerts, balls, and fireworks all attempted to give the resort the appearance of a fashionable abode. But the accounts of all observers accorded in emphasizing the dearth of activity by comparison with Brighton's splendour.

The Duchess of Berry, like the members of the royal family of England, had her own yacht. It bore the colour of the dynasty: white, trimmed with gold. The salon was gold and crimson, and the dining-room all white. On festive occasions, the ship was lit up. The nobles who stayed at Dieppe imitated the Duchess, taking walks along the harbour and going for outings on the sea. In so doing, they acquired that familiarity with the liquid element that characterizes the British aristocracy. One evening, the Princess of Béthune, Madame de Saint-Aldegonde, and the Duchess of Coigny took the moorings of Count Apponyi's boat into their own hands on the jetty, and hauled the boat into port.[93] By her presence, the Duchess of Berry encouraged charitable activity, patronized the offerings collected in the churches, and enlivened the charity bazaar. Anyone fashionable was invited to country parties like the one held on 24 July 1830 at the château in Eu.

Four days later, however, the Brighton steamer was packed, and the English suddenly rushed away. Chateaubriand, who had arrived from Paris, returned to the capital. In Dieppe, though, in the heat of the moment, the aristocrats misjudged the importance of the events that had occurred at the end of July. On 30 July, the day after the Three Glorious Days, Count Apponyi noted in his diary that 'the ladies of my acquaintance went to take their sea bathes'.[94]

The Revolution of 1830, far from destroying the symbolic association between the French nobility and the seaside holiday, tended to emphasize it. The new king abandoned Dieppe for his residence at Eu, which was not a seaside resort. The bourgeoisie, who tended to be represented by the Orléans branch, preferred to patronize the inland thermal spas. It was not until the arrival of Napoleon III, who already in 1812, aged 4, had bathed at Dieppe with his mother, Queen Hortense,[95] and that of Eugénie de Montijo, a frequent visitor to the Basque beaches in her childhood, that the seaside holiday was once again and very belatedly encouraged. And this effort was not accompanied by the symbolic purpose in favour of the Bourbon dynasty that guided the Duchess of Berry.

Nevertheless, after 1830, a small fraction of the legitimist nobility sought to manifest their discreet loyalty to a life-style developed under the last Bourbons, and one which they had come to enjoy: they took refuge in a few neglected coastal places. So it was that at Luc-sur-Mer and at Courseulles in 1832 and 1833, obscure nostalgic practices were played out as a prelude to internal exile. Loève-Veimars mentioned them the following year in the *Revue des deux mondes*:

The July Revolution had just dispersed what we call high society. The aristocracy's misfortunes were too recent; they could not yet go and be seen in the midst of parties and amusements at Bade,

Toeplitz, or Spa. The bourgeois society, which had already arisen so brightly on the ruins of the other, had invaded the Pyrenees, Aix, and Plombières that year. Dieppe recalled memories that were too sharp. So they took refuge on the melancholy coast at Luc. Women whose youth, beauty, and wit still revealed the advantages of a great name, famous great names such as *la fleur des pois* of the Faubourg Saint-Germain, a few fallen ministers ... met every evening in the hall of a miserable inn. In the morning, they could be seen heading out in little groups, the great lords on foot and the great ladies on donkeys, to walk sadly along the vast sea, the way the little court of James II walked after its defeat along the shore at La Hogue.[96]

In the German states, in Holland, and in Belgium, bathing establishments were generally founded by companies of merchants, civil servants, or physicians. At the same time, the sovereigns, enlightened despots with simple tastes, watched closely over the development of these resorts, and sometimes encouraged their foundation or their growth by contributing from their own fortunes. It was the assistance of the Grand Duke of Mecklenburg-Schwerin that enabled Dr Vogel to install the first sea bath at Doberan in 1794. A quarter of a century later, an anonymous bather declared that he was flattered to bathe there in the company of the monarch's heir.[97] In 1801, the Duchess of Oldenburg made a gift of a Bath chair to the resort that was being founded on the island of Wangerooge. In 1824, the King of Prussia, who had already encouraged the construction of a sea-bathing establishment in Colberg in 1820, appointed the directors of the resort at Swinemünde, granted them subsidies, and contributed to the embellishment of the club. Consequently, Swinemünde held a brilliant party each year on the sovereign's birthday.[98] The King of Hannover had a residence on the island of Norderney: the fact was attested in 1837. The governor of Iceland frequently visited Travemünde. The revival of Scheveningen after 1818 owed a great deal to the regular stays there of Count Gisjbert Karel Van Hogendorp, the main promoter of the restoration of the house of Orange-Nassau to the throne of the Netherlands.[99] Beginning in 1834, King Leopold I and his queen Louise-Marie spent the 'season' at Ostend;[100] the royal couple's presence encouraged efforts to equip the resort.

It would be a mistake, however, to imagine that only European nobility and members of the English gentry came together in the summer by the seaside, in the company of a few artists and the celebrities of the moment.[101] Of course the rare and more spontaneous activities of the working class or lower middle class described above were not enough to generate the finely codified sociability that characterized maritime holiday-making at that time. Over the decades, though, physicians' admonitions became more burdensome, the desire to imitate society's 'greats' increased, and means of transportation improved, facilitating the organization of pleasure parties near the major urban conglomerations. All this contributed to transmitting social practices and ensuring their penetration across class lines and their diverse reinterpretations in the process. On the basis of the clientele that was attracted, resorts underwent a shifting classification on a scale of quality, and in many places, visitors altered the dates of their stays.

The very detailed investigation published by Granville in 1841, the same year in which the railroad line to Brighton opened, indicates just how common the practice was becoming. Since 1754, the fashionable season had evolved.[102] Originally, the aristocracy, the gentry, and fashionable writers and artists came to Brighton in June, July, and August. This was still the pattern during the Regency. Only those who preferred colder waters, for therapeutic reasons, came to stay in the autumn. Gradually, London's proximity encouraged the arrival of other social classes; this incited many aristocrats to retreat to Ramsgate or Hastings or to turn to the new beaches of Devonshire. Those faithful to Brighton preferred to reside there in September and October, after the common folk had left. The summer was left to London merchants; in 1841 the month of September had become the domain of the legal profession. This is when the distinguished public developed the habit of staying there in November, December, and January. This made sea bathing increasingly difficult, and furthermore, the vogue for cold-water bathing was beginning to fade.

At the beginning of Victoria's reign, Brighton beach in the summer was often perceived – by the Queen herself – as a noisy spot, packed with tradesmen. As early as 1827, Prince Pückler-Muskau considered the resort to be primarily a winter residence for London high society.[103]

At the other end of the country, the calendar for Scarborough's season was also modified, if Granville is to be believed. The resort was still frequented by members of the landed aristocracy from the East and West Ridings, but they no longer arrived until after the great races that were run on the beach during August; then, writes Granville, the great mass of visitors of 'ignoble origin'[104] ceded the place to the members of the upper classes.

Around 1840, the visitors drawn to certain resorts remained working-class from one end of the season to the other. This is why the Tuggs family created by Charles Dickens[105] in 1836 refused to go to Gravesend or even to Margate, which they considered to be a rendezvous for shopkeepers. Granville, for his part, found Tynemouth overpopulated; the proximity of Newcastle and the crowds who frequented the resort made it discouraging, according to him, to jump into the water in the midst of so much poverty and ugliness.[106]

Many inhabitants of the great industrial towns sought to escape the summer heat of the city. Those from Manchester, eager, noted Granville, to cleanse their skin of industrial filth and to flee the smoke of the factories, went to Southport. The resort became the meeting-place for manufacturers and wealthy craftsmen, who could not afford to stay more than a week or two. The beach was even starting to attract the working class: one gaudy advertisement offered Manchester workers an outing to Southport for five hours for a modest sum.[107]

At Blackpool, further north, Lancashire manufacturers came together with the Preston middle class. At the boarders' table in Nixon's Hotel in 1840, tourists could see a forge master from Bradford or Halifax sitting next to a retired merchant from Liverpool. By the same token, Preston's upper classes disdained to visit the magnificent sands of Blackpool, preferring to stay in the more aristocratic resorts of Sussex.[108]

On the Continent, the same sort of divisions developed later on. The only difference was that the degree of regulation, often more strict and more attuned to social stratification, provided greater protection from the horrors of promiscuity, especially in Germany. On this point it is interesting to note that all the northern resorts made a point of separating the sexes, and implemented strict regulations on displays of nudity.[109] In this, they parted ways with the Basque model, discussed earlier, which was more carefree in this area, since the therapeutic aims and aristocratic sensibilities had not yet imposed such rules.

In 1827, the resort of Swinemünde[110] received 1,200 bathers, mainly aristocrats and members of the upper middle class. The beach, as described the following year, was divided into five zones. The middle zone, some 500 paces wide, was to remain empty. Its purpose was to separate the other two areas, one of which was reserved for men and the other for women. Little boys aged 4 and older were no longer allowed to accompany their mothers; instead, they were entrusted to a monitor on the men's beach. This classical division by sex, which was practised at Boulogne, at Granville, and at Dieppe, was extended here to include a division by social class. The men's sector was separated into two zones. In the first, members of the lower classes, deprived of changing cabins and bathing chairs, splashed in the waves. The second segment of beach was reserved for the wealthy; it offered twenty or thirty completely equipped changing cabins, bathing chairs, and a boardwalk to protect the gentlemen's feet from injury. The first of the zones reserved for women had the same type of installations available. These were not accessible to women from the lower classes, who were restricted to the last zone.

It is easy to understand that such regulation caused the spontaneous practices that had existed prior to the spread of the seaside holiday to disappear fairly quickly, be it the working-class paddling that was observed at Scheveningen or the 'free bathing' that visitors discovered at Ostend or Blankenberghe. [111] The common people of Rostock once had a tradition of organizing pleasure parties on the sand at Doberan; they went there to buy fish and to cook it on the beach. With the opening of the bathing establishment, this custom began to fade away, and by 1823, it had disappeared. [112]

The spread of seaside holiday-making to the masses, both where it had occurred and where it was foreshadowed, aroused criticism, scorn, and revulsion, even before 1840. Several works of fiction and fragments of diary writing illustrate the growing diatribe against the changes in this form of leisure. In 1817, Jane Austen wrote *Sanditon*. For the first time, to my knowledge, the sociability organized around sea bathing became the central focus of a novel. The scorn which the author reveals is aimed not only at the speech and the commonplace conversations concerning the picturesque quality of the sea, [113] but also at the lack of culture and morality and the stupidity of those who choose to make stays at Sanditon. She attacks this middle-class form of relaxation, born of the desire to imitate high society and of the spread of a new coenaesthetic sensibility.

Sanditon is the spitting image of an insignificant resort, a sort of anti-Brighton [114] founded by one Mr Parker, a lodger in search of guests and a pitiful profiteer taking advantage of those obsessed with their health. The setting for the novel's ridiculous scenes consists of a few lodging-houses christened 'Prospect House' and 'Cottage Bellevue', and above all a modest esplanade equipped with two benches, where the novel's heroes come to parade. The beach appears only in the background. The hilarious lower-middle-class gentleman, the handsomish dowry-hunter, a sort of sub-Lovelace, the parvenue lady with no moral sense, the rich heiress from the West Indies who has come to lose herself at Sanditon, and the lazy fat boy who cannot stop stuffing himself interact beneath the gaze of a lucid young girl from the Sussex middle class who is poor but ambitious orphan. Nothing in the novel, which was never finished, suggests the presence of a real working-class clientele.

Of a somewhat different tone were the criticisms levelled at Brighton by William Cobbett in 1822 and by John Constable in 1824.[115] They followed along the lines of those formulated earlier by Torrington. The painter chose to express the suffering arising from the contrast that was developing between the sea's magnificence and the tumult of a sort of Piccadilly-by-the-Sea, between the splendour of nature and the artificial rituals of the Marine Parade or the dandies on the jetty.

Still another tone appears in 'The Tuggses at Ramsgate', a short story by Dickens. This example, like the work of Henri Monnier and later Labiche in France, exudes derision in its purest form. The hero, a grocer who has come into a fortune, intends to become fashionable. To this end, he decides to stay at Ramsgate with his wife, son, and daughter. This provides an opportunity for a picturesque view of the 'sands':

> The ladies were employed in needlework, or watch-guard making, or knitting, or reading novels; the gentlemen were reading newspapers and magazines; the children were digging holes in the sand with wooden spades, and collecting water therein; the nursemaids, with their youngest charges in their arms, were running in after the waves, and then running back with the waves after them; and, now and then, a little sailing-boat either departed with a gay and talkative cargo of passengers, or returned with a very silent and particularly uncomfortable-looking one.[116]

The heroes, in yellow shoes, are seated on rush-bottomed chairs. The father eyes the young people who climb with their towels into the bathing-machines ready to head off towards the waves.

The Tuggses eat shrimp, then 'look at the crabs, the seaweed, and the eels'. In the evening, they go to a packed casino, and find themselves surrounded by swindlers, dandies, and mammas playing matchmakers for young things seeking husbands. For six weeks, one monotonous day follows another: 'Sands in the morning – donkeys at noon – pier in the afternoon – library at night – and the same people everywhere.'[117] The action is reduced to a ridiculous flirtation that leads to blackmail for presumed adultery.

Dickens's short story brings this analysis to a close. Ramsgate as seen by the Tuggses bears witness to the spread of a practice that became firmly established from Swinemünde to San Sebastian even before Victoria's arrival on the throne. In 1841 the railroad unleashed crowds on Brighton. This resort then definitively surpassed Bath, its former rival, but the function of the master of ceremonies had lost its meaning. The modern beach was born, out of the origins that have been retraced here.

Conclusion

Il aimait mieux l'écume que le miasme.
[He preferred the foam to the miasma.]
Victor Hugo, *Les Misérables*

The goal of this book has been to trace the emergence of a type of pleasure and how it was experienced. We settled ourselves on this coastal stage on which unconscious desires and obsolete emotions gradually took shape between 1750 and 1840; we watched carefully for fundamental and sometimes derisory quests, for pretentious posing, for customs long struggling to find their way, and for humble, obscure joys. Between the dunes and the water, we saw the primordial family circle form once again, and watched the beach being invented.

By looking through the eyes of these people from the past who were the first to be fascinated by black rocks, transparent waters, barren sands, rising viscosity, and unfurling waves, we have reached a better understanding of an age eager to read the archives of the Earth.

As images of time were shifting and the horror of stagnation in the tomb was taking deeper root, the shore, that place of longing where the elements converge, offered the sight of the restless sea to all those who feared the miasma and sought out the foam instead.

Considerations of Method

It is time for historians to call into question the idea of long-term prison and the out-of-step rhythms of Braudelian temporality. These rhythms have imposed an image of more or less wild rivers that move at different rates and whose waters never mix. This conviction makes it difficult to identify origins or trace genealogies, and even more so to detect any consistency among the representations that lend shape to a given period. In short, this model creates a virtually insurmountable obstacle to an authentic socio-cultural history.

It is important now to examine the ways and means whereby people in every age and, where possible, in every social category have interpreted ancient patterns and have reincorporated them into a consistent collection of representations and practices. This is how biblical exegesis, literary and aesthetic culture drawn from classical authors, medical science (which was also partially inspired by antiquity), and the experience of the great explorers from the dawn of the modern age all contribute to a whole system of ways of discussing and using the sea and its shores; they guide different types of behaviour which, taken together, comprise a historical phenomenon.

Although I have found it necessary, especially with regard to the Romantics, to refer to the intuitions of Gaston Bachelard and Gilbert Durand, the point is not to hone in on a belief in anthropological structures of imagination which are indifferent to the passage of time. Landscapes put out images which facilitate the movement from the conscious to the unconscious realm; topographical analysis provides symbols to which sensitivity responds; but to my mind, these operations occur as a function of mechanisms which can be dated.

This investigation crosses paths with the history of landscape evaluation. But with regard to the coasts, I have chosen not to construct grids like those developed by K. D. Fines[1] and Charles

Avocat[2] for interpreting landscapes. The method consisting in quantitatively measuring a landscape's attractiveness[3] has no place in the historian's perspective, as it would require developing a grid for each datable system of perception. In addition, it would be necessary to identify the stereotypes that encumber this discourse without offering any real perception or any real emotion of the senses. Finally, it would be important to take into account the history of discursive practices and the evolution of both the structures of descriptive style and the norms which determine what can be said and what is to remain unspoken. In short, such a method would tend to measure the changes in a literary genre rather than the evolution in the means of perceiving landscape. The effort is better spent concentrating on the history of desires, curiosities, and perceptual and discursive systems that shape these accounts.[4]

Notes

Notes to chapter 1

1 See below for the developments dedicated to the baroque poets, to Moisant de Brieux and Henry de Campion, and, concerning bathing, to Mediterranean practices.

2 [The *Shorter Oxford Dictionary* defines 'coenaesthesis' as 'the general sense of existence arising from the sum of bodily impressions; the vital sense' – trans.]

3 The bibliography, since the pioneering works of E. Cassirer, concerning the supposed sympathies between the microcosm and the macrocosm from the fifteenth to the seventeenth centuries is abundant; suffice it to mention Michel Foucault, *Les Mots et les choses* (Paris, Gallimard, 1966), pp. 32–57.

4 For more precise details, see 'Considerations of Method', pp. 283–4 [author's footnote in original].

5 We are talking here, of course, not of the Hebrews' vision of the sea, but of the interpretation that was dominant during the classical era. Indeed, for the men of the Bible, the Hebrew word *yâm* meant both what we call sea (body of salt-water) and the big freshwater bodies (lakes, immense rivers); see Jean-Paul Dufour, 'Étude lexicographique des paysages bibliques', in *Lire le paysage* (Presses de l'université de Saint-Étienne, 1984), pp. 71 ff.

6 See the commentary of Sébastien Munster, *Cosmographia universalis* (Basel, 1544), cited by Pierre de Latil and Jean Rivoire, *À la recherche du monde marin* (Paris, Plon, 1954), p. 50.

7 Genesis 1: 2; Exodus 20: 11.

8 Contemporaries frequently refer to this injunction of the Fathers of the Church; in particular Father Dominique Bouhours, *Les Entretiens d'Ariste et d'Eugène* (Amsterdam, 1671), p. 24; the Jesuit Father Georges Fournier, *Hydrographie contenant la théorie et la pratique de toutes les parties de la navigation* (1st edn, 1643; 2nd edn, Paris, Jean Dupuis, 1667; reprinted Grenoble, 1973, with comments by Jean Boudriot, René-Charles Duval *et al.*), p. 339. In his opinion, the phenomena of the tides will remain forever mysterious 'until we have the blessing of being in paradise'. Johann Albert Fabricius, *Hydrothéologie oder Versuch, durch aufmerksame Betrachtung der Eigenschaften, reichen Austheilung und Bewegung*

der Wasser die Menschen zur Liebe und Bewunderung ihres Schöpfers zu ermuntern (Hamburg, 1734), refers on this topic to Psalm 36: 6: 'Thy judgements are a great deep', and reminds us that God asks Job (Job 38: 16): 'Hast thou entered into the springs of the sea? or hast thou walked in the search of the depth?' [Throughout, translations of biblical texts are taken from the Authorized Version – trans.]

9 In Milton's paradise, the ocean is chafing under restraint.

10 On this subject, the work of Keith Thomas, *Man and the Natural World. A History of the Modern Sensibility* (New York, Pantheon Books, 1983) is an excellent source.

11 Which leads us to mention at this point the distinction made by the Greeks between the landscapes of the Oekoumene and those of the absolute. Whereas the image of the garden is organized according to the Dumézil triad (divine order: paradise; sovereign power: princely garden; working mass: productive garden), the desert, the inaccessible mountain, and, to a lesser extent, the forest, the marsh, and the lake do not constitute abodes for man; likewise the sea, from which of course, life emerged, but which, because of its passions and floods, constitutes a constant threat of engulfment. See Jacques Bethemont, 'Élément pour un dialogue: géographie et analyse du paysage', in *Lire le paysage*, pp. 102 ff.

12 [*Limes* is a Latin word which means limit, boundary, dividing line – trans.]

13 Thus Thomas Burnet (*The Theory of the Earth* (London, 1684) and William Whiston (*A New Theory of the Earth* (London, 1708)) lean towards the sources in the sky.

14 Not to mention Antonio Carrachio.

15 On this subject, Françoise Joukovsky, *Paysages de la Renaissance* (Paris, P.U.F., 1974), pp. 106–8, and Yvonne Bellenger, 'Les Paysages de la Création dans la *sepmaine* de du Bartas', in *Paysages baroques et paysages romantiques, Cahiers de l'association internationale des études françaises*, 29 (May 1977), p. 16.

16 Roy Porter (*The Making of Geology. Earth Science in Britain, 1660–1815* (Cambridge University Press, 1977), pp. 83 ff.), with good reason, emphasizes the importance of these theories and the debates they gave rise to. In assessing their importance, it is enough to consider how painstakingly Buffon tried to refute them. Moreover, Porter does a very good job of describing the religious, moral, and aesthetic implications of the debates over the Flood. The reader desirous of knowing the causes of the catastrophe as explained by various scholars, each interpreting the book of Genesis in his or her own way, is referred to this book and to the works concerned specifically with this subject. We should also add that the notion of the Flood, secularized under the rubric 'revolution of the globe', suggested by Leibniz in 1683 in his *Protogaea*, was to impose itself

as one of the major concepts of the history of geology. (On this
subject, Jean Ehrard, *L'Idée de Nature en France dans la première
moitié du XVIIIe siècle* (Paris, École pratique des hautes études,
1963), vol. 1, p. 202.)

17 The work was published in Latin in 1681; we cite here the English
edition of 1684, ch. 5, pp. 51 and 67.

18 *Ibid.*, p. 128.

19 *Ibid.*, p. 132. In France, the poet Louis Racine (*La Religion* (Paris,
J. B. Coignard, 1742), canto 3, p. 136), evokes the catastrophe
from the same perspective:

> Le ciel défigura l'objet de notre amour.
> La terre par ce coup jusqu'au centre ébranlée,
> Hideuse quelquefois, et toujours désolée,
> Vit sur son sein flétri les cavernes s'ouvrir,
> Des montagnes de sable en cent lieux la couvrirent.

> [The sky disfigured the object of our love.
> The earth by this blow, to the centre shaken,
> Hideous sometimes, and always desolate,
> Saw on her withered breast the caves open,
> Sand mountains in a hundred places covered her.]

20 The theory developed in 1695 by Woodward marks the genesis
of the Neptunian doctrine, which exerted an enormous influence
on geological thought; but the author presents a vision of the sea
which bears the mark of natural theology. It will be appropriate
to come back to it later.

21 The Ark, instrument and symbol of salvation, primary image of
navigation, constitutes a serious subject of reflection. In 1643, the
Jesuit Father Fournier devoted two chapters of his *Hydrograhie* to
it; late in his life (1675), Father Athanasius Kircher wrote a fascinating *summa* in which biology and architecture mix with the
science of domestic sanitary fittings and the art of home economics.
He retraces the hundred years of labour required for the preparation
of the work, describes the choosing of the wood and asphalts, the
installation of the ironworks and the timber yards, and the preliminary observations of the animals' habits; then he presents the
architecture, laden with symbols, of a three-levelled vessel; he shows
the inside layout imposed by promiscuity; he attributes the survival
of the passengers to the installation of a sanitary ventilation system.
The book, which reflects all the anxiety accompanying the urbanism
of the times, proposes an answer to a number of questions raised
by the reading of the Mosaic text. Kircher enumerates the animals
chosen by Noah, forcing him to distinguish between the primitive
species and the species which sprang from unrest and grafting and
were not worthy of being saved; and also forcing him to distinguish
between real and imaginary species. The author sets forth the ways

in which the stocks of meat and cheese, the barrels of fresh water and the biscuit required for the forty-day journey were collected. He describes the troughs and the tanks for the amphibians, as well as the various apparatuses, mangers, and cages used to assure peaceful coexistence among the passengers; not to mention the stocks of seeds and plants, the tools and clothing necessary for the reconstitution of agriculture after the drying out of the land.

It is hard to believe that problems as minute as those of the origins of the olive tree branch and the location of the only window in the Ark mentioned in the sacred text could have fuelled the polemic to the same extent as questions regarding the spreading of the catastrophe and the fate of the waters when they receded. Nevertheless, this was the case. Theoretical thought even led attempts at experimentation: at the beginning of the seventeenth century, declares Alexander Cattcott (*A Treatise on the Deluge* (reprinted, London, 1768), p. 39), Peter Jansen, a German Mennonite merchant, wanted to have built a ship 120 feet long, 20 feet wide, and 12 feet deep, according to the model of Noah's Ark.

22 Cattcott, *Treatise*, p. 33. He also goes through a calculation of the amount of food necessary for each animal species.

23 On this subject, Porter, *Making of Geology*, esp. pp. 164–5 and 197–8.

24 Richard Kirwan, *Geological Essays* (London, 1799), pp. 54–87.

25 [The *Shorter Oxford Dictionary* defines 'mephitism' as the 'poisoning of the air' due to a 'noisome or poisonous stench' – trans.]

26 Historians of geology, attentive to the consequences attributed to the Flood for the structure of the globe, too often disregard the persistent consequences for the composition of the air which are attributed to it.

27 Regarding the mountains, see Marjorie Hope Nicolson, *Mountain Gloom and Mountain Glory: The Development of the Aesthetics of the Infinite* (Ithaca, NY, Cornell University Press, 1959), as well as the recent book by Philippe Joutard, *L'Invention du Mont-Blanc* (Paris, Gallimard, 1986).

28 Thomas, *Man and the Natural World*, p. 259.

29 This conviction was expressed by the poet Jean-Antoine Roucher, who played a big role in the rise of descriptive poetry in France and who considers storms in themselves as remnants of the Flood (*Les Mois* (Paris, 1779), vol. 1, p. 88, and, concerning the magnificent description of the Flood, vol. 2, pp. 209–10). The author, who states that he has never seen the sea except in books, says he found his inspiration in scientific works. It is interesting to note that, regarding Roucher, Edouard Guitton (contribution to the Brest colloquium *La Mer au siècle des Encyclopédies*, 17–20 Sept. 1984), ponders the exact meaning of the expression 'see the sea', and wonders if it does not mean specifically the sea in a storm.

30 See the pages of Whiston (*New Theory of the Earth*, pp. 368 ff.) and the convictions of Louis Bourget analysed by Ehrard, *L'Idée de Nature*, vol. 1, p. 201.

31 See Jean Delumeau, *La Peur en Occident, XIVe–XVIIIe siècles* (Paris, Fayard, 1978), p. 37, and Alain Cabantous, 'Espace maritime et mentalités religieuses en France aux XVIIe et XVIIIe siècles', in *Mentalities/Mentalités*, ed. Hamilton, vol. 1, no. 1 (1982), p. 6.

32 Isaiah 27: 1.

33 This explains the location of the pilgrimage to the Mont, assigned to an ostentatious and symbolic fate in the France of Charles VII. See Colette Beaune, 'Les Sanctuaires royaux de Saint-Denis à Saint-Michel et Saint-Léonard', in *Les Lieux de mémoire*, ed. Pierre Nora (Paris, Gallimard, 1986), part 2: 'La Nation', vol. 1, pp. 75–80.

34 See Benedeit's text, recently published: *Le Voyage de Saint Brendan* (Paris, 10/·18, 1984), text established and trans. by Ian Short, annotated by Brian Merrilees.

35 Erich Pontoppidan, *The Natural History of Norway*, trans. from Danish (London, 1755), ch. 8, concerning certain marine monsters, pp. 183 ff. It is important to recognize, however, that the work is strongly imbued with the founding notions of natural theology (see pp. 23 ff.).

36 Regarding this point, but from a different perspective, see Gilbert Durand, *Les Structures anthropologiques de l'imaginaire* (Paris, Bordas, 1969), pp. 103 ff.

37 On this subject, see Jules Douady, *La Mer et les poètes anglais* (Paris, Hachette, 1912), pp. 73–4.

38 Corresponding images can be found in the Arab world, in particular in the work of the poet Mas'udi; but it is an entirely different mental universe.

39 Durand, *Les Structures anthropologiques*, p. 243.

40 On this subject, Numa Broc, *La Géographie de la Renaissance (1420–1620)* (Committee of historical and scientific studies, Paris, Bibliothèque nationale, 1980).

41 Joukovsky, *Paysages de la Renaissance*, p. 114.

42 In 1756, Madame du Bocage attributes to the Devil the unleashing of the storm in her *Colombiade*.

43 See Delumeau, *La Peur en Occident*, p. 40.

44 In this area, however, caution is necessary. The projection into the past of legends gathered by the folklorists of the nineteenth century from the coastal populations must be undertaken with care. Indeed, it is perilous to transpose into the early modern epoch these stories of ghost ships, sailors' hell, black knights, and elves of the shores, which apparently owe much to popular literature inspired by the romanticism of the sea. The works of Paul Sébillot thus warrant a textual analysis that distances itself somewhat with regard to

anthropology, takes into account literary history, and abandons the implicit denial of the social circulation of cultural objects.

45 Delumeau, *La Peur en Occident*, pp. 37–8.

46 Cabantous, 'Espace maritime', p. 9.

47 See A. Mandouze, 'Présence de la mer et ambivalence de la Méditerranée dans la conscience chrétienne et les relations ecclésiales à l'époque patristique', in *L'Homme méditerranéen et la mer*, Proceedings of the Third International Congress for the Study of Western Mediterranean Cultures, Jerba, April 1981 (Tunis, Éditions Salammbô, 1985), pp. 509–11.

48 Joukovsky, *Paysages de la Renaissance*, p. 114.

49 Jacques de Billy, quoted in ibid., p. 115; see also the quotes following, pp. 113–14. [There is an untranslatable alliteration in the French original: 'Mer très amère' – trans.]

50 E. Gombrich, 'Les Formes en mouvement de l'eau et de l'air dans les carnets de Léonard de Vinci', in *Écologie des images* (Paris, Flammarion, 1983), pp. 177 ff.

51 See George Shepard Keyes, *Cornelis Vroom, Marine and Landscape Artist* (Utrecht, 1975), vol. 1, pp. 26 ff., regarding Breughel's seascape-paintings. On this subject, see in particular J. Richard Judson, 'Marine Symbols of Salvation in the 16th Century', in *Essays in Memory of Karl Lehmann* (1964), pp. 136 ff.

52 See below, pp. 188–9.

53 In the Catholic schools of the seventeenth century, the spread of the techniques of Ignatian meditation, which emphasize daily labour and a disciplined imagination, rendered the 'composition of space' ordinary. This spiritual exercise consists in making the effort to concretely represent to oneself, on the basis of sensory experience, the scenes and landscapes which form the background of the meditation. It is evident that images of the shore might occasionally enter into these mental compositions, whose objective was to spiritualize the senses. On this subject, Alain Guillermou, *Saint Ignace de Loyola et la Compagnie de Jésus* (Paris, Le Seuil, 1960), esp. pp. 84–5.

54 This is the case in the work of Fournier cited earlier, *Hydrographie*, p. 676. Nevertheless, the sea is not proposed as a site for Purgatory. The hellishness of this distressing space, which is commonly situated in the deep, calls for images of fire, or, where a body of water is concerned, boiling water. See Jacques Le Goff, *The Birth of Purgatory*, trans. A. Goldhammer (Scolar, 1984).

55 Alan Ansen, *The Enchafed Flood, or the Romantic Iconography of the Sea* (University of Virginia Press, 1950), p. 12.

56 Thomas (*Man and the Natural World*, pp. 264–5) points out with good reason how highly literary the sensitivity of the time was; he notes as an example that, without referring to Horace or Virgil it is impossible to understand the English garden.

57 Joukovsky, *Paysages de la Renaissance*, p. 27. The author mentions a few favourable exceptions in the work of Rémi Belleau and in the first book of the *Franciade* by Pierre de Ronsard. It is of interest to note that these writers evoke even less often the ocean of the Greeks, without origin and without end, the circular river surrounding the relatively flat disc of the earth. (On this motif, see Jean-Pierre Vernant, *Mythe et pensée chez les Grecs. Études de psychologie historique* (Paris, Maspero, 1971), vol. 1, pp. 192 ff.). Okeanos for Homer is the origin of everything; it envelops the universe of which it is the boundary, without being limited or enveloped by anything.

58 On this subject, regarding the coherence of the pilgrims' system of appreciation of nature and their hatred of the spectacle of an agitated sea, Christiane Deluz, 'Sentiments de la nature dans quelques récits de pélerinage du XIVe siècle', in *Actes du 102e Congrès des sociétés savantes, Limoges, 1977*, section on philology and history, vol. 3, esp. p. 75, and above all, 'Pélerins et voyageurs face à la mer, XIIe–XVIe siècles', in Henri Dubois *et al.*, *Horizons marins et itinéraires spirituels V^e–XVIIIe siècles* (Paris, Publ. de la Sorbonne, 1987), vol. 2, pp. 277–88.

59 On this point, see the old but very well-documented work of Eugène de Saint-Denis, *Le Rôle de la mer dans la poésie latine* (Lyons, Bosc, 1935).

60 For examples of authors inspired by the model of the Virgilian storm, Dulard, *La grandeur de Dieu dans les merveilles de la nature* (Paris, 1749), canto 2, pp. 35–6; Fénelon, *Les Aventures de Télémaque* (Paris, Garnier, 1968), p. 127; Crébillon, *Électre*, act 2, scene 1; Delagrange, *Alceste*, act 2, scene 1 (speech of Tydée), (Paris, Ribou, 1704), pp. 16–17. (Heracles' speech), (Paris, Ribou, 1709), p. 18. Voltaire, *La Henriade*, canto 1.

61 On the storm in Thomson's work and the combined influence of the Psalmist and of Virgil, see Maurice Pla, *Les Saisons de James Thomson (1700–1748). Étude générique de la Géorgique* (thesis, University of Toulouse II, 1978), *passim*, and esp. pp. 478 ff.

62 See Monique Brosse, *La Littérature de la mer en France, en Grande-Bretagne et aux États-Unis (1829–1870)* (thesis, Paris, 1978, and University of Lille III, 1983), vol. 1, pp. 294–5, regarding the protocol of the storm.

63 Extensive developments on this subject are found in Saint-Denis, *Le Rôle de la mer*, pp. 285 ff. This work, laden with references, is unfortunately spoiled by the psychological anachronism into which the author sinks.

64 The influence of Lucretius on representations of the sea during the classical epoch has yet to be investigated. This task is, for the moment, outside our concerns here. The painter of the moving immensity of the seas could not find as great an echo as Horace, contemptuous

as he was of the monotony of the sea, in a time when the dominant aesthetic code discouraged appreciation of the immense, unlimited, infinite surface in which the gaze can lose itself. In the same way, the literature, at this time descriptive more than symbolic, dedicated, from the end of the first century A.D. to the limitless ocean (see Agnès Paulian, 'Paysages océaniques dans la littérature latine', *Caesarodunum*, no. 13 (1978), Proceedings of the colloquium *Archéologie du paysage* (Paris, E.N.S., May 1977), pp. 23 ff.) does not seem to have inspired the authors of modern times. We run into the influences of Lucretius and Lucian again when the sublime, then romantic, visions of the sea become so imposing.

65 Thus Bouhours affirms it peremptorily: *Les Entretiens*, p. 11.

66 On this subject see Margaret Deacon's fundamental book *Scientists and the Sea. 1650–1900. A Study of Marine Science.* (London and New York, Academic Press, 1971), p. 31.

67 A problem that Albert the Great resolved by appealing to a providential hypothesis: God miraculously uncovered part of the earth on which to install man and the animals; see Broc, *La Géographie de la Renaissance*, p. 68.

68 For developments, see Margaret Deacon's book *Scientists and the Sea*, pp. 5 ff.

69 A number of engravings of the sixteenth century represent these horrible places. Fournier mentions them at length in his *Hydrographie*, pp. 341 ff.

70 By virtue of its title, *A Submarine Voyage*, by Thomas Heyrick (1691).

71 On the assignment of places, of the limits which separate the cosmos from chaos, the sacred from the profane, the civilized from the savage in the culture of antiquity, and on the new secularization, the neutralization, and the indifference to place during the modern epoch, see the reflections of Françoise Paul-Lévy and Marion Ségaud in *Anthropologie de l'espace* (Paris, Centre Pompidou, 1983), pp. 9 ff. It is interesting to note that the sea, particularly for the populations of the archipelagos, does not always constitute a barrier.

72 Strabo, II, 5, 17, cited by Paul Pédech, 'Le Paysage marin dans la géographie grecque', *Caesarodunum*, no. 13 (1978), pp. 30–40.

73 See O. A. W. Dilke, 'Graeco-Roman Perception of the Mediterranean', in *L'Homme méditerranéen et la mer*, p. 54.

74 Pédech, 'Le Paysage marin'. On the poetic nature of this work, see Paul Schmidt, 'Avienus et le golfe Tartessien', *Caesarodunum*, no. 13 (1978), pp. 217 ff. It is appropriate, nevertheless, to remember that ancient geography owes much, in general, to fiction and imagination, and that the symbolic weighs heavily on descriptions at the same time as myths are being inscribed in the heavens – facts of which humanists of modern times were conscious. (On this aspect, see Christian Jacob and Frank Lestringant, *Arts et*

légendes d'espaces (Paris, Presses de l'E.N.S., 1981.) Regarding the discourse concerning the shore in the geography of antiquity, it is important to take into account: (1) the persistence of descriptive stereotypes; (2) the dissipation of the perception of spatial environment which characterizes landscape descriptions, in as much as this notion then means something; (3) the constraints of translating perception into discourse; (4) the desired effect on the public and the balance which installs itself between mimesis and fantasy (see Christian Jacob, 'Logiques du paysage dans les textes géographiques grecs. Quelques propositions méthodologiques', in *Lire le paysage*, pp. 159 ff.).

75 Fénelon, *Télémaque*, p. 493. The effort to climb the boulders in order to obtain an overall view introduces us here to another system of appreciation.

76 Racine, *Phèdre*, Act 1, Scene 3.

77 Fénelon, *Télémaque*, see shipwreck of Telemachus (p. 66), the farewells of Telemachus at Tyre (p. 118), the goodbyes of Philoctetes on the shores of his desert island (p. 341), Calypso's complaints on the shore (p. 65), the complaints of Philoctetes on the shore (p. 332); when Philocles is leaving, Protesilas rolls around on the sand of the beach (p. 317).

78 See Raymond Bloch, 'Les Dieux de la mer dans l'antiquité classique', in *L'Homme méditerranéen et la mer*, pp. 439–40.

79 Trans. De Latil and Rivoire, *À la recherche*, p. 16.

80 Fournier brings this up in his *Hydrographie*, p. 348.

81 Quoted by Paulian, 'Paysages océaniques', p. 28.

82 Bouhours, *Les Entretiens*, pp. 16 ff.; Fournier, *Hydrographie*, p. 341.

83 Regarding the *Néréides* by Diaper, Michèle S. Plaisant, *La Sensibilité dans la poésie anglaise au début du XVIIIᵉ siècle. Évolution et transformation* (thesis, Paris IV, 1974; Lille III, 1974), vol. 2, p. 519. It is appropriate here to mention the animal fantasy of the globe descended from the Pythagoreans and passed on by the Platonists and the Stoics. Democritus, for his part, saw the storm as the result of a fever of the terrestrial animal. 'Dans les profondeurs de la mer sont en quelque sorte les narines du monde,' writes the third-century geographer Solin; 'par leur respiration, elles enflent tantôt les mers, tantôt les abaissent' ('in the depths of the sea are, in a way, the nostrils of the world; by their breathing, they sometimes swell the seas and sometimes cave them in') (quoted by De Latil and Rivoire, *À la recherche*, p. 27). Before that, Pomponius Mela, geographer of the Claudian epoch, did not know whether to attribute the tides to the influence of the moon, to this animal breathing, or to the circulation which occurs through the caves hollowed out under the sea. These theories were well known during the seventeenth century (see Porter, *Making of Geology*, pp. 70–1); associated with the ideas of Paracelsus, presented in Kircher's work, they are also

associated with the alchemical, hermetic, and mystical trends; they thus appear to have been very pregnant before being gradually discredited at the end of the century. In France, Fournier mentions them at length, as does Bouhours; Brancas sets them out once again in his *Explication*.

84 See Jacob, 'Logiques du paysage', p. 165.

85 On nature as a strategic element, a support for battle, and on landscape as prey in the frame of reference of war construed by the ancients as predatory, see Claire Préaux, Simon Byl, and Georges Nachtergael, *Le Paysage grec* (Brussels, 1979), pp. 16–17.

86 A common theme in French poetry at the beginning of the seventeenth century; from *La Mort d'Hyppolyte* by Tristan to *L'Andromède* by Saint-Amant, see Jean-Pierre Chauveau, 'La Mer et l'imagination des poètes au XVIIe siècle', *XVIIe siècle* nos 86–7, (1970), pp. 107–34.

87 See Delumeau, *La Peur en Occident*, p. 31.

88 Nevertheless, it was necessary for the hero to circle his island by sea in order symbolically to take possession of his territory. On this subject, see Abraham Moles and Elisabeth Rohmer, *Labyrinthes du vécu. L'espace: matière d'actions* (Paris, Librairie des Méridiens, 1982), ch. 3: 'Nissonologie ou science des îles', pp. 55–7 (on the mastery of the shoreline).

89 Alain Corbin, *The Foul and the Fragrant: The Sense of Smell and its Social Image in Modern France*, trans. M. Kochan (Berg Publ., 1986).

90 Coleridge, 'The Rime of the Ancient Mariner'.

91 See Deluz, 'Pélerins et voyageurs'.

92 By this term we mean voyagers who travel through Europe according to the British model of the 'grand tour'.

93 Montesquieu, *Oeuvres complètes* (Paris, Nagel, 1950), vol. 2, p. 1061.

94 President de Brosses, *Journal du voyage en Italie. Lettres familières* (Grenoble, Roissard, 1972), vol. 1, p. 41, as well as the following quote.

95 *Ibid.*, p. 255. Thirty-nine years later (1778), Brissot, embarking at Boulogne for England, laments his inability to resist the sickness. 'Je me défendis du mal de mer, le plus longtemps qu'il me fut possible, en respirant des sels, ne mangeant point, me tenant toujours sur le pont' ('I struggled as long as I could against seasickness, smelling salts, not eating anything, staying always on the deck') (*Mémoires de Brissot ... Published by his Son* (Paris, Ladvocat, 1830), vol. 1, p. 297).

In 1789, the crossing accomplished by Meister becomes a real agony, lasting almost twelve hours (Jakob Meister, *Souvenirs de mes voyages en Angleterre* (Zurich, 1775), p. 2).

96 [A French economist and brother of the socialist and revolutionary Louis-Auguste Blanqui – trans.]

97 Adolphe Blanqui, *Voyage d'un jeune Français en Angleterre et en Écosse* (Paris, Dondey-Dupré, 1824), pp. 2–3.
98 Astolphe de Custine, *Mémoires et voyages ou Lettres écrites à diverses époques* (Paris, A. Vezard, 1830). Story concerning 28 Aug. 1822, vol. 2, pp. 297 ff.
99 Deacon (*Scientists and the Sea*, pp. 74–172) vigorously underscores the importance of this period during which oceanography progressed remarkably, thanks in particular to the impetus provided by the Royal Society during the years 1661 and 1662. This 'peak' from 1660 to 1675 was followed by a decline (1675–1700), which was interrupted by the awakening of oceanographical science in the eighteenth century.
100 A vast subject, well researched by Robert Mandrou and Robert Muchembled.

Notes to chapter 2

1 See, regarding them, Antoine Adam *Histoire de la littérature française au XVIIe siècle* (Paris, 1962), vol. 1, pp. 79–89 (Théophile de Viau), pp. 92–8 and 375–80 (Saint-Amant), pp. 369–75 (Tristan l'Hermite).
  More precisely, regarding the attraction they felt towards the sea-shore, Chauveau, 'La Mer'. Jacques Bailbé, 'Les Paysages chez Saint-Amant', in *Cahiers de l'association internationale des études françaises*, no. 29 (May 1977); 'Paysages baroques et paysages romantiques', ibid., pp. 25–44. It is interesting to note that these poets all came under the influence of the Italian Marino.
2 A number of seventeenth-century poets follow, as regards this subject, in the same vein as the authors of the preceding century mentioned earlier: e.g., Sponde (*Poésies*, (Geneva, 1949), p. 244); Gombauld (*Sonnets chrestiens*, xxiii, p. 291); André Mage de Fiefmelin, who lived in Oléron (an island on the Atlantic coast) (*Oeuvres*, 1601); Laurent Drelincourt (*Sonnets chrétiens* (1677), ii, 5). (Quoted by Chauveau, 'La Mer', pp. 113–15.)
3 The dangers of the sea in love, in accordance with the Petrarchan tradition, are evoked by Malherbe, Corneille, Marboeuf (Chauveau, 'La Mer', pp. 115–17).
4 Tristan l'Hermite, *La Mer*.
5 See Chauveau, 'La Mer', p. 125, concerning the description of the sea which Le Moyne develops in *Peintures morales* (1643), vol. 2, p. 121.
6 An essential aspect of this mode of appreciation; on the refraction of aerial worlds by the aquatic mirror in Saint-Amant's poetry, on the reversibility of the universe and of existence suggested by that of the metaphors of the bird and the fish, see Gérard Genette, 'L'Univers réversible', in *Figures* (Paris, Le Seuil, 1966), pp. 9–20 and 29–36; and Yvonne Bellenger, 'Les Paysages', p. 21.

7 Let us note, nevertheless, that Saint-Amant, who was not himself a pious man, was obliged to seek the good graces of the devout party.

8 The quotations are excerpts from the poem 'Le Contemplateur' (*Oeuvres* (Paris, Didier, 1971), vol. 1, pp. 49–69. This critical edition was established by Jacques Bailbé.

9 Jacques Thuillier, 'Le Paysage dans la peinture française du XVIIe siècle: de l'imitation de la nature à la rhétorique des belles idées', *Cahiers de l'association internationale des études françaises* (May 1977), p. 55. Let us note nevertheless that this little group of poets had close ties with the painters.

10 Marc Fumaroli, Introduction to the *Mémoires de Henri de Campion* (Paris, Mercure de France, 1967), p. 18.

11 Henri de Campion. *Mémoires*, pp. 172–3.

12 Fortin de la Hoguette, *Testament*, Part 2, ch. 32: 'De la Conversation', p. 182. Quoted by Fumaroli, *Mémoires*, p. 317.

13 Noémi Hepp, 'Moisant de Brieux devant l'antiquité classique', in *La Basse-Normandie et ses poètes à l'époque classique* (Caen, 1977), esp. pp. 218–19.

14 'L'on rêve bien doucement lorsqu'on se promène au bord de la mer' ['You dream very softly when you wander by the side of the sea'], the Grande Mademoiselle confided to Madame de Motteville, on 14 May 1660. (J. Rousset, *La Littérature de l'âge baroque en France* (Paris, José Corti, 1954), p. 279, n. 14.)

15 On this subject, concerning England, we owe much to Plaisant (*La Sensibilité*, esp. vol. 1, pp. 5–11, and vol. 2, pp. 972 ff.)

16 In this way, John Ray reintroduced in 1691 the role of Providence, and settled, in his own way, the crucial debate over God's mode of action. (*The Wisdom of God manifested in the Works of Creation* (Utrecht, 1714)). The author sets about, as of the introduction, refuting the Cartesian image of God the watch-maker.

17 A conviction expressed in particular by John Woodward (*An Essay toward a Natural History of the Earth, and Terrestrial Bodies, especially Minerals* (London, 1695)) and by Whiston (*New Theory of the Earth*, p. 271).

18 William Derham, *Physico-Theology: or a Demonstration of the Being and Attributes of God, from his Works of Creation* (London, 1703), *passim*. This series of qualifiers is expressed from p. 2. The author is the rector of Upminster in the county of Essex.

19 Woodward, *Essay*, pp. 53–4. After the Flood, God therefore left man in a state of happiness by giving him great occupations (pp. 56–7); the author thereby solves the problem of evil: sickness, death, hard labour are all stratagems used by the Creator to stop antediluvian man in the path of his folly (p. 53). Derham also considers the earth, as it is now, a ruin refashioned by a new creation.

20 Woodward, *Essay*, p. 35. The shapes of landscapes were, traditionally and up to this date, indeed commonly thought to be the

remnants of a perfect earth, which had been carved at the Creation and which, since then, had fallen into disrepair. Natural theology doubly refutes this theory: it suggests the idea of a refashioning the day after the Flood, and refuses the notion of a continuous decline. On this subject, see the enlightening pages of Porter, *Making of Geology*, esp. pp. 43–6.

21 Whiston, *New Theory of the Earth*, pp. 368 ff. 'The conflagration' will render the earth worthy of receiving the saints and the martyrs during the millennium until the Last Judgement and prior to the consummation of everything. This leads Whiston to separate the history of the earth into four successive phases: paradisiacal, antediluvian, primitive, and what will be his during the millennium; the latter will be characterized by the absence of seas and of any receptacle for bodies of water.

22 Derham, *Physico-Theology*, p. 602.

23 Abbé Antoine Pluche, *Le Spectacle de la nature ou entretiens sur les particularités de l'histoire naturelle* ... (Paris, 1732–50). The references are from the 7th edn (1739), vol. 3, p. 289.

24 Ehrard, *L'Idée de Nature*, vol. 1, pp. 186 ff.

25 Porter, *Making of Geology*, p. 102.

26 Plaisant, *La Sensibilité*, vol. 2, pp. 592 and 990.

27 Whose increasing attraction is associated, obviously, with the rise of the aesthetics of the sublime.

28 After the 1688 revolution, the Whigs found inspiration in the philosophy of Locke, which, in fact, agrees with the notions of order and harmony which rule in the Newtonian universe. See Plaisant, *La Sensibilité*, vol. 1, pp. 24–5.

29 Nieuwentijdt, *L'Existence de Dieu démontrée par les merveilles de la nature*. The first edition in English dates back to 1717–19, the first edition in French to 1725. The references are taken from the Amsterdam edition of 1760. The quotation comes from the introduction.

30 Paul Van Tieghem credits him with interesting developments in *Le Sentiment de la nature dans le préromantisme européen* (Paris, Nizet, 1960), pp. 14–16. In 1755, in Frankfurt, J. G. Sulzer published his *Unterredungen über die Schönheit der Natur*; in his turn, he also paraphrases Psalm 8 and sings of the riches hidden by the Lord in the waters of the sea (introduction). The author represents a late example of a traveller spurred by the desire to admire the beauties of nature willed by the Creator.

31 From 1633 to 1636, Father Yves of Paris published *La Théologie naturelle*. He enjoyed wandering out of doors and admiring the perpetual spectacle of the wonders of nature. Blessed with a very keen perception, he praises the fumes which rise on the horizon, the rising and setting sun. Each of his steps offers him a new joy; the bee, the slug, the ant please him no end. The smallest blade of grass

is, for him, a subject for contemplation leading to ecstasy (Henri Brémond, *Histoire littéraire du sentiment religieux en France, depuis la fin des guerres de religion jusqu'à nos jours* (Paris, Armand Colin, 1964), vol. 1, 'L'humanisme dévot', esp. pp. 431–3).

32 Brémond, *Histoire littéraire*, p. 336.
33 [This would appear to be a mistake. There is nothing to this effect in Psalm 52 in any English Bible – trans.]
34 Psalm 104: 6 and 7[−9], and Psalm 33: 7.
35 Jeremiah 5: 22.
36 Collections of texts and, in particular, the *Hexameron* of St Basil, were analysed from this perspective, as early as 1860, by Emile Gebhart in his *Histoire du sentiment poétique de la nature dans l'antiquité grecque et romaine* (Paris, Durand, 1860), pp. 199 ff.
37 Quoted by Bouhours, *Les Entretiens*, p. 27.
38 Thus, Racine, *La Religion*, canto I, p. 4:

> Et toi dont le courroux veut engloutir la terre,
> Mer terrible, en ton lit quelle main te ressere?
> Pour forcer ta prison tu fais de vains efforts;
> La rage de tes flots expire sur tes bords.

> [And you whose anger wants to engulf the earth,
> Terrible sea, in your bed which hand squeezes you?
> To escape your prison you make vain efforts;
> The rage of your waters expires on your edges.]

and Cardinal de Bernis, *La Religion vengée*, poem with ten cantos (Paris, Koenig, 1796) (but its text was known well before), canto III (p. 51), evoking the rage of the storm:

> Cependant l'onde approche et menace ses bords:
> Mais un bras invisible en soutient les efforts;
> La mer, à son aspect soumise et consternée,
> Abaisse devant lui son onde mutinée.

> [Nevertheless, the billows approach and threaten its edges:
> But an invisible arm supports the exertions;
> The sea, with its submissive and dismayed appearance,
> Lowers before him her mutinous waters.]

39 Dulard, *La Grandeur de Dieu*, p. 35. The commanding finger of the all-powerful God corresponds to the sensibility of the Catholic Reformation.
40 Richard Blackmore, *Creation: A Philosophical Poem which Proves the Existence and Providence of a God*, 5th edn (Dublin, 1727), book 1, pp. 20–1: 'The Sea'.
41 Ideas expressed in particular by Nieuwentijdt, *L'Existence de Dieu*, p. 273; by John Ray, *Wisdom of God*, pp. 82 ff.; by Fabricius, *Hydrothéologie*, p. 147; by Pluche, *Le Spectacle*, vol. 3, p. 195; and by Dulard, *La Grandeur de Dieu*, p. 38.

42 Pluche, *Le Spectacle*, vol. 3, p. 197.
43 Ibid., p. 194.
44 Nieuwentijdt, *L'Existence de Dieu*, p. 282.
45 See ibid., pp. 274–5; Derham, *Physico-Theology*, p. 66; and Fabricius, *Hydrothéologie*, p. 146.
46 Fabricius, *Hydrothéologie*, p. 339
47 Pontoppidan, *Natural History of Norway*, pp. 66 ff.
48 Bouhours, *Les Entretiens*, p. 41.
49 Thomson, *Seasons*.
50 On all these points, see Nieuwentijdt, *L'Existence de Dieu*, pp. 274 ff.
51 Ideas expressed by Nieuwentijdt, ibid., p. 284; by Blackmore, *Creation*, p. 21; by Pluche, *Le Spectacle*, vol. 3, p. 190; and by Dulard, *La Grandeur de Dieu*, p. 38.
52 Developed in particular by Pluche, *Le Spectacle*, vol. 3, pp. 274–5.
53 See Dulard, *La Grandeur de Dieu*, p. 65.
54 Fénelon, *Démonstration de l'existence de Dieu*, part 1, ch. 8, quoted by Ehrard, *L'Idée de Nature*, vol. 2, p. 622. Fénelon destroyed in this way an old argument favouring the inconsistency of the structure of the globe and the absurdity of the placing of the continents and seas.
55 Pluche, *Le Spectacle*, vol. 3, p. 211.
56 Bouhours, *Les Entretiens*, p. 25.
57 See Nieuwentijdt, *L'Existence de Dieu*, p. 278.
58 Especially those of the Flood and the boundaries established by God.
59 Ray, *Wisdom of God*, p. 82.
60 Fabricius, *Hydrothéologie*, p. 125. In so doing, this theologian makes somewhat abusive reference to Kircher, Boyle, Marsigli, and the Jesuit Louis Feuillée.
61 Bouhours puts these ideas into the mouth of Eugène in *Les Entretiens*, p. 31. The same idea about underwater species being replicas of those living on the Earth's surface is found in Pontoppidan, *Natural History of Norway*, p. 149.
62 Boufours, *Les Entretiens*, p. 31.
63 Dulard, *La Grandeur de Dieu*, p. 44.
64 Woodward, *Essay*, p. 149.
65 Bouhours, *Les Entretiens*, pp. 10–11.
66 It is important not to underestimate the importance of the concept of final cause for eighteenth-century and even early nineteenth-century scientists. Porter (*Making of Geology*, pp. 192–6) notes that even James Hutton was marked by it.
67 Cf. Daniel Mornet, *Le Sentiment de la nature en France de Jean-Jacques Rousseau à Bernardin de Saint-Pierre* (Paris, Hachette, 1907). The author makes some interesting comments (pp. 287–91) on the subject of the sea in literature.

68 Jacques-Henri Bernardin de Saint-Pierre. *Études de la nature*, in *Oeuvres complètes* (Paris, Méquignon-Marvis, 1818), vol. 3, pp. 150–1.

69 Ibid., pp. 154 and 213.

70 *Idem, Harmonies de la Nature* in *Oeuvres posthumes de J. H. Bernardin de Saint-Pierre* (Paris, Lefèvre, 1833), vol. 2, p. 183.

71 This was the case with Misson (*Nouveau Voyage d'Italie fait en l'année 1688* (Paris, 1691), vol. 1, pp. 2 ff.; the author also visits Holland) and with Diderot in 1773–4 (*Voyage en Hollande et dans les Pays-Bas autrichiens*, in *Oeuvres complètes* (Paris, Le Club français du livre, 1971), vol. 11, p. 365).

72 This link is encompassed in the broader alliance established by Max Weber between religion and economic development.

73 On this point, see Roelof Murris, *La Hollande et les Hollandais au XVIIe et XVIIIe siècles vus par les Français* (Paris, Champion, 1925), esp. pp. 30–1. On this subject, the author cites the assessment of François Janiçon (*État présent de la République des Provinces-Unies* (The Hague, 1729–30), an admirer of the Dutch, 'who set boundaries for the waves,' and of Daignan in 1777 (Murris, *La Hollande*, p. 37).

74 Diderot, *Voyage en Hollande*, p. 337

75 E.g., Abbé Coyer in 1769 (*Voyage d'Italie et de Hollande* (Paris, Duchesne, 1775), vol. 2, p. 220).

76 As did Carlo Pilati di Tassulo on 20 June 1778 (*Voyage de la Hollande ou Lettres sur ce pays* (2nd edn Haarlem, 1790), vol. 1, pp. 74 ff.).

77 Misson, *Nouveau Voyage*, vol. 1, p. 142.

78 The author is borrowing here from Wolfgang Stechow, *Dutch Landscape Painting of the 17th Century*, 2nd edn (London, Phaidon, 1968), esp. pp. 110–23; from F. C. Willis, *Die niederländische Marinemalerei* (Leipzig, 1911); from Lionel Preston, *Sea and River Painters of the Netherlands in the Seventeenth Century* (1937); from Laurens J. Bol, *Die Holländische Marinemalerei des 17. Jahrhunderts* (Braunschweig, 1973); and lastly, from Keyes, *Cornelis Vroom*, esp. pp. 17 ff.

79 Stechow, *Dutch Landscape Painting*, pp. 110–14.

80 In fact, it was when the journey to Holland became fashionable with the French, around 1745 according to Murris, that the new sensibility which I have undertaken to study began to emerge.

81 Murris, *La Hollande*, p. 25

82 Ibid., p. 23

83 Misson, *Nouveau Voyage*, vol. 1, p. 11, Pilati, *Voyage de la Hollande*, vol. 1, *passim*.

84 *Receuil des oeuvres de Madame de Bocage*, vol. 3: *Lettres sur l'Angleterre, la Hollande et l'Italie* (Lyon, 1764), p. 101 (letter to her sister, The Hague, 20 June 1750).

85 Charles Ogier, for one, enjoyed this sight greatly as early as 1636 (see Murris, *La Hollande*, p. 24). In 1719, Abbé Pierre Sartre raved during his visit to Holland about the effect produced by this bustle (cited in ibid., p. 21).

86 See the guides by Boussingault (*c.* 1660) and Lombard de Langres (1799) (ibid., p. 26).

87 Only in the middle of the sixteenth century – especially in northern Europe – did landscape painting become a genre, albeit a minor one, and specialists begin to produce for a fairly large market. But the first paintings of landscapes without any other clearly attributable subject appeared a century earlier. (See E. H. Gombrich, 'The Renaissance Theory of Art and the Rise of Landscape', in *Norm and Form* (London, 1966), pp. 107–21, and Svetlana Alpers, *The Art of Describing: Dutch Art in the XVIIth Century* (Chicago, 1983). The author of the latter (p. 128) underscores the role of topographical painting of ports in creating seascapes in northern Europe. In her opinion, painters from the North who made the trip to Italy in the sixteenth century were more attracted by discovering the world than by contemplating the vestiges of classical art.

88 Cited in Murris (*La Hollande*, p. 26).

89 Misson, *Nouveau Voyage*, vol. 1, p. 6.

90 David Hume, paraphrased by O. Brunet, *Philosophie et esthétique chez David Hume* (Paris, Nizet, 1965), p. 41. Other travellers analyse this impression, especially Madame du Bocage (*Receuil des oeuvres*, vol. 3, p. 82; The Hague, 20 June 1750); Coyer (*Voyage d'Italie et de Hollande en 1769*, vol. 2, p. 238); and later, Joseph Marshall (*Voyages dans la partie septentrionale de l'Europe, pendant les années 1768, 1769, et 1770* (Paris Dorez, 1776), pp. 11–12); and Samuel Ireland (*A Picturesque Tour through Holland, Brabant and Part of France made in the Autumn of 1789* (London, Egerton, 1789), vol. 1, p. 23). Still later, André Thouin claims to have experienced the same feeling at Buyskloot during a stay there in February 1795 (*Voyage dans la Belgique, la Hollande et l'Italie* (Paris, 40 rue Laffitte, 1841), vol. 1, p. 309).

91 Coyer, *Voyage d'Italie et de Hollande*, vol. 2, p. 259.

92 André Thouin says so, in *Voyage dans la Belgique*, vol. 1, p. 172.

93 Diderot declares that, for this reason, Amsterdam is a festering city (*Voyage en Hollande*, p. 429); Joseph Marshall (*Voyages*, p. 88), Dr Edward Rigby (*Voyage d'un Anglais en France en 1789* (Paris, Nouvelle Librairie nationale, 1910), p. 234), and Thouin (*Voyage dans la Belgique*, vol. 1, p. 282) all believe that this odour is intolerable to foreigners.

94 Misson (*Nouveau Voyage*, vol. 1, p. 11) and Aubry de La Motraye (cited by Murris, *La Hollande*, p. 33) therefore consider that the region around The Hague is the finest part of the country.

95 Misson, *Nouveau Voyage*, vol. 1, p. 11; Pilati, *Voyage de la Hollande*, vol. 1, p. 47.

96 Pilati, *Voyage de la Hollande*, vol. 1, p. 47.

97 Ireland, *Picturesque Tour*, vol. 1, pp. 72–3.

98 Samuel-François L'Honoré, *La Hollande au XVIIIᵉ siècle ou Nouvelles lettres contenant des remarques* ... (The Hague, 1779), p. 67. The author made this trip in 1776.

99 Diderot, *Voyage en Hollande*, pp. 425 ff. It is important, however, not to be fooled by the local colour found in *Voyage en Hollande*; Diderot delved heavily into earlier works, including that published in 1750 by A. de la Chesnaye du Bois. On this point, see G. Charlier, 'Diderot et la Hollande', *Revue de littérature comparée*, 82 (1947), pp. 190–229.

     Ann Radcliffe, *A Journey made in the Summer of 1794, through Holland and the West Frontier of Germany* (London, 1795), p. 48.

100 Misson, *Nouveau Voyage*, vol. 1, p. 11.

101 A practice described by Coyer (*Voyage d'Italie et de Hollande*, vol. 2, p. 266); by Pilati (*Voyage de la Hollande*, vol. 2, p. 174); and by Marshall (*Voyages*, p. 32).

102 Pilati, *Voyage de la Hollande*, vol. 2, p. 176.

103 Marshall, *Voyages*, p. 32.

104 J. de Luc visited Holland and mentally reviewed the works of landscape artists. Ireland (*Picturesque Tour*, vol. 1, p. 74) wrote of his trip to Scheveningen in 1789 that it seemed to him that little had changed since the time of Van Goyen, Simon de Vlieger, and other eminent painters who thought that this was a subject worthy of their crayons.

105 Taking into account the loss of a work by Dürer dating from 1598. See Stechow, *Dutch Landscape Painting*, p. 101, and Panofvsky, *Albert Dürer* (1943), p. 10.

106 Especially Aert Van der Neer.

107 As is indicated by his *Scheveningen* from 1658.

108 See S. de Vlieger, 1633.

109 And sometimes the same ones.

110 Stechow, *Dutch Landscape Painting*, p. 109.

111 On this subject, see David Cordingly, *Marine Painting in England, 1700–1900* (Studio Vista, 1974), esp. pp. 15–16.

112 This was the case with Ireland.

113 As Marshall did.

114 Diderot, *Voyage en Hollande*, p. 426.

115 Thouin, *Voyage dans la Belgique*, vol. 1, p. 227.

116 Numerous Latin writers in addition to Tacitus extolled the magnificent Campania coast (Tacitus, *History* iii, 60). On this subject, see D. Goguey, 'La Campanie dans la littérature latine: réalités géographiques', *Caesarodunum*, no. 13, esp. pp. 18–19, and R. F. Paget, 'From Baiae to Misenum', *Vergilius*, 17 (1971), pp. 22–38.

117 Tacitus, *History* i, 2.
118 Anne-Marie Taisne ('Peintures des villas chez Stace', *Caesarodunum*,
     no. 13, pp. 40 ff.) emphasizes how much the water's calm, despite
     the distant roar of the sea, played a role in the pleasure experienced
     inside the villa of Pollius Felix in Sorrento.
119 See Saint-Denis, *Le Rôle de la mer*, pp. 159–76. More generally,
     see also the classicist Henry Rushton Fairclough, *Love of Nature
     among the Greeks and Romans* (New York, reprint, 1963).
120 See below, pp. 148 ff.
121 Though obviously the visit to Venice and the trip along the shores
     of the Adriatic must not be forgotten this was sometimes considered
     a more comfortable itinerary.
122 See Elisabeth Chevallier, 'La Découverte des paysages de l'Italie du
     sud dans la seconde moitié du XVIIIe siècle: le voyage pittoresque
     de l'abbé de Saint-Non', *Caesarodunum*, no. 13, pp. 89–108. This
     makes it worth pointing out the importance, in this area, of the
     coastal trek made in 1778–9 by Vivant Denon and his group.
123 Misson, *Nouveau Voyage*, vol. 1, p. 273.
124 Cited by Elizabeth Wheeler Manwaring, *Italian Landscape in 18th
     Century England. A Study chiefly of the Influence of Claude Lorrain
     and Salvator Rosa on English Taste, 1700–1800* (reprint, New York,
     Russell, 1965), p. 9.
125 President de Brosses, *Journal*, vol. 1, pp. 242 and 227.
126 *Voyage en Italie de Monsieur l'abbé Barthélemy* (Paris, Buisson,
     1801), p. 55.
127 Coyer, *Voyage d'Italie et de Hollande*, vol. 1, p. 239.
128 Plaisant, *La Sensibilité*, vol. 1, p. 73, devotes some interesting re-
     flections to this subject. She cites the example of Lord Chesterfield,
     who registered his son at the Westminster school so that he would
     acquire 'the classical knowledge which, in this land, is necessary for
     a young man of quality' (see S. W. Brewer, *Design for a Gentleman*
     (London, Chapman and Hall, 1963), p. 64). The most famous
     writers, such as Addison, Gray, and Dr Johnson, wrote part of their
     work in Latin.
129 One English aristocrat requested that he be buried with the works
     of Horace; in 1745, Lord Lovat, condemned to be decapitated,
     quoted Horace on the scaffold. These examples are mentioned by
     Plaisant, *La Sensibilité*, vol. 1, p. 74.
130 See Georges Snyders, *La Pédagogie en France aux XVIIe et XVIIIe
     siècles* (Paris, 1965); Roger Chartier, Marie-Madeleine Compère,
     and Dominique Julia, *L'Éducation en France du XVIe au XVIIIe
     siècle* (Paris, S.E.D.E.S., 1976); and Jean de Viguerie, *L'Institution
     des enfants. L'éducation en France XVIe–XVIIIe siècle* (Paris,
     Calmann-Lévy, 1978), ch. 6: 'Les humanités', pp. 159–94. The
     article by Daniel Milo, 'Les Classiques Scolaires' was published in
     Les Lieux de mémoire, vol. 2: 'La Nation', vol. 3, pp. 517–62.

131 See P. A. Février, 'La Maison et la mer, réalité et imaginaire', in *L'Homme méditerranéen et la mer*, p. 342.

132 See Alain Schnapp, 'Archéologie et tradition académique en Europe aux XVIIIe et XIXe siècles', *Annales, économies, sociétés, civilisations*, vol. 37, nos 5–6 (Sept.–Dec. 1982), pp. 760–77. See also Francis Haskell and Nicholas Penny, *Taste and the Antique, The Lure of Classical Sculpture, 1500–1900* (Yale University Press, 1981); Kenneth Hudson, *A Social History of Archaeology: the British Experience* (London, Macmillan, 1981); and Glyn Daniel, *History of Archaeology* (London, Thames and Hudson, 1981).

133 J. Addison, *Remarks on Several Parts of Italy* (London, Jacob Tonson, Grays-Inn Gate, 1705).

134 Until well into the nineteenth century, classical archaeology continuously played a key role in the socio-cultural history of the West. See Schnapp, 'Archéologie'.

135 On the role of the *dilettanti* in rediscovering Greece and on the interest of collectors in Greece, the 'Antiquaries', and neoclassical architects, see Fani-Maria Tsigakou, *La Grèce retrouvée* (Paris, Seghers, 1984), pp. 18 ff.

136 On this public, see Daniel Roche, *Le Siècle des Lumières en province. Académies et académiciens provinciaux, 1680–1789* (Paris and The Hague, Mouton, 1978). As for Great Britain, the above-mentioned book by Manwaring offers a fine description of the rise of these amateur circles.

137 On this subject, E. Veryard, *An Account of Diverse Choice Remarks ... Taken in a Journey through the Low-Countries, France, Italy, and Part of Spain with the Isles of Sicily and Malta* (London, Smith and Walford, 1701), is particularly revealing. This traveller visited Sicily, which was rare at the time, and he notes the consistency and temperature of the sand at Baiae, and describes the grottoes and natural curiosities. Nevertheless, his way of seeing is dictated by the ancients, and Veryard is primarily interested in cities, antiquities, and the customs of the peoples.

138 On the grand tour, see also Geoffrey Trease, *The Grand Tour* (London, Heinemann, 1967), and Christopher Hibbert, *The Grand Tour* (London, Weidenfeld and Nicolson, 1969).

139 On the golden age and the codification of the journey to Rome in the eighteenth century, as well as the para-literary genre to which it gave rise, see Jean Rousset, 'Se promener dans Rome au XVIIIe siècle', in *Thèmes et figures du siècle des Lumières. Mélanges offerts à Roland Mortier* (Geneva, Droz, 1980), pp. 239–50.

140 Lord Shaftesbury, the distant initiator of the fashion for the sensitive soul (see Georges Gusdorf, *Naissance de la conscience romantique au siècle des Lumières* (Paris, Payot, 1976), pp. 219–44), settled in Naples at the end of his life, and contributed through his example to creating this system of appreciation which renders English neo-

classicism incomprehensible to anyone who neglects the importance of the grand tour.

141 This brings Montaigne to mind.

142 At the very end of the seventeenth century, Addison amused himself by cataloguing all the riches of Italy, a laboratory for different types of government, a great school of music, a land of famous collections and libraries, and offices full of curiosities, medallions, and 'Antiquities': 'One finds something more particular in the Face of the Country, and more astonishing in the Works of Nature, than can be met with in any other part of Europe' (*Remarks*).

143 Edward Gibbon, *Miscellaneous Works of E.G. ... with Memoirs of his Life and Writings; Composed by Himself ...* (London, 1796); from an essay of 13 Oct. 1763, p. 83.

144 On this point, it is worth emphasizing the influence of Misson's recommendations, especially those which were detailed in *Nouveau Voyage*, vol. 2, p. 290 ff.

145 One example is the accounts cited by President de Brosses and Abbé Coyer.

146 This was Goethe's case in 1786.

147 Michel Butor, 'Le Voyage et l'écriture', *Romantisme*, no. 4, (1972), p. 14.

148 A considerable bibliography has been devoted to the history of vedutism. See Chevallier, 'La Découverte', p. 106.

149 Brunet, *Philosophie*, p. 41

150 Addison, *Remarks*.

151 President de Brosses, *Journal*, vol. 1, p. 225.

152 Coyer, *Voyage d'Italie et de Hollande*, vol. 1, p. 223.

153 Jean-Baptiste Mercier Dupaty, *Lettres sur l'Italie en 1785* (Rome, 1788), vol. 2, p. 176.

154 Henry Swinburne, *Travels in the two Sicilies, in the years 1777, 1778, 1779 and 1780* (London, 1783–5), *passim*, esp. in April 1777 during the trip from Naples to Taranto.

155 A process easily decoded, given the prolix nature of the normative literature, which included such best-sellers as the works of Guyot de Merville, Abbé J. Richard, and especially Jérôme J. de Lalande, in addition to the main accounts mentioned here.

156 The young Bérenger, sensitive to the beauty of the coasts of his native Provence, was wandering along the beach in Marseilles in about 1770. He had just indulged in the voluptuous pleasure of rereading Theagenes and *Daphnis*, Gessner and Fénelon.

> Là, disais-je, à travers les eaux,
> des Grecs, pour fonder ma patrie,
> vinrent, du fond de l'Ionie,
> fixer l'ancre de leurs vaisseaux.
> Ici, ce peuple redoutable,
> ces fiers Romains ont respiré;

Ici Milon a soupiré;
César foulait ce même sable.

[There, I said, across the waters,
the Greeks came, to found
my country, from deepest Ionia,
to drop the anchors of their vessels.
Here, this fearsome people,
those proud Romans did breathe;
Here Milon did sigh;
Caesar trod this very sand.]

(*Les Soirées provençales ou Lettres de M. L. P. Bérenger écrites à ses amis pendant ses voyages dans sa patrie* (Paris, 1786; written in 1783), vol. 1, p. 99.)

Since 1978, a research unit of the C.N.R.S. (French National Centre for Scientific Research) has been systematically analysing the means of 'discovering' Provence in the seventeenth century. The intermediate conclusions emphasize a threefold sentiment in travellers: a taste for antiquities (they view the Crau through ancient texts); the attraction of images of primitive Christianity (here, the life of Mary Magdalene); and curiosity about urban business or the fertility of the gardens. This all combines to produce a literature of panegyrics and catalogues which corresponds to the image of the classical voyage. See Louis Godard de Donville, 'Présentations de l'ERA sur la découverte de la Provence au XVIIe siècle', in *La Découverte de la France au XVIIe* (Paris, C.N.R.S. publications, 1980), pp. 551–62.

157 Jean-Marie Roland de la Platière, *Lettres écrites de Suisse, d'Italie, de Sicile et de Malte* (Amsterdam, 1780), vol. 3, pp. 118–19.

158 Addison, *Remarks*, p. 186.

159 This is how Dupaty proceeded in 1785, *Lettres*, vol. 2, p. 176.

160 Misson, *Nouveau Voyage*, vol. 1, p. 316

161 Suggested in Suetonius's text; see Addison, *Remarks*, p. 185, on the subject of the 'brutal pleasures' of the Emperor; the site is also mentioned by Coyer, *Voyage de Hollande*, vol. 1, p. 239.

162 Including Misson, *Nouveau Voyage*, vol. 1, p. 317.

163 This is paradoxical in Addison. See Alain Bony, *Joseph Addison et la création littéraire. Essai périodique et modernité* (thesis, Paris III, 1979), pp. 395 or 566 ff.

164 With the exception of Dupaty, who in 1785 belatedly begins describing one which was of neoclassical inspiration and which revealed a heightened sensibility (*Lettres* vol. 2, p. 309): 'My thoughts fade away especially in this silence which spreads from one moment to another across these shores; from within it, the touching evening concert gradually rises, composed of the melancholy sound of oars cutting through the distant waves, the bleating of flocks scattered

in the mountains, the waves that murmur as they expire on the rocks, the trembling of leaves in the trees in which the breezes never rest, in short, all of these inaudible sounds, drifting distantly in the skies, over the waves, across the land, and forming together in this moment a sort of uncertain voice, a sort of melodic breathing of Nature asleep!'

165 On the subject of Virgil's influence on Thomson, see Pla, *Les Saisons, passim.*

166 With the exception of President de Brosses, who was fascinated by the pleasures of the Baiae of antiquity.

167 His incomprehension on this point is revealing: Addison cannot imagine that the Romans could delight, in the *summer*, in taking holidays by the sea along the shores of the Bay of Naples (*Remarks*, p. 162); he is persuaded that they go there in the winter.

168 Which does, however, become broader during the second half of the century, as is indicated by the text of Dupaty cited above.

169 Moisant de Brieux, *Œuvres choisies* (Caen, Le Blanc-Hardel, 1875), p. 211.

170 To my knowledge, it is not until the publication of *Cosmos* by Humboldt that the uselessness of looking for a topographical description of an itinerary in the *Aeneid* becomes accepted.

171 Addison, *Remarks*, preface.

172 As do Addison (*Remarks*, p. 183) regarding the bay seen from Capri and President de Brosses (*Journal*, vol. 1, p. 252), who attempts to reconstruct the beauty of the ancient shore in his imagination.

173 Swinburne, *Travels*, pp. 196 ff.

174 Johann Wolfgang von Goethe, *Italienische Reise* (Wiesbaden, 1959), p. 324. On the subject of this journey and the author's ability to read time through space, see Mikhaïl Bakhtin, *Esthétique de la création verbale* (Paris, Gallimard, 1979), 'Le Roman d'apprentissage', vol. 3, 'L'espace et le temps', pp. 232–57. More generally, see Humphry Trevelyan, *Goethe and the Greeks* (reprint, Cambridge, 1981).

175 Addison, *Remarks*, pp. 192 ff.

176 Jean Houel, *Voyage pittoresque des Isles de Sicile, de Malte et de Lipari* (...) (Paris, 1782), vol. 4, p. 115.

177 Fénelon, *Télémaque*, pp. 67 and 95.

178 Ibid., p. 102. In 1755, the same impression of motion is to be found in Fielding; see below, pp. 139 ff.

179 Consider this painter's influence on Shaftesbury; see Manwaring, *Italian Landscape*, p. 17.

180 This attitude is obvious, e.g., in Vivant Denon.

Notes to chapter 3

1 See especially Raymond Klibansky, Erwin Panofsky, and Fritz Saxl, *Saturn and Melancholy* (New York, Basic Books, 1964), and above

all, given the subject, Jean Starobinski, *Histoire du traitement de la mélancolie des origines à 1900*, Acta Psychosomatica, 3 (Basel, Geigy, 1960). Recently, the review *Le Débat* (no. 29 (Mar. 1984), pp. 44 ff.) devoted an interesting issue to this subject.

2 See Starobinski, *Histoire*, pp. 38 ff.

3 On the author and his work, Jean-Robert Simon, *Robert Burton (1577–1640) et l'anatomie de la mélancolie* (Paris, Didier, 1964), esp. pp. 278 ff. ('L'Air' and 'L'Hygiène de la motricité').

4 This would lead one to think that historians of descriptive literature do not pay sufficient attention to the influence of medical advice on the ways of appreciating landscapes.

5 Robert Burton, *The Anatomy of Melancholy* (London, George Bell and Sons, 1893; the 1st edn dates from 1621), vol. 2, p. 70.

6 Ibid., vol. 2, pp. 73 and 74.

7 Ibid.

8 Ibid., vol. 2, p. 70. On this point Burton is referring to Cardan's authority.

9 Ibid., vol. 2, p. 78, and for the development that follows.

10 One obvious contradiction emerges in reading Burton's book: in order to apply the principles there set forth, he should logically have recommended stays by the seaside. But as this was not a habitual practice, it does not occur to him.

11 See Thomas, *Man and the Natural World*, pp. 252–3, and, for the development that follows, ch. 6: 'The Human Dilemma', 'Town or Country', pp. 243 ff.

12 See Corbin, *The Foul and the Fragrant*, parts 1 and 2.

13 On this point, see Paul-Gabriel Boucé, *Les Romans de Smollett, Étude critique* (Paris, Publications de la Sorbonne, 1971), pp. 257 ff.

14 Tobias Smollett, *An Essay on the External Use of Water*.

15 Jean Deprun, *La Philosophie de l'inquiétude en France au XVIIIe siècle* (Paris, Vrin, 1979), p. 58.

16 Ibid., p. 90. On this subject, see also Gusdorf, *Naissance*, esp. pp. 137–45.

17 Deprun, *La Philosophie*, p. 88.

18 The above-mentioned book by Starobinski contains a very specific bibliography of works devoted at the time to this subject in France and in England, including the *Traité des affections vapoureuses des deux sexes*, published in 1763 by Dr Pomme, which constitutes a good example. More recently, Jean-Pierre Peter ('Entre femmes et médecins. Violence et singularités dans le discours du corps et sur le corps ... ', *Ethnologie française*, 1976); Catherine Fouquet and Yvonne Kniebielher (*La Femme et les médecins* (Paris, Hachette, 1982); and Jean-Marie Goulemot (Bienville presentation, *De la nymphomanie ...* (Paris, Le Sycomore, 1980) have worked on this fascinating subject, as has Thomas Laqueur, whose *Making Sex. Body and Gender from the Greeks to Freud* (Cambridge, Mass.,

Harvard University Press, 1990) summarizes English-language research – not to mention the fundamental book by Paul Hoffmann, *La Femme dans la pensée des Lumières* (Paris, Ophrys, 1977).

19 See Peter 'Entre femmes', and Goulemot, *De la nymphomanie.*

20 Jean Meyer, *La Noblesse bretonne au XVIIIe siècle* (Paris, S.E.V.P.E.N., 1966), vol. 2, pp. 1220–4.

21 Interesting thoughts on this subject can be found in Starobinski, *Histoire*, p. 30. The author analyses the distaste with living which is experienced by Werther as a failure to participate in nature's rhythms, and quotes Goethe, who wrote that 'All pleasure in life is founded on the regular return of external objects. The alternation between day and night, the changing seasons, flowers, and fruits ... these are the real driving forces of earthly life.'

22 See the successful work by Dr Samuel Tissot, *De la santé des gens de lettres* (Lausanne, Grasset; Paris, Didot, 1768; 1st edn, in Latin, dates from 1767).

23 A. P. Buchan, *Practical Observations Concerning Sea-bathing with Remarks on the Use of the Warm Bath* (London, 1804).

24 Buchan, *A Treatise on Sea Bathing; with Remarks on the Use of the Warm Bath*, 2nd edn (London, T. Cadell and W. Davies, Strand, 1810), pp. 146 and xii.

25 See ibid., p. 118.

26 This was the opinion of Dr Robert White expressed in *The Use and Abuse of Sea-water* (London, Flexney, 1775), p. 5. The shock of bathing in cold springs was actually much more harsh.

27 Cited by Buchan, *Treatise*, p. 118.

28 John Floyer, Ψυχρολουσια or the History of Cold-bathing both Ancient and Modern. We use the 1732 edition (S. Smith and Walford) to which an appendix by Dr Edward Baynard was added.

29 Ibid., p. 31.

30 Ibid., p. 30.

31 Ibid., p. 69.

32 See below, p. 110.

33 Jules Michelet, *La Mer* (Paris, Gallimard, 1983), preface by Jean Borie, book 4, ch. 1, p. 279.

34 Georges Vigarello (*Le Propre et le sale* (Paris, Le Seuil, 1985) has highlighted the importance of this trend, which he is a little hasty to attribute to a bourgeois mentality. It is important to see how it is rooted in the history of science and the evolution of sensibilities.

35 Dr Speed, *A Commentary on Sea-Water*, p. 154; appendix, translated from Latin, to the book by Russell, cited below.

36 It is also said that Louis XIV requested to take a young lady of the court who suffered from hydrophobia to the sea. This method was recommended by Van Helmont in relation to his theory on morbid ideas: the idea of the danger suggested by abrupt immersion was to erase the previous hydrophobic idea. Boërhaave approved of

the method in his aphorisms (see Joseph Monoyer, *Essai sur l'emploi thérapeutique de l'eau de mer* (thesis, Montpellier, 1818), p. 14).

37 See Edmund W. Gilbert, *Brighton, Old Ocean's Bauble* (London, Methuen, 1954), p. 12.

38 Ibid., p. 11, inspired by C. Morris, *The Journeys of Celia Fiennes* (1947).

39 Letter from Dr Richard Frewin to Dr Richard Russell, published in the work by Russell cited below, pp. 110–11.

40 We use Richard Russell, *A Dissertation on the Use of Seawater in the Diseases of the Glands, particularly, the Scurvy, Jaundice, King's Evil, Leprosy and the Glandular Consumption* (London, 1769).

41 See above, p. 28.

42 Russell, *Dissertation*, p. 32.

43 Which is why, as Russell stated in his aphorisms, a physician must always keep his eyes fixed on Nature as the surest of guides, and follow carefully in her footsteps.

44 Notably Théophile de Bordeu in France.

45 Russell, *Dissertation*, p. 126.

46 Ibid., p. 128.

47 Euripides, *Iphigenia in Tauris*, verse 1193.

48 This was the case with Dr Speed, whose text follows Russell's work in the 1769 edition we have used.

49 The memoir dates from 1767; it was published in Bordeaux in 1769.

50 Maret, *Mémoire*, p. 94; emphasis added.

51 Unless they choose to bathe in the morning.

52 Maret, *Mémoire*, p. 111.

53 On Dr John Awsiter and his *Thoughts on Brighthelmstone, Concerning Sea-Bathing and Drinking Sea-water with some Directions of their Use* (London, 1768), see Gilbert, *Brighton*, pp. 66–7.

54 Montesquieu, *L'Esprit des lois*, 'Des ports de mer'.

55 As was noted in particular by Dr Pierre J. B. Bertrand, *Précis de l'histoire physique, civile et politique de la ville de Boulogne-sur-Mer et de ses environs depuis les Morins jusqu'en 1814* (Boulogne, 1828), vol. 2, p. 405.

56 The corner-stone of the Brighton baths was laid in 1769.

57 Compare the beach at Cette, described in 1847 by Dr Jean Viel, *Bains de mer à Cette, de leur puissance hygiénique et thérapeutique, suivi de quelques observations cliniques par le docteur Viel* (Montpellier, Martel, 1847), p. 21.

58 This explains why the therapeutic value of the sea was quickly recognized in France by physicians at Montpellier, and especially by Professor Delpech. In Germany, Christoph Wilhelm Hufeland believed that the benefits of sea bathing were caused by the salinity, the impact of the waves, and also by electric and magnetic currents which stimulated vitality.

59 Dr James Currie, *Medical Reports, on the Effects of Water Cold and Warm as a Remedy in Fever and Other Diseases, whether Applied to the Surface of the Body, or Used Internally*, 3rd edn (London, 1805), 2 vols.

60 Cited by Gilbert, *Brighton*, pp. 13–14.

61 Studied by Gilbert, *Brighton*, p. 63. Relhan published a *Short History of Brighthelmstone, with Remarks on its Air and an Analysis of its Waters* in 1761, which comprised the first guide to the resort.

62 Jane Austen, *Sanditon, passim*. Wilbur Fisk, *Travels in Europe* ... (New York, Harper, 1838). He is astonished by how widespread seaside holidays have become in England, given that the whole country is covered by ocean air (p. 557). He notes that the fashion for various resorts often proves to be fleeting. Above all, he makes a mockery of the scientific distinctions observed by the English, who detect differences between the climatic qualities of places located very close to one another, something which an American would be unable to do.

63 See Gilbert, *Brighton*, p. 43.

64 As early as 1793, see the debate launched by Dr Woltmann and Drs Georg C. Lichtenberg, Hufenland, and Vogel. See below, p. 250.

65 Louis-Aimé Le François (*Coup d'oeil médical sur l'emploi externe et interne de l'eau de mer* (thesis, Paris, 26 Dec. 1812), for instance, vaunts the installation at Dieppe (p. 29). The desire to propagandize is even more obvious in Charles-Louis Mourgué, *Journal des bains de mer de Dieppe* (Paris, Seignot, 1823); see pp. 11–12. According to him, this is the only resort that deserves to play a national role, and it should become a model, no matter what the merits of Cette or Marseilles may be. The same opinion is given by Dr Gaudet, *Notice médicale sur l'établissement des bains de mer de Dieppe* (Paris, 1837). By contrast, Bertrand praised the advantages of the Boulogne beaches (*Précis*, vol. 2, pp. 531 ff.).

66 It is interesting, however, to note the premonitory opinion of Dr John Coakley Lettsom, who instigated the Royal Sea-Bathing Infirmary at Margate. This establishment included a solarium. According to Gilbert (*Brighton*, p. 18), this physician seems to have been the first to emphasize the role of air and light in the treatment of consumption.

67 Gilbert, *Brighton*, pp. 21 and 27.

68 Dr J. Le Coeur, *Des bains de mer, Guide médical et hygiénique du baigneur* (Paris, Labé, 1846), vol. 1, p. 31. This work, which contains no less than 870 pages, is an authority whose size indicates that the time of the genesis of seaside vacations is past.

69 Gilbert, *Brighton*, pp. 79–80, according to the book by Dr A. L. Wigan, *Brighton and its three Climates* ... (1834; 2nd edn, 1845).

70 Currie, *Medical Reports*, vol. 2.

71 Gaston Bachelard devotes some fascinating pages to the pleasure of confronting the raging waters (*L'Eau et les rêves* (Paris, José Corti, 1942), ch. 8: 'L'Eau violente', pp. 21 ff.).

72 Le Coeur (*Des bains*, vol. 1, pp. 387–8), emphasizes the importance of this 'moral support' for 'naturally fearful' individuals.

73 Notably Buchan and Mourgué (*Journal*, pp. 72 and 118). The latter notes that at Dieppe the patients are terrified by this practice, which is forced upon them by the bathing guides.

74 Le François (*Coup d'oeil* p. 20) notes in 1812 that this practice of drenching is very common in England, but is little used in the rest of Europe.

75 Bertrand, *Précis*, vol. 2, p. 558.

76 Le Coeur (*Des bains*, vol. 1, p. 339: After criticizing this 'stupid custom', he writes: 'For many female bathers especially, there is something original, something bizarre about being picked up off the ground by strong arms, waiting a few seconds for an unusual sensation, not knowing at exactly what moment they will feel it, experiencing it only against the will of their bodies, so to speak, and ... What do I know?'

77 Ibid., p. 17.

78 'I make beneficial use of these baths, either to hasten or to slow the menstrual function, by modifying the duration of the immersion, adjusting the impression of the water, and so forth' (Viel, *Bains de mer*, p. 87).

79 As was the case at a later date of Dr G. Hartwig (*Guide médical et topographique du baigneur à Ostende* (...) (Brussels, 1854), p. 98).

80 Bertrand, *Précis*, vol. 2, p. 576.

81 Léopoldine Hugo, *Corréspondance*, critical edition by Pierre Georgel (Paris, Klincksieck, 1976), pp. 218 and 220.

82 *Diary and Letters of Mrs. d'Arblay* (Fanny Burney) (London, Macmillan, 1904), vol. 2, p. 128.

83 In July 1828, the best 'bathers' in Dieppe were much in demand. See *Journal du comte Rodolphe Apponyi* (Paris, Plon, 1913), vol. 1, p. 125: 'The bather of the Duchess of Berry, named Cosvreau, is the bather in fashion; people fight to have him. For men, it's Fonvite, my bather ...'.

84 Anthony Dale, *The History and Architecture of Brighton* (Brighton, Bredon Heginbothom, 1950), p. 22.

85 In addition to Gaston Bachelard, see the interpretation of swimming in Maupassant by Elisabeth Roudinesco and her explanations of the correspondence between salt-water and the female sex (*La Bataille de cent ans. Histoire de la psychanalyse en France* (Paris, Ramsay, 1982), pp. 79–80).

86 Gilbert Andrieu, 'De l'art de surnager au XIXe siècle dans la Seine', *Revue des sciences et techniques des activités physiques et sportives* 5, no. 10 (Dec. 1984), pp. 64–74.

87 On this subject, see Georges Vigarello, *Le Corps redressé* (Paris, Delarge, 1978), *passim*.

88 Bachelard, *L'Eau*, p. 224. It should be noted, however, that this philosopher chooses very late examples and has only a limited knowledge of the history of somatic impressions.

89 This is not the place to raise the debate over this issue, which opposes the disciples of Norbert Elias to those who believe that though the configuration of this sentiment changes over the ages, its intensity does not vary, or that this is actually a subtle strategy of sensuality, as was suggested by Michel Foucault (*La Volonté de savoir* (Paris, Gallimard, 1976) and, from a completely different perspective, Peter Gay, *The Bourgeois Experience, Victoria to Freud*, vol. 2: *The Tender Passion* (Oxford University Press, 1986).

90 Anthony Pasquin (John Williams), *The New Brighton Guide* (1796), p. 6; cited by Gilbert, *Brighton*, p. 15.

91 C. Wright, *The Brighton Ambulator* (1818); cited by Gilbert, *Brighton*, p. 55

92 Cambry, *Voyage dans le Finistère* (1835 edn), p. 205.

93 See below, p. 279.

94 The account by Torrington is very revealing on this subject (see below).

95 And if they did not choose to lace up the wool boots with buffalo soles worn by certain ticklish individuals. See Le Coeur, *Des bains*, vol. 1, p. 323.

96 Thomas Pennant, *A Tour in Scotland* (1769; 3rd edn, Warrington. 1774), p. 19.

97 At Boulogne, insists Bertrand, 'Bathers may place their feet anywhere with confidence' (*Précis*, vol. 2, p. 535).

98 And more generally, of their intense sense of place.

99 Le Coeur, *Des bains*, vol. 1, p. 390.

100 Dr Charles Londe, *Nouveaux Éléments d'hygiène* (Paris, 1838), vol. 2, p. 296.

101 According to Gilbert, *Brighton*, p. 14.

102 Buchan, *Treatise*, p. 71, on the beaches of the Thanet peninsula.

103 The respective comfort of the bathing-machines was an element in evaluating resorts. A German traveller in 1822, for instance, noted that at Putbus, unlike at Doberan, the bather was provided with a pair of slippers (*Reise eines Gesunden in die Seebäder Swinemünde, Putbus und Doberan* (Berlin, 1823) (an account of trips made in 1822)).

104 These tents were used at Boulogne in the 1820s (Bertrand, *Précis*, vol. 2, p. 552). They were also observed later at Ostend.

105 This subject, more closely related to the history of dress, is not within the scope of this work. We will therefore limit our comments to these few indications on the origins of these practices.

106 Louis Garneray, *Vues des côtes de France dans l'océan et dans la*

Méditerranée (text by E. Jouy) (Paris, Panckoucke, 1823), p. 16.

107 This is when the production of bathing trunks began at Caen. Here and for the following quotations, see Le Coeur, *Des bains*, vol. 1, p. 307.

108 Ibid., vol. 1, p. 317.

109 Note that such activities may take place on the beach without bathing; this was the case at Scheveningen (see above, p. 37) and also along the German coasts; see comments about Doberan in *Reise eines Gesunden.*

110 Gilbert, *Brighton*, p. 12.

111 Le François, *Coup d'oeil*, p. 26.

112 Auguste Bouet, 'Les Bains de Biarritz', in *La France maritime*, vol. 3 (1837), pp. 318–19. 'Here there is no etiquette and no uncomfortable clothing; the women are barely covered by light dresses in striped cotton, and as for the men, nature alone takes charge of their dress. The atmosphere is full of shouts, songs, and untroubled, unbridled pleasure.' On this resort, see Pierre Laborde, *Biarritz, huit siècles d'histoire, 200 ans de vie balnéaire* (Biarritz, Ferrus, 1984), 'Les Bains de Biarritz, 1814–1854', pp. 27 ff. and from a somewhat broader point of view, the author's thesis, *Pays basques et pays landais de l'extrême Sud-Ouest de la France, Étude d'organisation d'un espace géographique* (Bordeaux, 1979).

113 Cited by J. Laborde, 'Biarritz et ses origines à la fin du Second Empire', *Bulletin de la Société des sciences, lettres et arts de Bayonne* (Oct. 1965–Jan. 1966). On bathing at Biarritz in 1818, see Louis Garneray, *Voyage pittoresque et maritime sur les côtes de la France, dans l'océan et dans la Méditerranée* (Paris, 1823), comment on 'Vue de Biarritz'.

114 ['*Cacolettes*' is the name given to the seats placed on either side of a mule and used to carry passengers or wounded persons – trans.] By extension, it also designates the young girls who use these seats on the mules and who guide tourists.

115 Bertrand, *Précis*, vol. 2, p. 531.

116 E. Jouy, in Garneray, *Vues des côtes*, p. 28: 'Sea bathing in Le Havre is yet another distraction for all classes of society ... as unlike Dieppe, no major establishment there offers a central meeting-place for bathers; the humble huts or shelters covered with canvas are used as dressing-rooms for doing one's toilet, which is basically the same for everyone who enjoys the pleasure of bathing.'

117 Letter from Subdelegate Moracin reproduced in Laborde, 'Biarritz'.

118 Pilati, *Voyage de la Hollande*, vol. 2, p. 176.

119 Dr Thoré (*Promenade sur les côtes du golfe de Gascogne* (Bordeaux, Brossier, 1810), p. 297) criticized the nudity practised so openly on the beach; this was an old debate in the region over the indecency of bathing. See the fine article by Josette Pontet, 'Morale et ordre

public à Bayonne au XVIIIe siècle', *Bulletin de la Société des sciences, lettres et arts de Bayonne* (1974), pp. 127 ff.

120 See Le Coeur, *Des bains*, vol. 1, p. 299. As an example, see the rules set by the mayor F. Vallée concerning nudity on the beach at Granville, dated 1 July 1837. Three sectors were established: the women's bathing area, the area for 'bathing by clothed men', and 'the bathing area for unclothed men'. (catalogue of the exhibition *La Vie balnéaire et les bains de mer à Granville, 1840–1940*, Musée du Vieux Granville, 5 July–5 Oct. 1987, p. 9).

121 Hartwig, *Guide médical*, p. 18.

122 Paquet-Syphorien, *Voyage historique et pittoresque* ... (Paris, Didot, 1831), vol. 2, p. 136.

123 See Hélène Tuzet, *La Sicile au XVIIIe siècle vue par les voyageurs étrangers* (Strasbourg, Heitz, 1955), p. 447; Joseph Hager published his account in Vienna in 1795; it was translated into English in 1800.

124 Bérenger, *Les Soirées provençales*, vol. 2, p. 131.

125 Coyer, *Voyage d'Italie et de Hollande*, vol. 2, p. 126.

126 Monoyer, *Essai*, p. 16. The author was from Saint-Tropez.

127 Goethe, *Italienische Reise*, p. 326.

128 Patrick Brydone, *A Tour through Sicily and Malta* (London, 1773), vol. 1, pp. 13 ff.

129 Ibid.

130 Monoyer, *Essai*, is in large part devoted to Professor Delpech's experiments.

131 He did so after Dr Lettsom, but long before the Lyons school. Adopting a vitalist point of view, Viel also recommended baths of burning sand: 'the heat of the sand helps us to recover our vital heat' (Viel, *Bains de mer*, p. 94, and on the subject of his praises of the beneficial effects of sunshine, p. 37). Note that in Germany, Johann Georg Krünitz was proclaiming the benefits of exposure to sunlight as early as 1801.

132 The pages written by Maine de Biran in a diary from 1816 during a stay at Barèges in the Pyrenees mountains clearly demonstrate this relationship.

133 See Pierre-Jean Georges Cabanis, *Rapports du physique et du moral de l'homme* (Paris, 1802).

134 Notably under the influence of Baron von Haller.

135 [Hyperaesthesia is defined by the *Shorter Oxford Dictionary* as the 'excessive and morbid sensitiveness of the nerves or nerve-centres' – trans.]

136 Viel, *Bains de mer*, p. 42. For this reason, the therapy 'consists in being alert to the responses of the individual nature of each patient' to these demands.

137 Le Coeur, *Des bains*, vol. 1, p. 214.

138 Bertrand, *Précis*, vol. 2, pp. 579–83. In May 1827, for a 55-year-old patient suffering from muscular asthenia and imminent chorea,

'we prescribed receiving the wave in the midst of the foam'. In July 1827, for a young twenty-eight-year-old man suffering from 'circulatory neurosis' which caused 'bizarre sensations', Bertrand prescribed seven or eight thirty-second immersions, 'and after each one, a rub-down with flannel'. The patient continued the bathes until mid-October, but by then they had become 'merely a source of pleasure' for him: a revealing shift from the therapeutic purpose to a hedonistic one.

139 Le Coeur, *Des bains*, vol. 1, p. 411. On this subject, the practitioner gave the following tips: choose hot weather, without wind, at noon, when the sun is shining. Walk on the beach, sit down, roll in the hot sand, dive into the water, and repeat the sequence four or five times. 'I recommend this procedure to bathers,' added the author; 'I have often employed it.'

140 See the correspondence of young Flaubert, *passim*.

141 Bertrand, *Précis*, vol. 2, pp. 578–9.

142 Viel, *Bains de mer*, pp. 69, 90, and 92.

143 Richard Smollett, *Travels through France and Italy* (London, 1766).

144 *The Torrington Diaries, Containing the Tours through England and Wales of the bar. ... John Byng, between the Years 1781 and 1794* (London, Eyre and Spottiswoode, 1934), vol. 1, p. 90.

145 Townley, *Journal* (Whitehaven, J. Ware and Son, 1791), 2 vols. This journal covers the period from April 1789 to April 1790.

146 Ibid., vol. 1, p. 30.

147 Ibid., 19 July 1789, vol. 1, p. 116.

148 Ibid., 4 Aug. 1789, vol. 1, p. 38.

149 Ibid., 12 Dec. 1789, vol. 1, p. 293.

150 'More desirable places for privacy and retirement could not be wished for' (ibid., 11 Dec. 1789, vol. 1, p. 293).

151 Ibid., 27 Nov. 1789, vol. 1, p. 277.

152 On this subject, see Thomas, *Man and the Natural World*, pp. 281 ff.

153 Townley, *Journal*, 15 Aug. 1789, vol. 1, p. 152.

154 Cowper, 'Retirement' (1782).

Notes to chapter 4

1 On this subject, see Porter, *Making of Geology*, pp. 116–18.

2 Interesting discussion of these issues in ibid., pp. 104 ff., 197 ff.

3 See above, pp. 287–8.

4 See Rhoda Rappaport, 'Geology and Orthodoxy: The Case of Noah's Flood in 18th Century Thought', *The British Journal for the History of Science*, 11 (1978), pp. 1–18, esp. pp. 14–15.

5 The late Nicolas Boulanger, *L'Antiquité dévoilée par ses usages, ou Examen critique des principales opinions, cérémonies et institutions religieuses et politiques des différents peuples de la terre* (Amsterdam, Rey, 1766), p. 382.

6 See Rappaport, 'Geology and Orthodoxy', p. 15.

7 Benoît de Maillet, *Telliamed ou entretiens d'un philosophe indien avec un missionnaire français* (Amsterdam, 1748), vol. 1, pp. 110–28. There is an abundant literature on this work. See Albert V. Carozzi, 'De Maillet's *Telliamed* (1748): An Ultra-Neptunian Theory of the Earth', in *Toward a History of Geology. Proceedings of the New Hampshire Inter-Disciplinary Conference on the History of Geology, Sept. 7–12, 1967*, ed. Cecil J. Schneer (Cambridge, Mass., MIT Press, 1969), pp. 80–100. More recently. Miguel Benitez, 'Benoît de Maillet et l'origine de la vie dans la mer: conjecture amusante ou hypothèse scientifique?', *Revue de synthèse*, 3rd series, nos 113–14 (Jan.–June 1984), pp. 37–54. By the same author, 'Benoît de Maillet et la littérature clandestine: étude de sa correspondance avec l'abbé Le Mascrier', *Studies on Voltaire and the Eighteenth Century*, 183 (1980), pp. 133–59. Lastly, Jacques Roger, *Les Sciences de la vie dans la pensée française du XVIIIe siècle* (Paris, Armand Colin, 1963), pp. 520 ff.

8 Voltaire, *Dictionnaire philosophique*, 'Déluge universel' and 'Inondation' entries.

9 Buffon, 'Histoire et théorie de la Terre', speech made at Montbard 3 Oct. 1744, in Georges-Louis Leclerc, Count de Buffon, *Oeuvres complètes* (Paris, Levasseur, 1884), p. 94.

10 See Reyer Hooykaas, *Natural Law and Divine Miracle, A Historical-Critical Study of the Principle of Uniformity in Geology, Biology and Theology* (Leiden, E. J. Brill, 1959), esp. 'The Uniformitarians'.

11 See also Porter, *Making of Geology*, pp. 109 and 192–6.

12 On the primitive ocean, starting-point for 'Neptunian' theory, see Martin Guntau, 'The Emergence of Geology as a Scientific Discipline', *History of Science*, 16 (1978), p. 285.

13 Carozzi, 'De Maillet's *Telliamed*', p. 81.

14 On this aspect of Buffon's thought, see *Les Époques de la Nature* (Paris, Ed. du Muséum, 1962), introduction by Jacques Roger, esp. pp. xvii ff. See also Ehrard, *L'Idée de Nature*, vol. 1, pp. 208 ff., and Numa Broc, *La Géographie des philosophes* (thesis, University of Lille III, 1972), pp. 269 ff.

15 'It seems that our earth was once at the bottom of the sea,' states Buffon, and he deduces that 'therefore, to discover what happened earlier on earth, let us see what is happening today on the bottom of the sea' (*Histoire*, p. 43). This way of thinking is very distant from that expressed by the Plutonian Buffon of *Époques*.

16 Rappaport, 'Geology and Orthodoxy', *passim*.

17 On this distinction, Alexander M. Ospovat, 'Reflections on A. G. Werner's Kurze Klassifikation', in Schneer, *Toward a History*, pp. 242 ff., and the article by Leroy E. Page cited below.

18 On this conflict, see Porter, *Making of Geology*, pp. 196 ff. It should be noted, however, that Leroy E. Page ('Diluvialism and its Critics in Great Britain in the Early Nineteenth Century', in

Toward a History, ed. Schneer, pp. 257 ff.) considers that the impact of religious ideas on early nineteenth-century geologists has been exaggerated.

19 See Richard J. Chorley, Antony J. Dunn, and Robert P. Beckinsale, *The History of the Study of Landforms on the Development of Geomorphology* (London, Methuen, 1964), vol. 1, pp. 54–6. This work also contains important observations on Werner and Neptunism.

20 Porter, *Making of Geology*, pp. 201–2.

21 On this point, see also Hooykaas, *Natural Law*.

22 On the subject of Cuvier's diluvian theory, see Page, 'Diluvialism', pp. 261–2. The knowledge about this scientist has recently been considerably expanded by Dorinda Outram's book *Georges Cuvier. Vocation, Science and Authority in Post-Revolutionary France* (Manchester University Press, 1984).

23 According to Cuvier, the sea first abandoned its bed in order to occupy the continents. The sea bottom thereby uncovered became the present-day earth. Furthermore, a similar shift had already taken place before the creation of man. Consequently, the Flood resulted in bringing the sea back into the sea bed it had left earlier. This set of hypotheses could be reconciled with Scripture.

24 D. R. Oldroyd, 'Historicism and the Rise of Historical Geology', *History of Science*, 17 (Mar. 1979), p. 192.

25 See Broc, *La Géographie des philosophes*, p. 603. It should be noted that Abbé Le Mascrier had vaguely suggested that the earth could be several million years old (cf. Rappaport, 'Geology and Orthodoxy', p. 4, which quotes the abbé, and 'Essai sur la chronologie', in *Le Monde, son origine et son antiquité* (J. B. Mirabaud, 1778).

26 Broc, *La Géographie des philosophes*, p. 620.

27 See the very interesting article by François Ellenberger, 'De l'influence de l'environnement sur les concepts: l'exemple des théories géodynamiques au XVIIIe siècle en France', *Revue d'histoire des sciences*, 33, no. 1 (1980), esp. pp. 66 ff. on the 'neo-catastrophist revenge', the end of belief in 'slow causes', and the isolation of Giraud-Soulavie. At the beginning of the nineteenth century, belief in a long chronology of the earth was still 'tainted by libertine subversion'. This counter-offensive ran out of steam between 1830 and 1840 as awareness of glacier-age phenomena grew.

28 I will say little about shells, which were considered early on as prizes in the history of the earth and which fascinated scientists as early as the seventeenth century (see Roger, Introduction to Buffon, *Époques*, p. xvi). They continued to cause excitement as long as the diluvian problem was addressed in its classic form. What was most astonishing was the presence of shells on mountains and within continents. It seemed to support the biblical texts, before discrediting them (see the work done by Réaumur in Touraine).

None the less, the coastal cliffs remained one of the sites where it was easily possible to develop a collection of 'diluvian' shells. More important for the purposes of this book, these observations tended to emphasize the mobile nature of the shore.

29 On this episode, see Eugène Wegmann, 'André Celsii: remarques sur la diminution de l'eau, aussi bien dans la Baltique que dans l'Atlantique', *Sciences de la terre*, 21 (1977), pp. 39–52.

30 More precisely, at the beginning of a period that corresponded to one-third of the time that had passed since the Creation (ibid., p. 136).

31 More specifically, between 1765 and 1769.

32 See Eugène Wegmann, 'Évolution des idées sur le déplacement des lignes de rivage, origines en Fennoscandie', *Mémoires de la Société vaudoise des sciences naturelles*, 14, no. 4 (1967), no. 88, pp. 129–90, esp. pp. 137 ff.

33 See Ellenberger, 'De l'influence', pp. 50 ff. The author emphasizes the role of the environment in the origin of scientific theory.

34 De Maillet, *Telliamed*, vol. 1, pp. 8 and 9.

35 See Deacon, *Scientists and the Sea*, pp. 176–80.

36 On this aspect of Philippe Buache's work, see J. Thoulet, 'L'Étude de la mer au XVIIIᵉ siècle. De Maillet, Buache et Buffon', in *Mémoires de l'Académie de Stanislas, 1908–1909*, pp. 214–56.

37 R. P. de Dainville. 'De la profondeur à l'altitude. Des origines marines de l'expression cartographique du relief terrestre par côtes et courbes de niveau', in *Le Navire et l'économie maritime du Moyen Âge au XVIIIᵉ siècle, principalement en Méditerranée*, ed. Michel Mollat du Jourdin (Paris, S.E.V.P.E.N, 1958).

38 This is merely one example intended to demonstrate the range of interpretations possible.

39 Jean-André de Luc, *Lettres sur l'histoire physique de la terre …, renfermant de nouvelles preuves géologiques et historiques de la mission divine de Moyse* (Paris, 1798), lettre 5: 'Naissance de nos continents', 1 July 1792, pp. 223 ff. Peremptorily, de Luc states that it is obvious that the sea has 'no visible effect' on the hard rocks along the coast which are exposed to the waves (p. 259). Erosion therefore occurs only on beaches. Like many scientists of his age, he distinguishes between alluvium and 'diluvium'.

40 The cliff, note François Dagognet, François Guéry, and Odile Marcel (colloquium: *Mort du paysage, philosophie et esthétique du paysage* (Paris, Champvallon, 1982), preface p. 27) does not designate a thing, but rather 'an accumulation of tragedies'. 'The landscape incorporates the drama from which it is derived … It [is] a sacred function of an unconscious telluric memory; it contains and conserves the forces that find rest in it.' The historicity of this interpretation is exactly what we want to show here. It is obvious that this goes along with the calling into question in the political arena of

the immobile legitimacy of a monarchy based on divine right; this contributed to heightening the passions aroused by the sharp debates over geology.

41 On this process, see Alexander M. Ospovat, 'Reflections', and Chorley *et al.*, *History*, pp. 85–6.

42 Think of the fascination that the depths of the Earth exerted over the German Romantics, as revealed notably in *Heinrich von Ofterdingen* by Novalis, who was himself a mining engineer and a disciple of Werner.

43 Porter, *Making of Geology*, p. 142.

44 Ibid., pp. 180–1.

45 On the subject of geological cross-sections, Martin J. S. Rudwick speaks of the 'artificial cliff' (cf. Martin J. S. Rudwick, 'The Emergence of a Visual Language for Geological Science, 1760–1840', *History of Science*, 14, (1976), pp. 149–95).

46 Henry C. Englefield, *A Description of the Principal Picturesque Beauties, Antiquities, and Geological Phenomena of the Isle of Wight* (London, 1816).

47 At the same time, the concept of the documentary landscape was developing.

48 Rudwick, 'Emergence', p. 176.

49 On these workings, see Maurice Lévy, *Le Roman gothique anglais, 1764–1824* (Toulouse, Faculté des Lettres, 1968).

50 See Benitez, 'Benoît de Maillet et l'origine', pp. 44 and 48.

51 Broc, *La Géographie des philosophes* pp. 292–8.

52 It is important, however, not to fall into the too frequent confusion between 'terrestrialization' of marine species as Benoît de Maillet imagined it and the procedures of evolutionism, of which the author of *Telliamed* cannot be considered a forerunner (see Benitez, 'Benoît de Maillet et l'origine', p. 42).

53 Sandor Ferenczi, *Thalassa. Psychanalyse des origines de la vie sexuelle* (Paris, Payot, 1977), *passim*.

54 This is clearly explained by Benitez, 'Benoît de Maillet et l'origine', p. 42.

55 Emphasized by Roger, *Les Sciences de la vie*, p. 524.

56 Anecdote recounted in De Maillet, *Telliamed*, vol. 2, pp. 160–1.

57 See the text by Montesquieu mentioned above, p. 69.

58 De Maillet, *Telliamed*, vol. 2, pp. 197 ff. and 215.

59 Buffon, *Époques*, Roger, Introduction to pp. lxx–lxxi and text by Buffon, p. 155.

60 See below, pp. 199 ff.

61 On this point, Porter, *Making of Geology*, pp. 90 ff.

62 More specifically, beginning in 1730–5 (ibid., pp. 93 ff.). Porter calls into question the notion that there was a decline in investigative science in England during this period. On all these points, see also Thomas, *Man and the Natural World*, *passim*, and especially on

the subject of botany, pp. 281 ff. and the now classic work by John Harold Plumb, *Georgian Delights* (Boston, 1980); a pleasant overview is also offered by Roy Porter, *English Society in the Eighteenth Century* (Penguin Books, 1982).

63 Roche, *Le Siècle des Lumières*, vol. 1, pp. 125–6 and 155–76.

64 Ibid., p. 125.

65 François Ellenberger, 'Aux sources de la géologie française. Guide de voyage à l'usage de l'historien des sciences de la terre sur l'itinéraire Paris–Auvergne–Marseille', *Histoire et Nature. Cahiers de l'association pour l'histoire des sciences de la nature*, 15 (1979), pp. 3–29; here, p. 28.

66 On this evolution, see Porter, *Making of Geology*, p. 173.

67 Cf. the fascination exercised in France by the natural site called l'Enfer de Plogoff (Plogoff's hell-hole) and in Scotland by Buchan's Buller.

68 Barthélemy Faujas de Saint-Fond, *Voyage en Angleterre, en Écosse et aux îles Hébrides ayant pour objet les sciences, les arts, l'histoire naturelle et les moeurs* (Paris, 1797), vol. 1, p. 370.

69 Ibid., vol. 2, pp. 42 and 44.

70 B. Faujas de Saint-Fond, *Description de l'île de Staffa, l'une des Hébrides et de la grotte de Fingal* (Paris, 1805), p. 13.

71 L. A. Necker de Saussure, *Voyage en Écosse et aux îles Hébrides* (Geneva and Paris, 1821), vol. 1, pp. 137–8.

72 Ibid., vol. 2, pp. 492 and 498.

73 Ibid., vol. 3, p. 39.

74 Thomas Pennant, *Le Nord du globe ou tableau de la nature dans les contrées septentrionales* ... (Paris, Barrois, 1789), vol. 1, p. 34. This scientist played an important role in the history of travellers' sensibility to impressions of wild nature.

75 See Thomas, *Man and the Natural World*, p. 283. This work clearly presents the connections that develop at this time between botany, zoology, and the means of appreciating nature.

76 See the examples set forth in Margaret Deacon's book *Scientists and the Sea*, pp. 251 ff.

77 See François Carré, *Les Océans* (Paris, P.U.F., 1983).

78 Goethe, *Italienische Reise*, pp. 92 ff.

79 This same feeling accounts for the powerful attraction then exercised by fish markets.

80 Thomas, *Man and the Natural World*, p. 283. On this point, see Peter S. Dance, *Shell Collecting, a History* (Berkeley, University of California Press, 1966). As early as the decade of the 1730s, ladies such as the Duchess of Beaufort and the Duchess of Portland devoted themselves passionately to this hobby. In 1784, the *Universal Conchologist* offered its readers a list of the most important collectors. At the same time, on the Continent, scholars were fascinated by coral, zoophytes, and lithophytes (see Ehrard, *L'Idée de Nature*, vol. 1, pp. 192 ff.).

81 See Jean-Victor Audouin and Henri Milne-Edwards, *Recherches pour servir à l'histoire naturelle du littoral de la France. Voyage à Granville, aux îles Chausey et à Saint-Malo* (Paris, Crochard, 1832). The year 1815 was indicated by Milne-Edwards in *Observations sur les ascidies des côtes de la Manche*, Académie des Sciences, 11 Nov. 1839.

82 Armand de Quatrefages de Bréau. 'Souvenirs d'un naturaliste, L'Île de Bréhat, Le Phare des Héhaux', *Revue des deux mondes*, 15 Feb. 1844, p. 619.

83 Audouin and Milne-Edwards, *Recherches*, p. ii.

84 The two scientists, and later Armand de Quatrefages, stayed between Granville and the Île de Bréhat. The first two explorers did toy for some time, however, with the impossible idea of 'successively travelling across all of our coasts' in order to undertake a systematic study of their animal species (Audouin and Milne-Edwards, *Recherches*, p. iii).

85 Ibid., pp. 86 and 89.

86 De Quatrefages, 'Souvenirs', p. 613.

87 See Chorley, *et al.*, *History*, p. 183.

88 [The Citizen King was Louis-Philippe, and his reign was also known as the Bourgeois Monarchy, or *monarchie censitaire*. That is, it was a monarchy with limited suffrage, with the right to vote limited to landowners and bourgeois who could afford to pay the electoral *cens* – trans.]

89 De Quatrefages, *Souvenirs*, p. 607.

Notes to chapter 5

1 On this point see Théodore A. Litman, *Le Sublime en France, 1660–1714* (Paris, Nizet, 1971), ch. 7: 'Saint-Évremond: l'aspect négatif du sublime et du vaste', which quotes the 'Dissertation sur le mot vaste', 1685, esp. p. 153.

2 The need for boundaries is not only a visual one. It involves the 'complicity of vision and the whole body with the landscape'. The horizon is a border in relation to which the whole system of appreciating a landscape is built up. It defines the landscape as the viewer's territory. 'The path followed by vision merely anticipates the body's movements' (Michel Collot, 'L'Horizon du paysage', in *Lire le paysage*, p. 122).

3 See Plaisant, *La Sensibilité*, *passim*.

4 At the end of his life, when he retired to the foothills of Mount Vesuvius, Sannazaro described the life and work of fishermen in his *Eclogues*.

5 Cited by Plaisant, *La Sensibilité*, vol. 2, p. 517. Tickell's critique appeared in the *Guardian*, no. 28 (13 Apr. 1713).

6 Praised by Milton in the portrait of that of Adam and Eve in *Paradise Lost*.

7 William Diaper, *Nereides or Sea Eclogues* (1712), quoted by Plaisant, *La Sensibilité*, vol. 2, p. 473.

8 See Bony *Joseph Addison*, p. 565.

9 Quoted by Plaisant, *La Sensibilité*, vol. 1, p. 303. Note that as early as 1671, Bouhours (*Les Entretiens*, p. 5) made his character Ariste speak of the tempest and say: 'All that inspires an indescribable horror, combined with pleasure.'

10 It would go well beyond the present subject to discuss appropriately the emergence of the concept of the rhetorical sublime, then the natural sublime, within the framework of aesthetic debates. Let us simply note that the range of sublime objects listed by pseudo-Longinus, translated in 1674 by Boileau, includes the sea: 'We do not naturally admire small streams ... but we are truly surprised when we look at the Danube, the Nile, the Rhine, and especially the Ocean.' Boileau was solely concerned with the rhetorical sublime. It should be admitted, however, that this is a common, though fairly irrelevant distinction. See also the translation by William Smith (Dublin, 1740). It is obvious that delighting in reading the *Odyssey* is a first step towards appreciating the Mediterranean and its shores. Boileau's Longinus made its way to England as early as 1674. A few decades earlier, in 1652, John Hall had also translated the Greek work. His *Peri Hupsous* went through several editions during the following fifty years. On that side of the Channel, however, the influence of the Cambridge Platonists, especially Henry More (see Plaisant, *La Sensibilité*, vol. 1, pp. 143 and 146), prepared the new aesthetics, whose 'essential criteria were immensity, greatness, and terror'. Henry More identifies the idea of God with space, immense, eternal, uncorruptible, omnipresent.

11 *Spectator*, vol. 7 (Sat., 20 Sept. 1712), pp. 58–9.

12 See Bony, *Joseph Addison*, pp. 427–8, including the following quotation.

13 Which, it should be noted, contradicts his manner of appreciating the coastline in Campania (see above, p. 47).

14 See Louis Marin, 'Le Sublime classique: les tempêtes dans quelques paysages de Poussin', in *Lire le paysage*, p. 201. It should be remembered that at the end of the century, Kant in his *Critique of Judgement* was to make the sublime into a movement of the spirit, rather than a quality inherent in an object.

15 On this subject, Bony, *Joseph Addison*, pp. 566 ff. The author analyses the way in which Addison extols primitive pleasures. At the same time, this represents the disappearance of a belief in a chaotic nature; since if vision could behold only a fallen nature, the position of the spectator would be discredited.

16 Or in 1730, if one takes into account the publication of the complete edition of *The Seasons*.

17 Plaisant, *La Sensibilité*, vol. 2, pp. 625 ff.
18 Nicolas Grimaldi, 'L'Esthétique de la belle nature. Problèmes d'une esthétique du paysage', in *Mort du paysage?*, p. 120.
19 Especially if the English landscape garden is considered as a way of putting desire in a cage through obsessive shaping and structuring; within this structure, there are 'no more surprises, no more happiness or unhappiness, no more action'. Consequently, it is the antithesis of the wandering of the grand tour (see Michel Cusin, 'Le Jardin anglais au XVIIIe siècle: aménagement imaginaire et déménagement signifiant', in *Lire le paysage*, pp. 225–6). On this subject, see John Dixon-Hunt and Peter Willis, *The Genius of Place. The English Landscape Garden. 1620–1820* (London, Elek, 1975).
20 D. Mallet, *The Excursion* (London, J. Walthoe, 1731). The poem contains a description of the 'horrible day' of the storm (pp. 17–19), the polar sea (p. 28), and its icebergs, with an evocation of

> Tartar's cruel coast,
> By utmost Ocean Wash'd,
> On whose last wave
> The blue sky leans her Breast.
> diffus'd immense
> In solitary Length the Desert lies.

The description of the terrible solitude leads to an image of the central abyss.
21 Cited by Plaisant, *La Sensibilité*, vol. 2, p. 742.
22 See John Sunderland, 'The Legend and Influence of Salvator Rosa in England in the XVIIIth Century', *Burlington Magazine*, 115, no. 849 (Dec. 1973), pp. 785–9, which qualifies the theses of Manwaring, *Italian Landscape*, pp. 40 ff.
23 See below, pp. 234 ff.
24 Edmund Burke, *The Works of Edmund Burke* (London, George Bell and Sons, 1894). The French reprint edition (Vrin, 1973) includes an introduction by Baldine Saint-Girons which emphasizes (p. 21) that the major role of Burke was to liberate the sublime from its original bond with rhetoric and to seek to identify its features in nature itself. The author also reveals the bond that connects this notion to the development of pathos in poetic and Romanesque literature, especially in Richardson's *Pamela* and in the funeral elegies of Gray and Young. On this subject, see the classic work by Samuel H. Monk, *The Sublime, a Study of Critical Theories in XVIIIth century England* (New York, 1935), and that of Theodore E. B. Wood, *The Word 'Sublime' and its Context. 1650–1760* (Paris, Mouton, 1972).
25 Burke, *Works*, vol. 1, 1900 edn, p. 88.
26 The word is borrowed from Richard Payne Knight, *An Analytical Inquiry into the Principles of Taste*, 4th edn, (1808) p. 370.

27 Burke, *Works*, vol. 1, p. 92.
28 Ibid.
29 What Hume calls the passion of surprise. On the lines that follow, see the beautiful book by Barbara Maria Stafford, *Voyage into Substance, Art, Science, Nature and the Illustrated Travel Account. 1740–1840* (Cambridge, Mass., MIT Press, 1984), esp. pp. 403–9. The author notes that eighteenth-century aesthetics are marked by the search for a cure against visual boredom (p. 414).
30 Ibid., p. 421.
31 Burke, *Works*, vol. 1, p. 150.
32 Ibid., p. 151.
33 Ibid., p. 148.
34 Ibid., p. 149.
35 See Robert Sauzet, *Contre-réforme et réforme catholique en Bas-Languedoc au XVIIe siècle. Le diocèse de Nîmes de 1598 à 1694* (dissertation, Paris, IV, 1976), vol. 1, p. 302.
36 See Olivier Michel, 'Adrien Manglard, peintre et collectionneur. 1695–1760', *Mélanges de l'École française de Rome*, 93 (1981), 2: 'Moyen Âge, Temps Modernes', pp. 823–926. Adrien Manglard, who arrived in Rome in 1715 at the age of twenty, had already developed a flourishing clientele by 1726, essentially comprising Roman aristocrats. On this painter, see also André Rostand, 'Adrien Manglard et la peinture de marines au XVIIIe siècle', *Gazette des beaux-arts* (July–Dec. 1934), pp. 263–72.
37 *Mémoires de Marmontel*, published by Maurice Tourneux (Paris, Librairie de bibliophiles, 1981), vol. 2, p. 179. About Toulon he says: 'Here, what should have impressed me the most was what surprised me the least. One of my desires was to see the open sea. I saw it, but when calm; and Vernet's paintings had depicted it so faithfully that the reality did not evoke any emotion in my mind; my eyes had becomed accustomed to it as though I had been born on its shores.'
38 Simon Pelloutier, *Histoire des Celtes* (Paris, Coustellier, 1741). Paul-Henri Mallet, *Monuments de la mythologie et de la poésie des anciens peuples du Nord* (Copenhagen, Philibert, 1756).
39 Some time ago, Paul Van Tieghem painstakingly established the chronology of the spread of Ossianism throughout Europe, first of all in the Germany of Klopstock and then Friedrich von Stolberg, then later in France, long after the fashion for England which dominated in France between 1750 and 1770 had faded away, and beyond the neoclassicism of the last years of the eighteenth century, in Imperial and Restoration France. See Paul Van Tieghem, *Ossian en France* (Slatkine reprints, 1967), 2 vols; and by the same author, *Ossian et l'Ossianisme dans la littérature du XVIIIe siècle* (Groningen, 1920).
40 For the extent that fog implies a loss of awareness of time and space in exterior phenomena, see Stafford, *Voyage into Substance*, p. 400.

Generally speaking, says the author, the way eighteenth-century figures insisted on invoking the concept of reverie shows their concern with the fundamental distinction between linear time and immemorial time.

41 See Patrick Rafroidi, *L'Irlande et le Romantisme* (dissertation, University of Lille III, 1973), *passim*. This work shows (p. 300) that it was actually Irish songs which the forger MacPherson revived on the Romantic Continent.

42 James Beattie, *The Minstrel* (Edinburgh, 1771).

43 Ibid., p. 146.

44 Pennant, *Tour in Scotland*; Dr Samuel Johnson, *A Journey to the Western Islands of Scotland* (London, Humphrey Milford, 1924). He took this trip in 1773 accompanied by Boswell.

45 Jean Meyer ('Quelques directions de recherche sur les marines de guerre du XVIIIe siècle', *Bulletin de la Société d'histoire moderne*, no. 1 (1985), pp. 6–19), showed that at this time naval warfare was primarily based on massive use of troops.

46 A tremendous amount of literature is devoted to this discovery of Britain, and we merely skim the surface here. We cite only the important work of Esther Moir, *The Discovery of Britain; the English Tourists, 1540 to 1840* (London, Routledge and Kegan Paul, 1964). At that time, a Scottish tour was developing, and its features were of course very different from those of the grand tour.

47 See Margaret Isabel Bain, *Les Voyageurs français en Écosse, 1770–1830, et leurs curiosités intellectuelles* (Paris, Champion, 1931).

48 Martin Martin, *A Description of the Western Islands of Scotland*, 2nd edn (London, 1716).

49 Johnson, *Journey*, p. 21, on the subject of his visit of Buchan's Buller.

50 Charles Nodier, *Promenade de Dieppe aux montagnes d'Écosse* (Paris, Barba, 1821), p. 327.

51 Pennant, *Tour in Scotland*, vol. 1, p. 19.

52 M. Uno de Troil, bishop of Linkoeping. The French edition of his letters, *Lettres sur l'Islande* (Paris, 1781), pp. 373 ff., includes an appendix (pp. 392 ff.) containing a description of the isle of Staffa by Mr Joseph Banks, who accompanied the bishop on the trip he undertook in 1772.

53 On this subject, see the revealing work by Edouard de Montulé, *Voyage en Angleterre et en Russie, pendant les années 1821, 1822 et 1823* (Paris, 1825).

54 In Scottish Gaelic, Staffa is said to mean 'the melodious grotto'.

55 See Faujas de Saint-Fond, above. p. 113.

56 For this and the following quotation, see Jean-Didier Urbain, 'Sémiotiques comparées du touriste et du voyageur', *Semiotica*, 58, nos 3–4 (1986), pp. 269–70.

57 These practices took over once the original heroic model had been forgotten by second-generation tourists – our generation, which is happy to follow the beaten path and returns rather curiously to some features of the classical model, being satisfied simply to recognize sites.

58 Cruttwell, *The New Universal Gazetteer; or Geographical Dictionary* (London, 1798).

59 Blanqui, *Voyage d'un jeune Français.*

60 B. Ducos, *Itinéraire et souvenirs d'Angleterre et d'Écosse, 1814–1826* (Paris, Dondey-Dupré, 1834), 4 vols.

61 [A reference to Labiche's play *Le Voyage de Monsieur Perrichon*, in which a prosperous tradesman thinks it his duty to take his whole family on a trip to the Alps. They meet with a few amusing misadventures – trans.]

62 Ducos, *Itinéraire*, vol. 3, p. 278.

63 Ibid., pp. 282–3.

64 Ibid., p. 285. It should be noted on this point that the images and feelings are similar to those which prevail during the Restoration, especially during missions.

65 Denise Delouche, *Les Peintres de la Bretagne avant Gauguin* (dissertation, University of Rennes II, 1978), vol. 1, pp. 2 ff.

66 *Voyages de Guibert dans diverses parties de la France et en Suisse faits en 1775, 1778, 1784 et 1785* (Paris, D'hautel, 1806), pp. 35–6.

67 Although the text was published only in 1806.

68 Cambry, *Voyage*, p. 34. This gap in literature explains why he was so alone on the shores.

69 Ibid., p. 37.

70 Ibid., p. 111.

71 Ibid., p. 158

72 Ibid., p. 167.

73 On this point, see François Guillet, *Curiosité et comportements touristiques dans les régions côtières de l'Atlantique et de la Manche sous la monarchie censitaire d'après les guides de voyage*, Mémoire de Maîtrise (Tours, 1984).

74 See the exhaustive studies of Plaisant (*La Sensibilité*, esp. vol. 1, pp. 374 ff.) on the rise of topographic and loco-descriptive poetry and (vol. 1, pp. 291 ff.) on what it owes to the sensualist theory of imagination, to the laws of sensory association, and consequently to the influence of Addison and Shaftesbury. On the social penetration of the taste for pastoral retreats among the Whig middle class, see ibid., vol. 1, pp. 370 ff.

75 See Broc, *Géographie de la Renaissance*, pp. 211–18.

76 Edouard Guitton, *Jacques Delille (1738–1813) et le poème de la nature en France de 1750 à 1820*. The author mentions (p. 187) the necessity which the poet felt to travel, especially from Honfleur to Le Havre, sometimes on horseback and sometimes by sea. By

contrast, he was overwhelmed by weariness with the open sea on his trip to the Orient in 1784–5 (ibid., p. 387).

77 Bachelard, *L'Eau*, pp. 206–11.

78 Quoted by Plaisant, *La Sensibilité*, vol. 2, p. 563.

79 *The Works of Henry Needler*, p. 15; cited in ibid., p. 630.

80 John Gay, 'Rural Sports, a Georgic', in *Poems on Several Occasions* (Dublin, 1730), p. 4.

81 Quite to the contrary (see above. p. 14), since the shores in Diaper's poem stand out as a repugnant and nauseating place.

82 This poem should be considered together with the translation of the *Halieuticks* by Oppien and the translation of Sannazaro's works, a distant Italian model for the seaside pastoral. See Plaisant, *La Sensibilité*, vol. 2, p. 518.

83 An idea to which Michel Vovelle is very much attached. See *Ideologies and Mentalities*, trans. E. O'Flaherty (Polity Press, 1990), ch. 5.

84 Henry Fielding, *The Journal of a Voyage to Lisbon* (London, 1755), p. 160.

85 See Roger Martin, *Essai sur Thomas Gray* (Paris, P.U.F., 1934), p. 176.

86 'The oak trees grow right up to the edge of the shore. The sea forms small and large bays which can be seen, sparkling, amongst dense groves. In addition, there is the fleet (...) ships passing incessantly (...). There are thousands of similar sights and delightful spots to be enjoyed along the coast between Fareham and Southampton. You have to climb Portsdown if you want a view of the whole area.'

87 As is demonstrated, for instance, by Thomson's silence on the matter.

88 See Carl Paul Barbier, *William Gilpin, his Drawings, Teaching, and Theory of the Picturesque* (Oxford, Clarendon Press, 1963), p. 7.

89 There is an abundant bibliography on the subject; in addition to the previously mentioned works by Barbier and Stafford, see Christopher Hussey, *The Picturesque. Studies in a Point of View* (London, Frank Cass, 1967; 1st edn, 1927), ch. 4: 'Picturesque Travel', pp. 83–127; Michel Conan, 'Le Pittoresque: une culture poétique', postface to William Gilpin, *Three Essays: On Picturesque Beauty, on Sketching Landscape, on Picturesque Travel* (London, Blamire, 1794).

90 William Gilpin, *Observations on the Mountains and Lakes of Cumberland and Westmoreland* (London, R. Blamire, 1786).

91 See Odile Morel, 'Les Aveux d'un amateur de paysage', in Dagognet et al., *Mort du Paysage*, pp. 197 ff., and Barbier, *William Gilpin*, p. 104.

92 See Stafford, *Voyage into Substance*, pp. 403–4.

93 Guitton, *Jacques Delille*, p. 573.

94 On loco-descriptive poetry, see the now classic work by John Barrell, *The Ideas of Landscape and the Sense of Place, 1730–1840, an Approach to the Poetry of John Clare* (Cambridge, 1977).

95 P. H. Valenciennes, *Éléments de perspective pratique à l'usage des artistes* (Paris, 1800), p. 340.

96 In particular, see the above-mentioned work by Barbier, *William Gilpin*, pp. 121–47.

97 On this point, it is radically different from the classical voyage.

98 Gilpin, *Three Essays*.

99 And in the substitution of the association of ideas and emotions for the emblematic. See Marie-Madeleine Martinet, *Art et nature en Grande-Bretagne au XVIIIe siècle, de l'harmonie classique au pittoresque du premier romantisme* (Paris, Aubier, 1980), pp. 6–7.

100 See Guillet, *Curiosité, passim*.

101 From which Vernet saw the sea for the first time: its beauty was praised by Bérenger.

102 In the corpus that we have established, this site is celebrated by Amédée Pichot, *Voyage historique et littéraire en Angleterre et en Écosse* (Paris, Ladvocat, 1825), vol. 3, p. 174. About Arthur's Seat, he writes that this is 'the most extraordinary panorama that any city in Europe could offer'. Saussure, *Voyage en Écosse*, vol. 1, p. 102. La Tocnaye, *Promenade d'un Français dans la Grande-Bretagne*, 2nd edn (1801), p. 126. The view to be seen from Calton Hill is surpassed only by that of Constantinople. Louis Simond, *Voyage d'un Français en Angleterre pendant les années 1810 et 1811* (Paris, Treuttel, 1816), vol. 1, p. 371. Ducos (*Itinéraire*, vol. 2, p. 352 ff.) gives evidence of how heavily visited the site was. Marc-Auguste Pictet, *Voyage de trois mois en Angleterre, en Écosse et en Irlande pendant l'été de l'an IX (1801)* (Geneva, 1802), p. 63. From the top of Calton Hill, 'One enjoys one of the most beautiful views that can possibly be imagined.' And the anonymous author of *Tournée faite, en 1788, dans la Grande-Bretagne, par un Français ...* (Paris, 1790, B.N. 8° N 150), p. 164 states, 'We went to Calton Hill first ...'. For the British, climbing the hill was one of the obligations on the Scottish tour.

103 Whose admirers included Ann Radcliffe, *Journey*, vol. 2, p. 197.

104 See Thomas Frognall Dibdin, *A Bibliographical, Antiquarian and Picturesque Tour in France and Germany* (London, 1821), vol. 1, p. 139. In 1818, he walked up to the 'wonderful summit' in order to enjoy the famous panorama of Rouen and its environs, something which no English tourist whose itinerary passed via Dieppe would want to miss.

105 See Leopold von Buch, *Reise durch Norwegen und Lappland* (Berlin, 1810), vol. 1, p. 27, on the subject of the view from Elsinore: 'The effect surpassed my expectations. It is impossible to see anything more beautiful or more striking ... In this highly varied and lively scene, there is something almost magical.' See also P. H. Mallet, 'Voyage en Norvège', appendix to the *Voyage en Pologne, Russie, Suède, Danemark, etc., par M. William Coxe* (Geneva, 1786), p. 309; on 8 June 1755 he took a walk along the banks of the magnificent strait.

106 At the time, remember, the traces of bustling activity were still quite accepted by lovers of solitude. The time had not yet come for nostalgia caused by the destruction of the landscape.

107 See Guillet, *Curiosité, passim.*

108 See Armand Narcisse Masson de Saint-Amand, *Lettres d'un voyageur à l'embouchure de la Seine* (Paris, Guibert, 1828), p. 133. '"Monsieur seems to be an enthusiast for beautiful sites", my coachman said to me, as he had heard me rave ecstatically all along the way'; the driver took the liberty of suggesting a viewpoint to the traveller. In January 1820, the Duke of Kent became a victim of his love of viewpoints when he died on the way back from a tiring excursion on a hill near Sidmouth in Devonshire. See Gilbert, *Brighton*, p. 16.

109 Gilpin, *Three Essays.*

110 Cf. Torrington during his stay at Weymouth, above, p. 90.

111 William Gilpin, *Observations on the Coasts of Hampshire, Sussex and Kent, Relative chiefly to Picturesque Beauty, Made in 1774* (London, 1804).

112 This is why it is important to pause over the visual analysis of this coast, since Gilpin recognizes that it deserves special attention.

113 *Ibid.*, vol. 1, p. 207.

114 John Hassell, *A Tour of the Isle of Wight* (London, 1790).

115 Englefield, *Description*, preface.

116 *Ibid.*, p. 66.

117 On this subject, compare the travel style adopted by the companions of the anonymous author of *Tournée faite en 1788, passim.*

118 S. B. J. Noël de la Morinière, *Premier essai sur le département de la Seine-inférieure* (Rouen, 1795), 2 vols. In the neoclassical invocation that concludes this essay, the author speaks to the beaches and cliffs of the Saint-Valéry-en-Caux region. These found their place in the literature in the same year as the Armorican coasts described by Cambry:

Pardon me, solitary places, I have troubled the silence of tombs and the asylum of the dead (an allusion to young girls crushed beneath the cliffs); mossy terraces, covered with marine algae, fissured cliffs from which odourless thistles and wild stock hang, dwellings of the rasping crow, nature's silent rest, ancient of ages that saw this line of marly mountains arise which now form the borders of the vast basin in which the waters of the sea rock, riparian fish, mussels, nereides, polyps, vegetable-animals, you who seem to confound the series of levels in nature, always the same, always varied in its productions, I will long remember the diverse sensations that I have experienced on the banks of the Channel, long will the memory of the rocks of Életot drift into my thoughts when they are left to themselves to meditate upon the past (*Voyage*, p. 244).

119 On this subject, Saussure's way of travelling is revealing: he strives
to visit a great many viewpoints, in order to sum up his emotions
and cultivate the apprehension of these pleasures. For instance,
above the belvedere of Ben Lomond: 'I spent a good part of the
day at the top; I could not tire of contemplating this magnificent
view; I enjoyed turning my gaze toward the sites I already knew;
a thousand interesting memories were attached to the view of the
places I had been through; I also liked to observe those I was to
visit, and the view of the Hebrides, to which I was going gave me
new zeal and a premonition of a host of new pleasures' (*Voyage en
Écosse*, vol. 2, pp. 215–16).

120 See Pierre Goubert, *Jane Austen, étude psychologique de la roman-
cière* (Paris, P.U.F., 1975), pp. 75–82. The description of the environs of Lyme in *Persuasion* reveals Jane Austen's belated conversion.

121 And through the works of Dari, Baglivi, and Pascali.

122 Dr Pierre Thouvenel, *Traité sur le climat de l'Italie considéré sous
ses rapports physiques, météorologiques et médicaux* (Verona, 1797).

123 This vision endured even after Lavoisier's discoveries in chemistry.

124 Thouvenel, *Traité*, vol. 1, p. 175.

125 Ibid., p. 177

126 Ibid., pp. 175–6, including the citations that follow.

127 This is difficult for the modern observer to understand, as we are
primarily aware of the constant dryness of the summer season.

128 Thouvenel, *Traité*, pp. 195 and 200.

129 La Platière, *Lettres*, vol. 2, pp. 159 ff.

130 Thouvenel, *Traité*, vol. 1, p. 178, including the citation that
follows.

131 Ibid., p. 207.

132 This set of expressions is to be found in ibid., pp. 178–9.

133 See Bakhtin, *Esthétique*, p. 239.

134 G. Burnet, *Some Letters, Containing an Account of What Seemed
Most Remarkable in Switzerland, Italy, Some Parts of Germany etc.
in the Years 1685 and 1686* (Rotterdam, 1687), book 2, pp. 101–2.

135 Président de Brosses, *Journal*, vol. 1, p. 215 (2 Nov. 1739).

136 Choiseul-Gouffier, *Voyage pittoresque de la Grèce* (1782), vol. 1,
commentary on plate 8.

137 Creuzé de Lesser, member of the legislature, *Voyage en Italie et
en Sicile, fait en 1801 et 1802* (Paris, Didot, 1806), p. 161.

138 Charles-Victor de Bonstetten, *Voyage sur la scène des six derniers
livres de l'Énéide, suivi de quelques observations sur le Latium
moderne* (Geneva, Paschoud, 1805).

139 Lesser, *Voyage en Italie*, pp. 351 and 353.

140 Fielding, *Journal*, pp. 235–6.

141 Johann Georg Sulzer, *Journal d'un voyage fait en 1775 et 1776 dans
les pays méridionaux de l'Europe* (The Hague, Plaat, 1781), pp. 151,
161, and 228. This system of appreciation endured into the heart

of the nineteenth century. Honoré-Zénon Gensollen (*Essai historique, topographique et médical sur la ville d'Hyères* (Paris, 1820), p. 54) praises the walks around Hyères, notes that travellers admire the orange-trees, the natural fragrances, and the cheerful groves, and enjoy the gentle climate; but he never says anything about the possible beauty of the coast, and he does not even allude to the seascapes.

142 Aubin-Louis Millin, *Voyage dans les départements du Midi de la France* (Paris, Imprimerie impériale, 1807), vol. 2, pp. 384 and 435.

143 As an example, see Edward Rigby, *Dr Rigby's letters from France* (London, 1807). On the subject of Nice, he notes that one morning his party had begun to walk towards the seaside, but the heat soon became so great that they could not continue their walk.

144 Valenciennes, *Éléments*, p. 435.

145 On this point, see Tuzet, *La Sicile*.

146 Valenciennes, *Éléments*, pp. 417 and 629–30.

147 Ibid., pp. 417–18 and 427. The journey of tourist-photographers today respects many of Valenciennes' requirements.

148 Ibid., p. 494.

149 Goethe, *Italienische Reise*, pp. 232 and 253.

150 See Stafford, *Voyage into Substance*, p. 406.

151 Valenciennes, *Éléments*, pp. 272 and 273.

152 Ibid., p. 493

153 See Stafford, *Voyage into Substance*, pp. 331, 356, and 405–6.

154 About his point of view, see Chevallier, 'La Découverte', pp. 89–91.

155 Houel, *Voyage pittoresque*, vol. 1, p. 54 (the journey began on 16 Mar. 1776).

156 Ibid., vol. 4, pp. 87–8.

157 [Gilliatt is a character in Victor Hugo's novel *Les Travailleurs de la mer* who tries to salvage a boat that is wrecked off the coast of Guernsey – trans.]

158 [*L'Homme qui rit* is a novel by Victor Hugo in which, among other adventures, a tramp rescues a boy and a baby girl who had been abandoned on a lonely sea-shore – trans.]

159 Jacques Milbert reports that naturalists and artists got into the habit of sitting next to the porthole of the main cabin each evening and 'spending hours watching the gold and silver mass moving in all directions at the bottom of the waves, which proved to be particularly active when the night was especially dark and the waters choppy'. With the phosphorescence, it was possible then to watch the animals, especially the swordfish (Jacques Milbert, *Voyage à l'île de France*, vol. 1, pp. 111–12, cited by Stafford, *Voyage into Substance*, p. 332). Hartwig, *Guide médical*, pp. 31 ff., an inspiration to Michelet, speaks of the attraction this phenomenon represented for tourists in the Ostend region.

160 Forster, *Reise um die Welt*, p. 170. About this character, see Marita Gilli, *Georg Forster: l'oeuvre d'un penseur allemand réaliste et*

révolutionnaire, 1754–1794 (University of Lille III), (Paris, Champion, 1975).

161 La Platière, *Lettres*, vol. 2, pp. 310 ff.
162 Thouin, *Voyage dans la Belgique*, vol. 2, pp. 463–4.
163 Houel, *Voyage pittoresque*, vol. 1, p. 109.
164 Ibid., pp. 118–19.
165 La Platière, *Lettres*, vol. 3, pp. 7–10.
166 Lucien Febvre has emphasized the importance of studying the limits of the thinkable; the same goes for the limits of affects.
167 Bernardin de Saint-Pierre, *Harmonies de la Nature*, p. 194.
168 Ibid.

Notes to chapter 6

1 See Dorothy Wordsworth, *Journal of a Tour in the Isle of Man, 1828*, in *Journals of Dorothy Wordsworth* (London, Macmillan, 1952), vol. 2, pp. 400–19. A fascinating text which responds to that of Townley analysed above.
2 Brosse, *La Littérature*, vol. 1, p. 9.
3 It is evident that the sensibility of the Romantics where nature is concerned is a never ending topic; among the most recent syntheses, see Georges Gusdorf, *Les Sciences humaines et la pensée occidentale*, vol. 11, 'L'Homme romantique' (Paris, Payot, 1984), and Michel Le Bris, *Journal du Romantisme* (Geneva, Skira, 1981); these works contain rich bibliographies. We present here only a brief exposé in order to clarify the evolution of the practices.
4 Considered, above all, as a hothead or dare-devil.
5 Immense bibliography, on this subject also; see the works by Pierre Miquel and David Cordingly mentioned below.
6 This is also true of Turner, of Bonington, and of Constable, as well as of Eugène Isabey and Paul Huet.
7 Cordingly, *Marine Painting*, pp. 117–18, regarding Turner.
8 Jacques Carré, 'Couleur et paysage dans la peinture romantique anglaise', in *L'Imaginaire du romantisme anglais. Romantisme*, 49 (1985), p. 103.
9 See the analysis of the structure of this scene in Hubert Damisch, *La Théorie du nuage* (Paris, Le Seuil, 1972), pp. 257 ff.: 'Les Merveilleux Nuages'.
10 At the same time, to the Romantics, the ocean's boundless waters constitute an open space, a symbol of freedom. In this regard, the Atlantic becomes the absolute ocean, as well as the English Channel which prolongs it, though it is a smaller, more intimate space, rendered all the more fascinating by the fact that one can easily roam about its opposite shores; see Brosse, *La Littérature*, vol. 1, pp. 174 and 177; see also *idem*, 'Byron et la mer', *Romantisme*, 7 (1974), p. 66. Regarding the English Channel, Turner planned for a while to execute a joint study of the two

opposite shores, project of the *English Channel or Manche*; towards that end that, around 1829–32, he studied the coasts of Picardy and Normandy. See nos 195 ff. of the catalogue of the J. M. W. Turner exhibition, National Galleries of the Grand Palais, 14 Oct. 1983–10 Jan. 1984 (Éditions des Musées nationaux, 1983).

11 See on this subject, pages of Mircea Eliade and of Gilbert Durand, *Les Structures anthropologiques*, p. 261.

12 Between the years 1801 and 1805, the artist adds a number of studies of breaking waves to his sketch-book; see Luke Herrmann, 'Turner and the Sea', *Turner Studies*, 1, no. 1 (1981), pp. 6–9, and more generally, Andrew Wilton, *Turner and the Sublime* (British Museum Publications, 1980), esp. pp. 37, 39, 46. Ansen (*Enchafed Flood*, pp. 17–20) underscores how important the absence of historical change in the sea and the desert was to the Romantics.

13 In this way, Turner establishes as essential the manner of portraying the air, the wind, the cloud; see Wilton, *Turner*, p. 46, and pp. 78 ff. concerning the dynamical perspective.

14 Jean-Pierre Richard, *Paysage de Chateaubriand* (Paris, Le Seuil, 1967), p. 64.

15 See Jean Perrin, *Les Structures de l'imaginaire shelleyen* (Grenoble, P.U.G., 1973), p. 49.

16 Richard, *Paysage*, p. 65. Shelley, like Chateaubriand, never failed to mention the quality of the wind.

17 On this subject, see the work of Turner analysed by Wilton, *Turner, passim*.

18 Like other Romantic painters, he is probably trying to demonstrate, by so doing, the frailty of visual perception, and trying also to break away from the optimism induced by Newtonian perspective.

19 Impression probably even clearer in his paintings of Brighton made in 1824, see Cordingly, *Marine Painting*, p. 123.

20 After 1796, remarks Wilton (*Turner*, p. 46), Turner 'explores methodically the possibilities of representing the sea and the interrelations between the sea and man'.

21 Perrin, *Les Structures*, p. 173.

22 Chateaubriand, *Les Natchez* (Paris, Ladvocat, 1826), vol. 1, p. 134.

23 A pietist, Friedrich was subject to the influence of Kant, Schiller, and Boehme. He was an admirer of Tieck, Novalis, and the brothers Schlegel, as well as of Herder and Klopstock. In short, he was situated at the confluence of a number of German cultural currents.

24 In the fashion of the *Sturmers* of 1770, he emphatically prefers the hours of nightfall which favour introspection and which impose the chthonian presence of the dark water, associated with the cold femininity of the moonlight, on anyone walking the beach. One is reminded, by this subject, of the 'Stanzas to the Moon', composed by Wordsworth on the shores of Cumberland; see Douady, *La Mer*, p. 218.

25 See plate 19.

26 On this subject, in French, Marcel Brion, 'Caspar David Friedrich, inventeur du paysage tragique', in *Caspar David Friedrich, le tracé et la transparence* (Paris, Centre culturel du Marais, 1983).

27 Obviously, as an example, in Lamartine's poetry, as demonstrated by Van Tieghem, *Ossian en France*, vol. 2, pp. 313–28. The poet confesses that he owes Ossian the melancholy of his brushes, which is the sadness of the sea itself.

28 Durand, *Les Structures anthropologiques*, pp. 103–22.

29 See the theme of underwater caves in Shelley's poetry as analysed by Perrin, *Les Structures*, pp. 111–12.

30 C. G. Jung, *Symbols of Transformation*, trans. R. F. C. Hull (Routledge, 1956).

31 Durand, *Les Structures anthropologiques*, p. 227.

32 Thus well before Victor Hugo's sea/mother (mer/mère in French) association and sea-water fascination, analysed by Charles Baudouin, *Psychanalyse de Victor Hugo* (Paris, A. Colin, 1972; 1st edn, 1943), pp. 180–1.

33 Christian Lacassagnère, 'Image picturale et image littéraire dans le nocturne romantique. Essai de poétique inter-textuelle', in *L'Imaginaire du romantisme anglais*, pp. 57–9.

34 Durand, *Les Structures anthropologiques*, p. 256.

35 e.g., think of the *Narrative of Arthur Gordon Pym of Nantucket* by Poe or of the storm that assails the boat in *L'Homme qui rit* by Victor Hugo.

36 See Brosse and her analysis of the progressive enrichment of the motive of the abyss, *La Littérature*, vol. 1, p. 286.

37 See Jung, *Symbols of Transformation*.

38 Perrin, *Les Structures*, p. 205. Hélène Lemaître also writes that in Shelley's world the sea and the ocean symbolize simultaneously the human crowd and the universal unconscious (*Shelley, poète des éléments* (Caen, Caron, 1962), pp. 65–80: 'Sea and Ocean').

39 Jean Bousquet, *Les Thèmes du rêve dans la littérature romantique (France, Angleterre, Allemagne)* (Paris, Didier, 1964), esp. pp. 95–6 and 148.

40 Emphasized in the English language by the resemblance between the words 'dream' and 'stream'; see Perrin, *Les Structures*, p. 116.

41 Heinrich Heine, *Die Nordsee*, no. 3 (1826), in *Reisebilder* (Berlin and Weimar, 1979), pp. 86 and 90. Same tone in poem no. 10 of the first cycle of the *Nordsee*. On this subject and on the 'pathetic fallacy' in Heine's texts, see Jeffrey L. Sammons, *Heinrich Heine, a Modern Biography* (Princeton University Press, 1979), p. 117.

42 'To the Sea'.

43 Novalis, *Disciples at Saïs*, in *Pollen and Fragments: Selected Poetry and Prose of Novalis*, trans. A. Versluis (Phanes, 1990).

44 Perrin, *Les Structures*, p. 30.

45 Richard, *Paysage*, pp. 112–13.
46 Custine, *Mémoires*, pp. 348–9.
47 Richard, *Paysage*, p. 114; who showed the importance of navigation in Chateaubriand's work.
48 Jules Michelet, *Journal* (Paris, Gallimard, 1959), vol. 1, Fri., 7 Aug. 1831, p. 83.
49 Nodier, *Promenade de Dieppe*, p. 18. Victor Hugo, letter to Adèle, Saint-Malo, 25 June 1836, in *Oeuvres complètes* (Club français du Livre, 1967, vol. 5, p. 1093.
50 [Or 'la mer/mère', the 'sea/mother' in the French original – trans.]
51 See the ambiguity about the femininity of the beach in Shelley's 'West Wind'.
52 Let us underline, on this subject, the hiatus which separates not only the libertine and *galant* works, but also the literature of the eighteenth century, which is not afraid to mention coitus in nature (see *Dolbreuse* of Loaisel de Tréogaste), and the texts of the novelists of the early nineteenth century who refuse to talk of such a thing.
53 E.g. in the novella of Émile Souvestre, 'Le Traîneur de grèves', in *Scènes des rives et des côtes* (Brussels, Lebègue, 1852), vol. 1, pp. 5–81; see also the texts which describe the legend of the 'Chambre d'amour' [room of love] at Biarritz. Near Piriac, says Edouard Richer, are natural water pools where women bathe nude; the inhabitants of the area named the most attended of the ponds the 'Trou du moine fou' [Crazy monk's hole].
54 See plate 25, taken from 'La Normandie', vol. 1, in the *Voyages pittoresques* by Baron Taylor and Charles Nodier.
55 See the death of Virginia or that of 'La Jeune Tarentine' by André Chenier.
56 See Brosse, 'Byron et la mer', p. 62.
57 See Pierre Miquel, *Eugène Isabey, 1803–1886, la marine au XIXe siècle*, ed. de la Martinelle (Maurs-la-Jolie, 1980), vol. 2, pp. 21–3.
58 This was particularly the case with Turner. This fashion of painting the shore led to a proliferation of students in the studios of the marine painters Gudin and Garneray, to an increase in that type of subject in the Salon between 1823 and 1836, as well as to the success of the Navalorama installed by Louis Gamain in Le Havre (1830) and then on the Champs-Elysées (1838); see Miquel, *Eugène Isabey*, vol. 2, p. 30.
    In France, this increase in paintings of coastlines was in response to the spectacle of a marine fleet decimated, for a time, by the Revolution and the War of the Empire and of which only a few coasting vessels were left; during the Restoration it was a way of extolling the end of the *ancien régime*, particularly happy on the sea. Lastly, it corresponds to a will to idealize the misery of the people; we will come back to it later.

59 Cordingly, *Marine Painting*, p. 96. The practice was not new to England, but it was becoming more widespread. In 1732, Samuel Scott accomplished a trip around the island of Sheppey. For five days, William Hogarth, accompanying him that year on a little sailboat, had bitten into the region of Gravesend and Sheerness. Charles Brooking also inspected the coastline around the island of Sheppey and along the coast of Kent in 1752 (see ibid., pp. 76 and 81. In 1803, Constable travelled along the coast from London to Deal. In 1814, 1816, 1824, and 1825 he alternated between the coastal excursion and the seaside resorts required for the health of his wife. The spectacle of the beach at Margate was familiar to Turner as of the age of thirteen (see Herrmann, 'Turner and the Sea', p. 5). At the end of the 1780s, he was already knowledgeable about the estuaries of the south-east and south of Great Britain. In 1813, he made a long coastal trip. Other trips inspired the *Picturesque Views of the Southern Coast of England* and then, in 1811, his *Views of Sussex*, while waiting for the coastal trip which he took around Brittany in 1826 (see Delouche, *Les Peintres de la Bretagne*, vol. 1, pp. 93 ff.). In 1807–8, John Sell Cottman sailed with his family along the coast, reconciling his artist's work with his desire for the sea. (On all these points also, see Cordingly, *Marine Painting*, passim. On William Daniell's trip mentioned below, see pp. 97–9.

60 It would be appropriate, on this subject, to ponder the exaltation by the Romantics of row-boat trips, a practice already well anchored in the seaside resort ritual. 'Promise of a test', path of heroization (see Brosse, *La Littérature*, vol. 1, p. 274, and Durand, *Les Structures anthropologiques*, pp. 286–7), this practice actually constitutes more a theme of lacustrine poetry. The 'nacelle', a substitute for the crib, womb-like haven, or miniature island, gives a maximum of pleasure on circumscribed bodies of water.

61 See above, p. 90. Torrington's experiment.

62 On the correspondence between the horse and the piano in Romantic music, see Danièle Pistone's beautiful pages, *Le Piano dans la littérature française des origines jusque vers 1900* (Lille III), (Paris, Champion, 1975).

63 On this subject, see Douady, *La Mer*, p. 221.

64 Hermann Fürst von Pückler-Muskau, *Briefe eines Verstorbenen* (reprint, New York and London, 1968, of Stuttgart, 1836); extracted from a letter of 10 Feb. 1827; vol. 1, p. 354. Other examples of such rides can be found in this work, esp. in a letter of 22 Sept. 1827. In the region surrounding Scarborough, the foreshore is 'soft like velvet'. (See also vol. 3.) The Prince represents only one of many proponents of horse rides on the beach.

65 The best example: Honoré de Balzac, *Un drame au bord de la mer* (1834), inspired by a 1830 trip to Guérande.

66 On this extension which includes the beach as well as the heath in the catalogue of wild beauties, see Yves Luginbuhl. 'Paysage sauvage, paysage cultive. L'ordre social de l'harmonie des paysages', in *La Nature et le rural*, colloquium of the Association des ruralistes français, Strasbourg, no page numbers.

67 Chateaubriand, *Mémoires d'Outre-Tombe* (Paris, Flammarion, 1949), vol. 1, p. 61. On this subject, Paul Viallaneix, 'Chateaubriand voyageur', in *Approches des Lumières, Mélanges offerts à Jean Fabre* (Paris, Klincksieck, 1974), p. 567.

68 Van Tieghem, *Ossian en France*, vol. 2, p. 194.

69 A. de Chateaubriand, *René*, p. 187, and *Le Génie du christianisme*, part I, book 6, ch. 6, p. 311, of Ladvocat's edn (1826).

70 'Hellebek' (1776).

71 It should be emphasized specifically, on this subject, that we are talking about poems and not a private diary.

72 It is interesting to compare this impression with that of inside emptiness and reverie aroused by the auditive sensation produced by the lapping of the waters of the lake of Bienne in Rousseau's works.

73 Concerning him, see Douady, *La Mer*, p. 226, and esp. Brosse, *La Littérature*, vol. 1, pp. 35–46.

74 Reminiscence of the resounding sea in Homer's work.

75 See *Childe Harold*, canto 67; see Brosse, 'Byron et la mer', p. 61, on the new familiarity between the poet and the sea.

76 Pierre Arnaud, *Ann Radcliffe et le fantastique, Essai de psycho-biographie* (Paris, Aubier, 1976), pp. 268 and 311. The novelist and her husband, on 23 July started on a trip to Eastbourne, planning to pass through Beachy Head. Exhausted by the heat and the effort, Ann sat down on the beach, while her husband walked on. The young woman started to feel anxiety, then fear, take hold of her mind: 'The tide was low; there was nothing in front of me but the sea; the white cliffs rose above me ...; the beach all around [was nothing but] a chaos of rocks and land slides from the cliffs, moving far into the waves; the sea-gulls flew around screeching.' Throughout the entire journey to Bexhill, then Hastings, Ann tried to stay off the beaten path. She adored long walks on the beach. In October 1822, at Ramsgate, she wrote: 'I walked twelve miles back and forth to Broadstairs, almost without sitting, but felt tired because of lack of sleep.'

    Pierre Arnaud quotes in particular Talfourd (Sir Thomas N.), 'Memoir of the Life and Writings of Mrs Radcliffe', in *Gaston Blondeville*, 4 vols (London, H. Colburn, 1826), vol. 1, pp. 41–2.

77 Custine, *Mémoires*, p. 300.

78 Delouche, *Les Peintres de la Bretagne*, vol. 1, pp. 14–17.

79 In 1806, the painter Turpin displayed at the Salon René wandering on the beach at night.

80 It is interesting to note that, on this subject, Delouche and Catherine Bertho ('L'Invention de la Bretagne. Genèse sociale d'un stéréotype', *Actes de la recherche en sciences sociales*, no. 35 (Nov. 1980): 'L'Identité' do not agree on the chronology, since the latter puts the emergence of the image of the Arcadia of Brittany as 1830.

81 Ducos, *Itinéraire*, vol. 2, pp. 327–8. He evokes 'the noise of the waters that were taken to be the grumbling of an irritated people' (10 July 1826); during the Restoration, once again, the perception of the sea is subject to the images of the revolutionary crowd. On this late correspondence between the crowd and the sea in Victor Hugo's work, see Walter Benjamin, *Charles Baudelaire. A Lyric Poet in the Era of High Capitalism* (New Left Books, 1977). This theme has been re-evaluated by all specialists in Victor Hugo since the first publication of this book in 1955.

82 Indeed, think about the endless journeys of the young Gavarni in the Pyrenees.

83 See Miquel, *Eugène Isabey*, vol. 1, *passim* (esp. pp. 33–4). In 1831, Eugène Le Poittevin had a studio built on the cliff of Étretat, so as to be able to see the sea at all moments of the day.

84 Paul Huet (1803–69), according to his notes, his correspondence, and his contemporaries, published by René-Paul Huet (Paris, Laurens, 1911), *passim*. On 10 Aug. 1828, he wrote to his sister, (p. 105): 'I'm awaiting Jadin at Fécamp to return to Paris on foot with him ... Here I am, then, trotting, a cardboard folder under my arm, my pockets full of pencils.' These habits are in fact the doing of a group of friends united by the fraternity displayed at the Saint-Siméon farm. This reunion of young artists, supported by the network established around Queen Hortensia, around Gérard, Gudin, and Nodier, contributed to the popularity of sea-shores in the cultivated milieux of the capital. These painters produced so many beach scenes that by 1824 Thiers considered them to be a worn-out cliché (see Miquel, *Eugène Isabey*, vol. 1).

85 In Richer, *Voyage pittoresque*, letter 7, pp. 112–13.

86 Here are the principal ones: 8 Aug. 1834, at Brest, he wets his feet in the ocean; 6 Aug. 1835, stay at Tréport, on the beach. Walk on the cliff. Then he follows the seaside from Tréport to Le Havre; ten-mile walk at Étretat, then from Étretat to Montivilliers, at low tide, through the slippery piles of seaweed, the puddles of water, the large shingles. June 1836, trip in Normandy and Brittany accompanied by Juliette Drouet and Célestin Nanteuil. He walks fifteen miles between Dol and Saint-Malo. He swims in the sea, and describes the way he confronts the waves. 'As I've walked sixty or so miles in the sun over the past four days, end to end, my face is peeling, I am red and horrible' (letter to Adèle, 25 June 1836, ibid., p. 1094). 17 July at Saint-Valéry-en-Caux, he says he spent eight hours contemplating the angry sea.

Aug.–Sept. 1837: trip to Northern France and to Belgium. Victor Hugo wanders for two full hours along the sea at Ostend. 'It is while wandering in the dunes that one feels the profound harmony tying the earth to the sea' (ibid., p. 1292). (See Claude Gély, ' Notices sur les voyages de 1834 et 1835', ibid., vol. 5, pp. 1048–50.)

87 See Brosse, *La Littérature*, vol. 1, pp. 164–7.

88 See the lines dedicated to this place by Nodier in *La Normandie*. See also Victor Hugo's excursion mentioned above, 17 July 1836.

In Scotland, the rocks of Dunbar provide this dominating position. Simond (*Voyage d'un Français*, vol. 2, pp. 66–8), in Feb. 1811, spent a whole day enjoying the horrors of the sea from the viewpoint of the 'magnificent ruin of the rocks'. It is true that he is more attentive to the sublime spectacle than to the confrontation of the elements. Note that he names this section of chapters 'Rocks and shores'.

89 A later literary example: the visit to the lighthouse in Fromentin's *Dominique*.

90 Prince Pückler-Muskau, *Briefe*, vol. 3, p. 92 (July 1828). The author is nourished by the work of Byron, which he takes with him on trips.

91 See Durand, *Les Structures anthropologiques*, pp. 266 and 289.

92 Analysed in ibid., p. 264.

93 On this subject, see Richard, *Paysage*, p. 33.

94 See Shelley, *The Revolt of Islam*, first canto, xx.

95 Balzac, *Un drame*, in *L'Oeuvre de Balzac* (Paris, Le Club français du Livre), vol. 5, p. 686. It is interesting to note that the repulsion still caused by the brilliance of sunlight definitively separates this eroticism from the dominant coenaesthetic impressions of our twentieth century. This noon, notes the Pauline described by Balzac, 'throws a devouring colour on these three expressions of the infinite [the shining yellow of the sand, the azure of the sky, and the unified green of the sea] ... I see in it despair, responds the narrator, ... I looked at Pauline and asked her if she felt she had the courage to bear the pounding brilliance of the sun and the strength to walk on the sand – I have little boots, let's go, she said to me' (ibid., pp. 693 and 694).

96 'To go towards the joy of your waves and dive into you and refresh myself, revel in you, strengthens me' (Pour aller vers la joie de tes vagues et me plonge en toi et me rafraîchis, me délecte, me renforce) ('A la mer', 1777).

97 'Les Mers', 1777.

98 Cited by Robert Escarpit, *Lord Byron, un tempérament littéraire* (Paris, Cercle du Livre, 1955), vol. 1, pp. 154–5.

99 See Brosse, 'Byron et la mer', esp. pp. 60 and 62. In 1818, in *Childe Harold*, the poet unites the two Romantic models of bathing: 'From a boy / I wantoned with thy breakers – they to me were a

delight – & if the freshening sea made thou a terror – 'twas a pleasing fear – For I was as it were a child of thee / And trusted to thy waters far & near / And laid my hand upon thy Mane – as I do here' (*Childe Harold's Pilgrimage* (New York, Garland Publishing, 1991), p. 329, canto 4).

100 Brosse, *La Littérature*, vol. 1, pp. 406 ff.

101 In this perspective, Stolberg's appreciation of straits is associated with the fascination with insularity.

102 Richer, *Voyages pittoresques*, p. 108.

103 See Jones quoted by Durand, *Les Structures anthropologiques*, p. 274; see Moles and Rohmer, *Labyrinthes* pp. 63–4.

104 From this perspective, the hiding of Eugénie de Guérin in her poor Cayla represents more this feeling than the classic desire for withdrawal.

105 Durkheim emphasized the weight of these phenomena in anomic suicide, a form of withdrawal, like the private diary or the Crusoe-like adventure, from the new realities of the social struggle.

106 Which is found in Baculard d'Arnaud's *Makin* (see below, and Brosse, 'Byron et la mer', p. 63). In practice, the Romantics give a new meaning to the classic voyage to the island of Capri.

107 Marcel Brion, *Casper David Friedrich*, p. 110. On these marine painters (Dahl, Carus, Runge), see Hans Jürgen Hansen, *Deutsche Marinemalerei* (Oldenburg, Stalling, 1977).

108 See Sammons, *Heinrich Heine*, p. 116. The author emphasizes the homology between the *Nordsee* verses and the Homeric periods which record the majestic movements of the sea. He considers that this is the first major work of German poetry dedicated to the sea, which is debatable. He bases his argument on this subject on Gerhard Hoppe, *Das Meer in der deutschen Dichtung von Friedrich L. Graf zu Stolberg bis Heinrich Heine* (dissertation, Marburg, 1929).

109 Heinrich Heine, *Die Nordsee*, p. 90.

110 Often, it is in fact a composite practice, as shown by the Lisztian pilgrimage.

111 'Travel is ordinarily simply a commentary on what I felt while crossing the border', confides Custine (*Mémoires*, p. 84). A feeling which will lead to the uselessness of travel, see des Esseintes' trip to England (Huysmans, *A rebours*), which boils down to a departure.

112 See Nodier, *Promenade de Dieppe*, p. 329.

113 Note, nevertheless, that in practice, most travel accounts reveal themselves to be composite. The authors change intentions, and change appreciation systems at several different points in the works.

114 This explains why historians are blinded when they take these works as paintings of reality.

115 Nodier, *Promenade de Dieppe*, preface and p. 21.

116 Ibid., p. 149.

117 Custine, *Mémoires*, p. 75.
118 Ibid., p. 80.
119 Guillet, *Curiosité*, *passim*.
120 Who as early as 1833 published an interesting article on Cornwall in the *Revue des deux mondes* and in 1836, *Le Finistère*, not forgetting his fiction set in Brittany.

Notes to chapter 7

1 It seems that classical Greece did not establish the harbour before Alexander's reign as a subject of landscape painting, despite the existence of a very old marine painting tradition, in Mycenae, e.g.; see C. Picard, 'Pouzzoles et le paysage portuaire', *Latomus*, 18 (1959), pp. 23–51. On the following, see also Reddé, *Mare nostrum. Les infrastructures, le dispositif et l'histoire de la marine militaire sous l'Empire romain* (École française de Rome, 1986), esp. part 2.
2 Chevalier de Jaucourt, *Encyclopédie*, vol. 13, pp. 129–31.
3 See Cordingly, *Marine Painting*, p. 53.
4 A confluence already praised by Descartes in a letter to Guez de Balzac. (Chauveau, 'La Mer', p. 128) and by Saint-Amant who calls Amsterdam the 'miracle of the world' (Bailbé, 'Les Paysages').
5 Nevertheless it is true that the behaviour and the postures of Vernet's characters apparently became models and stereotypes for later tourists.
6 See *Le Port de Marseille* in the Louvre. On this subject, Léon Lagrange, *Joseph Vernet et la peinture au XVIIIᵉ siècle* (Paris, Didier, 1863–4), pp. 69 ff.
7 In particular, the fame of the views of Rouen, of Plymouth, of Dover, and of the shores of the Sund.
8 Particularly evident in the *Port de la Rochelle*.
9 Lagrange, *Joseph Vernet*, pp. 193–4.
10 On this subject and particularly concerning the canvases dedicated to the harbour of Bayonne; see our communication to the annual colloquium of the French Historical Society, Los Angeles, Mar. 1985.
11 Since the Middle Ages (see Michel Mollat du Jourdin, 'Sentiments et pratiques religieuses des gens de mer en France, du XIIIᵉ au XVIᵉ siècle', in *Revue d'histoire de l'Église de France*, July–Dec. 1984, p. 306), the corruption of the place was denounced by clerics. On the violence, see Alain Cabantous, *Les Populations maritimes françaises de la mer du Nord et de la Manche orientale (vers 1660–1794)* (thesis, Lille III, Dec. 1987), p. 693. We were unfortunately unable to use this fine work with which we became acquainted when correcting the proofs; but it generally confirms our theory.
12 Alain Cabantous (*La Mer et les hommes. Pêcheurs et matelots dunkerquois de Louis XV à la Révolution* (Dunkirk, Westhoek,

1980), p. 71) emphasizes the number of foreigners in the harbours of the English Channel and of the North Sea, in particular the importance of the migrations of sailors towards London and the harbours of the United Provinces.

13 Emphasized by Vice-Admiral Antoine Thévenard in the year 1800. (*Mémoires relatifs à la marine* (Paris, Laurens), vol. 2, p. 58).

14 Quoted by Lagrange, *Joseph Vernet*, pp. 76–7.

15 Jean-Michel Deveau, 'Le Port de La Rochelle au XVIIIᵉ siècle', in Comité des Travaux Historiques et Scientifiques, *111ᵉ Congrès des Sociétés savantes* (Poitiers, 1986).

16 Up until this date, notes Delouche, harbour painting perpetuated the essential characteristics of Vernet's work. On this subject, see *Les Peintres de la Bretagne*, vol. 1, pp. 105–9.

17 Since the collection is composed of 60 plates, not counting 16 others dating from after 1786.

18 Already, the lighthouse of Cordouan, built around 1600 at the mouth of the Gironde, symbolized the 'lights' of the royalty fighting against the storms (see J. Guillaume, 'Le Phare de Cordouan, merveille du monde et monument monarchique'. *Revue de l'art*, 8 (1970), pp. 33–52); and the goal of the painting of the harbour of Misenum, which is displayed at Versailles in the Apollo salon, is to extol the construction of the harbour of Rochefort. (See Edouard Pommier, 'Versailles, l'image du souverain', in *Les Lieux de mémoire*, vol. 2, p. 197.)

19 On this subject, the bombarding of Copenhagen by an English squadron in 1807 had important repercussions.

20 An ambiguity particularly clear under the Revolution.

21 See Florence Ingersoll-Smouse, *Joseph Vernet, peintre de marine. Étude critique et catalogue raisonné* (Paris, Étienne Bignou, 1926), vol. 1, p. 22.

22 *Mémoires du duc de Croÿ sur les cours de Louis XV et Louis XVI (1727–1784), Extraits de la nouvelle revue rétrospective*, 1895–1896, p. 146.

23 Montesquieu, *Oeuvres complètes*, vol. 2, pp. 1059, 1062, and 1073–5.

24 Président de Brosses, *Journal*, vol. 1, p. 32. 'The quay of the harbour, which is paved with bricks, to facilitate walking, is perpetually covered with all sorts of people, all sorts of nations and all sorts of sexes.' Grosley, *Nouveaux mémoires ou observations sur l'Italie et sur les Italiens par deux gentilshommes suédois* (London, 1764), vol. 2, p. 158, and Marmontel, *Mémoires*, p. 179.

25 Marmontel, *Mémoires*, vol. 2, p. 179.

26 Sulzer, *Journal*, p. 111.

27 Dupaty, *Lettres*, vol. 1, p. 38.

28 Bérenger, *Les Soirées*, vol. 1, p. 92.

29 Alexandre Grimod de la Reynière, *Lettre d'un voyageur à son ami*

ou réflexions philosophiques sur la ville de Marseille, 2nd edn (Geneva, 1792), p. 10. Thomas Pennant, *A Journey from London to the Isle of Wight*, (Edward Harding, 1801); quoted heavily by Margaret J. Hoad in *Portsmouth, as Others Have Seen It*, part 2, 1790–1900 (Portsmouth, Arthur Coomer, 1973), pp. 4–5.

30 Thouin, *Voyage dans la Belgique*, vol. 1, pp. 328–9.

31 Millin, *Voyage*, vol. 3, pp. 249–50.

32 See J. Gury, 'Images du port', in *La Mer au siècle des Encyclopédies*, p. 56.

33 Pilati, *Voyage de la Hollande*, vol. 1, pp. 213 ff.

34 [According to the Republican calendar set up by the French Revolution (and abolished by Napoleon in 1805), 1793 was Year I. Thus, Years VII and VIII are 1799 and 1800 – trans.]

35 Citizen Barbault-Royer, *Voyage dans les départements du Nord, de la Lys, de l'Escaut, etc., pendant les années VII et VIII* (Paris, Lepetit, 1800), pp. 158, 187, and 189. The desolation of the English trading ports was then much less in evidence, if we are to believe François Crouzet. It would be appropriate in particular, according to this historian, to correct what contemporaries continually say about the disastrous effects of the Continental blockade. See François Crouzet, *L'Économie britannique et le blocus continental (1806–1813)* (Paris, P.U.F., 1958), vol. 2, pp. 764–6. The brutal decline in activity during the 1811 recession was mostly a consequence of foreign ships, and if Liverpool and Hull were hit rather hard, this was not the case for other big ports.

36 Delouche, *Les Peintres de la Bretagne*, vol. 1, p. 52.

37 Victor Hugo, *Oeuvres complètes*, vol. 5, p. 1108.

Notes to chapter 8

1 Here, the author is talking almost exclusively about the French coasts; on the other side of the Channel, the spectacle of the shore, too quickly disturbed by seaside resort practices, does not make for as easy an account of the evolution of the attention paid to the people of the strands.

2 Concerning this notion, see Jacques Revel, 'Une France sauvage', in Michel de Certeau, Dominique Julia, and Jacques Revel, *Une politique de la langue. La Révolution française et les patois* (Gallimard, 1975), p. 49.

3 The work, in particular, of J. D. Cassini and his son, of Father Feuillée, of Lahire, and of Laval. As an example, J. M. Homet, 'Les Astronomes et la découverte du littoral méditerranéen', in *La Découverte de la France au XVIIe siècle*, pp 319–27.

4 See Meyer, Quelques directions', and Cabantous, *Les Populations maritimes*, pp. 51 ff. As Madeleine Pinault ('Diderot et les enquêtes de Le Masson du Parc', in *La Mer au siècle des Encyclopédies*, pp. 344–5) emphasizes, among the goals fixed by Le Masson du

Parc during the extensive investigation he led, is the wish to portray the role of fishermen on foot.

5 François Tiphaigne, *Essai sur l'histoire économique des mers occidentales de France* (Paris, Bauche, 1760), p. 117.

6 Documents collected by Duhamel du Monceau for the preparation of his work; in particular, accounts of the investigation made in execution of a decree issued 26 May 1720 and of that made by Le Masson du Parc (1729); Archives nationales 127 AP 21: answers to the questionnaires concerning Brest (1720), Oléron (1717), and the coast of Poitou (n.d.).

7 John Knox, *A Tour through the Highlands of Scotland and the Hebrides Isles, in 1786* (J. Walter, esq., London, 1787).

8 Johann Christian Fabricius, *Reise nach Norwegen, mit Bemerkungen aus der Naturhistorie einer Ökonomie* (Hamburg, 1779).

9 Geneviève Delbos, 'De la Nature des uns et des autres. À propos du dépeuplement des eaux', in *La Nature et le rural*, Strasbourg colloquium, no page numbers.

10 Tiphaigne, *Essai*, p. iv.

11 Menc, *Memoir Composed by the Reverend Father Menc, Dominican ... Quelles sont les causes de la diminution de la pêche sur les côtes de Provence* (Marseilles, Sibié, 1769).

12 See La Morinière, *Premier Essai*, pp. vi–vii. Only the first volume of the treaty was published, in 1815.

13 Tiphaigne expresses that opinion, *Essai*, pp. 117–18.

14 Fabricius, *Reise nach Norwegen*, p. 285.

15 This is what comes out of the reading of the previously mentioned file in the Archives Nationales.

16 Concerning the aggressiveness of the glaciers at the beginning of the eighteenth century and the great glacier tides of the following decades, see Emmanuel Le Roy Ladurie, *Le Climat depuis l'an mille* (Paris, Flammarion, Collection 'Champs', 1983), vol. 1, pp. 245 ff.

17 Tiphaigne, *Essai*, p. 121.

18 Menc, *Memoir*, pp. 6 and 37.

19 A whole series of regulations had been implemented on the coast of the English Channel in order to ensure the furnishing of the King's table.

20 Menc, *Memoir*, p. 23.

21 It is necessary here to add a few details concerning the French case: since 1544, by royal ordinance, the coasts were part of the royal domain, untransferable and indefeasible. The 1681 Colbert ordinance, which reaffirms these dispositions, annuls the attributions prior to the 1544 edict. Nevertheless, short of being able to hold fast to this principle, the royal power grants and concedes fishing privileges, while multiplying regulations. Natural shoals, in particular oyster-beds, were part of the royal domain since the Edict of Moulins in 1566, confirmed by an edict in March 1584 and

then by the previously mentioned ordinance of 1681. Usually, the fishing products were reserved for the sailors of the Royal Navy. But it was necessary very early to regulate their use, see Delbos, 'De la Nature'.

Already, an edict of Henry III in March 1584 (art. 83) prohibited fishing with dragging nets (tow-nets), and tolerated the drag-net only for the service of the King's table. The August 1681 ordinance concerning the merchant marine renewed the ban on the tow-net and the drag-net. It remained unenforced. Following thorough investigations at the beginning of the 1720s, the declarations of 22 Apr. 1726 and of 18 Mar. 1727 extended the prohibitions. They were detailed by the declaration of 10 Dec. 1729 and by the ordinances of 16 Apr. and 31 Oct. 1744 which focused on the protection of the breeding season. Nevertheless, as in many other domains, the instructions were not followed. On the local level, the regulations were often extremely detailed. A decree issued by the Rennes Parliament on 17 Oct. 1775 specifically regulated the gathering of oysters in the Tréguier region. The warden of the oyster-beds who, as in other domains, seemingly based his policies on prior customs emanating from the community of fishermen, codified the places, the calendar, the techniques, and the gestures of the operation (see Michelle Salitot, 'Formes de l'activité huîtrière à Cancale depuis le XVIIIᵉ siècle', in *La Nature et le rural*). In England, this warden answered to the local authorities.

Despite the theoretical impossibility of transferring the maritime domain, a process of monopolization set in as the decades passed; thus, in addition to the development of all sorts of fishing, during the second half of the eighteenth century in the region of Cancale a private oyster-breeding industry appeared slowly in the 'étalages' [shop windows], stores destined to regulate sales (see ibid.).

22 Delbos, 'De la Nature'.

23 Menc, *Memoir*, pp. 7–9.

24 A practice about which the drawings found in Le Masson du Parc's investigation already provide information.

25 Denounced in a memoir of M. de Gasville, Administrator of Rouen, dated from 21 Aug. 1720 (Archives nationales, 127 AP 21).

26 Tillet, *Observations faites par ordre du Roi sur les côtes de Normandie au sujet des effets pernicieux qui sont attribués, dans le pays de Caux, à la fumée du varech* (1771). As well as for the details that follow.

27 A similar, but later, debate developed on this subject along the coast of Norway, around the year 1802; see Fabricius, *Reise nach Norwegen*.

28 See Jean-Claude Perrot, *L'Âge d'or de la statistique régionale française. An IV–1804* (Paris, Société des études robespierristes, 1977). Already, Le Masson du Parc had visited in detail, between

1723 and 1737, the shores of Picardy, Normandy, and the Atlantic coast to Guienne, on horseback, accompanied by a guide and a local administrator, visiting each village, sometimes house by house. See Madeleine Pinault, 'Diderot'; A. Cabantous mentions the activity of inspector Sicard.

29 Tillet, *Observations*, p. 20.

30 See our introduction to Parent-Duchâtelet, *La Prostitution à Paris au XIXe siècle* (Paris, Le Seuil, 1981).

31 Tillet, *Observations*, pp. 25 and 26.

32 From Le Masson du Parc to Tillet or Knox. Sent in 1786 by the British Society to increase the fishing and to better the coasts of the kingdom, to study fishing and to develop activity there based on the model of New Scotland, this last journey, mostly on foot, covered 3,000 miles in six months along the Scottish coasts.

33 According to Habasque, who quotes a report by the marshall of Brancas, in 1739, there were more than 90,000 coast guards. In theory, a single signal was enough to alert and gather them (see President François Habasque, *Notions historiques, géographiques, statistiques et agronomiques sur le littoral des Côtes-du-Nord* (Saint-Brieuc, Guyon, 1832), vol. 1, p. 363).

34 Which inspired Diderot (see what Madeleine Pinault ('Diderot') says of the influence of the texts and plates of Le Masson du Parc's study). It is through the plates of the anonymous sketcher who illustrated the work of Duhamel du Monceau that Diderot, well before he ever set eyes on the fishermen of Scheveningen, visually encountered the things of the sea. We know, on the other hand, of the mediocrity of the *Encyclopédie* on this subject. The articles on marine biology are simply a collection of references to Pliny, Boyle, and Rondelet, without any reference to marine biotope through diving (see Pierre Niaussat, 'Regards actuels sur la biologie marine dans *l'Encyclopédie*', in *La Mer au siècle des Encyclopédies*, pp. 223–41). In the same way, the plates of the *Encyclopédie* devoted to the subject, apart from a certain social realism which manifests itself in the presence of clothing in rags and of the descriptions of the nets, the sails, and the fishing objects which the sketcher could have seen, represent absolute fiction; the monsters of the Middle Ages were still present, and the amassing of ruins already indicates a desire to privilege the picturesque (see J. F. Pahun, 'Précision et fantaisie dans les planches marines de *l'Encyclopédie*', in *La Mer au siècle des Encyclopédies*, pp. 333–42).

35 See Habasque, *Notions*, vol. 1, p. 68. The author says that such rafts were mentioned in 1618 at Lézardrieux. A decree of 12 Prairial of the year XI [1803] very badly enforced, bans the use of dromes.

36 Before jurists came to differentiate the floating seaweed, always free, shore seaweed, often used by local farmers to fertilize their fields, and seaweed from the depths reserved for men enrolled for

naval service (see Charles Le Goffic, 'Les Faucheurs de la mer', *Revue des deux mondes*, 1, (1906), p. 364).

37 The inhabitants of these islands also used clumps of earth. On these practices, Vice-Admiral Thévenard, *Mémoires relatifs*, vol. 2, p. 58.

38 Tillet, *Observations*, *passim*.

39 La Rochelle, memoir, 1723, Archives nationales, 127 AP 2 1.

40 Michel Darluc, *Histoire naturelle de la Provence* (Avignon, 1782), vol. 1, p. 419.

41 Knox, *Tour*, vol. 1, p. 267.

42 Tillet, *Observations*, p. 37.

43 Ibid., p. 21.

44 The following description is based on the previously mentioned works by Duhamel du Monceau, by Dr Tiphaigne, and by Knox, and especially on documents found in the Archives nationales under the call number 127 AP 2.

45 Crustacea were then liked very little by gourmets and were even disdained by the fishermen of certain regions, at La Rochelle, e.g., or in Scotland. The spiny lobster, in particular, considered indigestible, was reserved for the indigenous population.

46 In particular that of the scoter duck. This animal was hunted in the Médoc, along the coast of the La Rochelle area, as well as in Scotland, though, according to Knox, this activity was getting rarer.

47 E.g. in the Côtes-du-Nord, on the beaches of Langueux, of Yffiniac, or of Hillion (Habasque, *Notions*, vol. 2, pp. 331–41).

48 In 1554, these privileges had been forbidden for the future!

49 See Tiphaigne's description, *Essai*, p. 289.

50 Reports concerning Belle-Île and the area around Brest. The fishermen of the Loire estuary roasted the lampreys before covering them 'with a wine and white vinegar sauce seasoned with salt, pepper, clove and a few bay leaves' (Report concerning the area around Nantes, Archives nationales 127 AP 2).

51 Report previously mentioned concerning Belle-Île.

52 Knox, *Tour*, vol. 1, p. 362.

53 E.g. to Tiphaigne, *Essai*, p. 101; the author separates them from the riverside fishermen who work at the mouths of rivers.

54 Report concerning fishing on the strands of Oléron (1717) and in the Blaye region (n. d.), Archives nationales, 127 AP 2.

55 Report concerning the areas around Nantes and Royan; ibid. The questionnaire strives to make a global study of the dressing habits of the fishermen of the kingdom.

56 As an example, Darluc, *Histoire naturelle*, pp. 6 and 7; emphasis added.

57 We will content ourselves with the works in which a portrait, endlessly reiterated into the middle of the nineteenth century, emerges.

58 Lépecq de la Cloture, *Collection d'observations sur les maladies et les constitutions épidémiques* (Rouen, 1778), esp. the pages dealing with the fishermen of the Polet, near Dieppe (pp. 172 ff.) and with the cantons of the shores of the pays de Caux.

59 Desmars, *De l'air, de la Terre et des eaux de Boulogne-sur-Mer* (Paris, Pierres, 1761).

60 Initially, he, like Bigot de Morogues, thought that the sea 'absorbed the floating parts in the air' and that the air which covered its surface 'was the most healthy that one could breathe, and the most free from heterogeneous parts'; but in 1761 he changed his mind because the sea forms around itself an atmosphere composed of the substance of the 'huge quantity of animals and plants which live and die in its midst' (*De l'air*, pp. 24–5). The correspondence between this atmosphere and the moods of man breaks down. As for H. Tully, he incriminates the salinity of the fumes from the sea.

61 J. A. Dulaure, *Description des principaux lieux de France* (Paris, 1788), vol. 1, preface, p. 34. On this fundamental conviction, which led to the cultural policy of the Revolution, see Certeau *et al.*, *Une Politique*, *passim*.

62 Desmars, *De l'air*, p. 21.

63 We see here the very old scientific roots on which the hardening technique is founded, a technique whose appearance is emphasized by Michel Foucault's disciples, esp. Georges Vigarello.

64 Darluc, *Histoire naturelle*, vol. 1, p. 6. They are opposed in this way to the 'class of men who are softened and drained of energy by the lax climate of the sea-shore' (ibid.).

65 Cambry, *Voyage*, p. 68. This great longevity could coexist with a strong and precocious morality, according to a veritable natural selection process. This longevity of coastal populations is also noted by Dr Souquet, *Essai sur l'histoire topographique, physico-médicinale du district de Boulogne-sur-Mer* (Boulogne, 1794), pp. 49–52.

66 It is necessary, in this perspective, to assign a lot of importance to this notion of 'weakening', tied to the humidity of the air which determines the fluidity of the humours; it implies a weak moral vigour and an inclination to habits that are also weakened.

67 On this type of nourishment, its images and its supposed virtues, see Thomas, *Man and the Natural World*, chapter entitled 'Meat or Mercy', pp. 289 ff.

68 Desmars, *De l'air*, p. 57.

69 On this subject, see Du Jourdin, 'Sentiments'.

70 In a word, as close reflections of the elements, they seem to have escaped the process of restraining and interiorizing standards dear to Norbert Elias; more generally, see *Images du peuple au XVIIIe siècle* (Centre aixois d'Études et de recherches sur le XVIIIe siècle, Paris, 1973).

71 A classic subject of investigation for the clergy in the seventeenth and eighteenth centuries; see that of Abbot Jean-Baptiste Thiers, *Traité des superstitions* (Paris, Dezallier, 1679). Let us note, however, that accounts of the investigation made *c.* 1720 propose another image: rare are the fishermen on foot and 'small-time fishermen' who believe in the existence of marine monsters washed up on the beach (see the negative answers in the reports on the regions of Rochefort, Royan, Oléron, and Blaye). During the same time period, some scientists were convinced of the existence of marine men, dating back to before the 'terrestrialization' process; see above, pp. 109 ff.

72 Desmars, *De l'air*, p. 22.

73 E.g. Tiphaigne, *Essai*, p. 262. The author becomes lyrical when talking of these products: Nature 'sows some in the sand, ties some to the surface of rocks, locks some inside rocks; she even piles them one on top of another quite often. On a shell, she will attach plants, on these plants more shells, and on these still more plants; it is as if she fears that she will run out of space.'

74 An egalitarian virtue underlined by Tiphaigne (*Essai*, p. 9) and attested by the memoir concerning the coast of the Rousillon, Archives Nationales, previously mentioned call number.

75 Tiphaigne, *Essai*, pp. 51 and 150.

76 Desmars, *De l'air*, p. 50.

77 Reports of the previously mentioned investigation.

78 Emphasized concerning the coast of Brittany; previously mentioned investigation.

79 Diderot, *Voyage en Hollande*, p. 426.

80 Tiphaigne, *Essai*, p. 252.

81 La Cloture writes (*Collection*, p. 37): 'The people of Normandy are generally prolific and inclined to ensure the reproduction of the species. Authentic proofs of this can be found especially in the towns bordering the sea, along a coast stretching more than 250 miles.' Regarding the inhabitants of Polet, the author underlines the paradox: 'They rarely see their wives, and are prolific' (p. 172). In 1775, Lemoyne, ex-mayor of Dieppe, wrote a manuscript report extolling the rusticity and fertility of the rural fisherman of upper Normandy. Two years later the theme was taken up again by a merchant of Granville (Cabantous, *La Mer et les hommes*, pp. 689 ff.).

Cabantous (*Les Populations maritimes*, pp. 193–201) demonstrated scientifically, in the region of Dunkirk, the existence of the high fertility rate.

82 On this subject, Du Jourdin, 'Sentiments', Cabantous, 'Espace maritime, *passim*, and, as a source, Fournier, *Hydrographie*, pp. 674–7.

83 On this subject, reflections of Marie-Nöel Bourguet, 'Race et folklore. L'image officielle de la France en 1800', *Annales, économies,*

sociétés, civilisations, July–Aug. 1976, esp. pp. 815–18. The knowledge of man is founded on the feeling of distance.

84 An area rendered particularly erogenous by the religious sculptures of the seventeenth century, and even more so those of the eighteenth century, which was fond of sensuality in the legs of its angels.

85 See above p. 37.

86 Indeed, the descriptions of the Mediterranean coasts, integral part of the classic journey, fulfilled different purposes, as we have seen. In the preface to his *Description des principaux lieux de France*, published in 1788 (vol. 1, pp. 33–4), Dulaure, inspired by neo-Hippocratic views, established a strict dichotomy between the people of the coasts and the inland people. 'Those who inhabit the sea-shores, the arid and marsh-like countries (the beach); who live off the products of fishing, have neither the habits, nor the costume of the farmer living more happily inland.'

87 Cambry, *Voyage*, pp. 203 ff.

88 See Bertho, 'L'Invention', pp. 46 and 51. References, esp. to the exaltation of the island of Ouessant by L. Billardon de Sauvigny (*L'Innocence des premiers âges en France* (Paris, 1768)). Furthermore, in 1792, Joseph Lavallée thought he had found the virtuous island in the island of Groix, and Cambry, two years later, in the island of Batz.

89 It is appropriate to highlight the role of the journey to the Highlands in the spread of this feeling. Already in 1773 the goal of the travels of Johnson and Boswell was to observe habits and superstitions, and to note speech patterns and traditions. Johnson's book is completed by a critique, on the spot, of the recording procedures, in particular those guiding the oral investigations.

90 If we think, e.g., of the comparisons used by Arthur Young during his travels.

91 Cambry, *Voyage*, p. 181.

92 Ibid., p. 38.

Notes to chapter 9

1 Jacques Léonard highlights well this ongoing presence of neo-Hippocratic thought among Western physicians of the nineteenth century; in a more general way, he demonstrates how this presence underlay public hygiene (see *La France médicale au XIXe siècle* (Paris, Gallimard, 1978), pp. 173 ff.). As an example, Gigot, *Essai sur la topographie physique et médicale de Dunkerque* (Paris, 1815).

2 Christophe de Villeneuve-Bargemon, *Statistique du département des Bouches-du-Rhône*, 4 vols (Marseilles, Picard, 1821–9), vol. 1, pp. 895–8.

3 Bertrand, *Précis*.

4 See Corbin, *The Foul and the Fragrant*.

5 De Quatrefages de Bréau, 'Souvenirs', p. 606.

6 See Bertrand, *Précis*, vol. 2, p. 299.

7 Note, however, that article 538 of the Civil Code prescribes that 'the shores, the silt of the sea, the harbours, the rocks underneath the level of the tides, which are part of the public domain, are not subject to private possession'.

8 E.g. Audoin and Milne-Edwards, *Recherches*, vol. 1, pp. 173–80.

9 It is poverty that then inspired the attempts at fish-breeding, and later, the ambitious restocking project for the waters of the sea. On all these subjects see Delbos, 'De la Nature'.

10 See Villeneuve-Bargemon, *Statistique*, vol. 1, pp. 896 ff.

11 Ibid., vol. 2, pp. 1130 ff.

12 Joseph Lavallée (according to L. F. Cassas's itinerary), *Voyage pittoresque et historique de l'Istrie et de la Dalmatie* (Paris, Year X), p. ii.

13 Ibid.

14 Ibid., p. 108. Concerning the concept of degeneration in the anthropology of the Enlightenment, in particular in Buffon, see Michèle Duchet, *Anthropologie et histoire au siècle des Lumières* (Paris, Flammarion, 1977), pp. 202 ff.

15 Hence, once again, the special attention we give here to those of the Continent.

16 Concerning the age-old suggestiveness of the figure of the Gaul, see Corrado Vivanti, 'Les Recherches de la France d'Étienne Pasquier. L'invention des Gaulois', in *Les Lieux de mémoire*, part 2: 'La Nation', vol. 1, pp. 215–45, and Pierre Ronzeaud, *Les Représentations du peuple dans la littérature politique en France sous le règne de Louis XIV* (thèse d'État, Tours, 1985); also the older works of Arlette Jouanna and of Claude-Gilbert Dubois.

17 On this subject, Mona Ozouf, 'L'Invention de l'ethnographie française: le questionnaire de l'Académie celtique', *Annales, économies, sociétés, civilisations*, 36th year, no. 2 (Mar.–Apr. 1981), pp. 210–30.

18 See Bertho, 'L'Invention', p. 47; a fear manifested already by Samuel Johnson during his trip to the Highlands in 1773.

19 Habasque, *Notions*, vol. 1, pp. 279–81.

20 Fulgence Girard, 'Le Mont-Saint-Michel', *La France maritime*, 1834–7, 2nd edn (1852), vol. 1, p. 165.

21 See Vérusmor, 'Le Mont Tombelène' (Tumba Beleni), *La France maritime*, vol. 3, pp. 262–4.

22 See, belatedly, A. Dauvin, 'Les Îles des Saints et d'Ouessant', *La France maritime*, 2nd edn, vol. 2 (1852), pp. 321–3.

23 On this subject, Delouche, *Les Peintres de la Bretagne*, vol. 1, p. 21.

24 See Michel de Certeau, *La Culture au pluriel* (Bourgois, 1980), ch. 3: 'La Beauté du mort' (in collaboration with Dominique Julia and Jacques Revel).

25 Girard, 'Le Mont-Saint-Michel', p. 166.

26 Thus Charles Nodier criticizes Dulaure for his direct plunge into the world of Druids and his indifference to the intermediary influence of multiple epochs (see Ozouf, 'L'Invention' p. 225).

27 Heinrich Heine, *Die Nordsee*, pp. 76–9.

28 Nodier *et al.*, *Voyages pittoresques*, vol. 1: 'La Normandie Ancienne' (1820), pp. 4–5.

29 Ibid., pp. 5 and 94.

30 La Cloture, *Collection*, p. 35. La Morinière, *Premier Essai*, vol. 2, p. 106.

31 Richer, *Voyage*, letter 7, pp. 58–9. J. Morlent, *Le Havre ancien et moderne* (Le Havre, Chapelle, 1825), vol. 2, p. 2.

32 See M. Demonet, P. Dumont, and E. Le Roy Ladurie, 'Anthropologie de la jeunesse masculine en France au niveau d'une cartographie cantonale (1819–1830)', *Annales, économies, sociétés, civilisations*, July–Aug. 1976.

33 Philarète Chasles on the other hand, to cite only one example, notes the presence along the English coast of the North Sea of twelfth-century Normans: 'One feels as though one has gone back five to six hundred years in European history' (Philarète Chasles, 'Scènes d'un village maritime en Angleterre', *Revue de Paris*, 7 (1829), p. 82). The scenes mentioned by the author take place prior to 1816. Another example: E. Jouy, in Garneray, *Vues des côtes*, concerning the inhabitants of the Pollet, suburb of Dieppe inhabited by fishermen (p. 40).

34 On the condition, however, that nothing spoils the game of the imagination. For this reason Jouy hates hybrids, and it pains him to see a manufacturing plant installed inside a ruined monument, whose import is then lost (Garneray, *Vues des côtes*, p. 44).

35 Girard, 'Le Mont-Saint-Michel', p. 163.

36 On all these problems, but set against the backdrop of the moors, see Marcel Calvez's fine article, 'La Dimension naturelle de Brocéliande. Analyse à partir de récits d'espace du XIXe siècle', in *La Nature et le rural*. It is in this perspective that one can reread Paul Sebillot's works, in particular *Contes de terre et de mer* (Paris, Charpentier, 1883); *Légendes, croyances et superstitions de la mer* (Paris, Charpentier, 1886), and especially relevant to our subject, *Contes des landes et des grèves* (Rennes, Caillère, 1900).

37 [A fantastic character in the Arthurian cycle who appears as a good fairy with healing powers in the songs of Merlin and Ogier the Dane – trans.]

38 Habasque, *Notions*, vol. 1, p. 281.

39 E.g. Alexandre Bouet, 'La Baie d'Audierne', *La France maritime*, 3, pp. 155 ff., and Souvestre, 'La Cornouaille', *Revue des deux mondes*, 3 (1833), p. 690.

40 E.g. Xavier Marmier, 'Souvenirs de voyages', *Revue de Paris*, June 1837. An investigation into the popular readings in Limousin led me to establish that, in that region, Marmier was one of the most widely read authors of the second third of the century.

41 We are thinking of course about the wonderful 'Archipel de la Manche' which precedes *Les Travailleurs de la mer* by Victor Hugo.

42 Heinrich Heine, *Die Nordsee*, p. 86.

43 In this way, Catherine Bertho underlines the change which then occurred concerning the characters of the Breton and of Brittany. As of 1830, according to her, the stereotypes establishing the black image of the region were established, imposed by a Parisian elite; and the great Romantic travellers served only to legitimize this solidly constructed image. But a Breton elite, aided by the retreat of the legitimists to the country and by the political success of the principle of nationalities, then started to rebel against the stereotypes issuing from the capital ('L'Invention' p. 51).

44 E.g. Souvestre.

45 A number of scenes in Ovid's *Métamorphoses* are set on sea-shores.

46 Ozouf, 'L'Invention' p. 227.

47 Balzac, *Un drame*, pp. 699–700.

48 Dauvin, 'Les Îles des Saints', pp. 321 and 322.

49 See below, regarding Eugène Isabey's work.

50 Crabbe, *Peter Grimes*; Balzac, description of the salt-marsh worker of Guérande, in *Un drame*. See also Bertrand, *Précis*, vol. 2, p. 51, regarding the 'dens' of the fishermen of Equihen.

51 Chasles, 'Scènes', p. 85.

52 A curious note about the significance of the boulder and of lichen in the travel accounts of Prince Pückler-Muskau; he tells (*Briefe*, vol. 3, p. 290) of his visit to O'Connell and evokes the path: 'In particular I noticed a boulder of great beauty ... three kinds of moss, yellow, red and purple, grew in its crevices and marked the black lines in a most unusual manner.' It is noteworthy that, for the Prince, this impression was worthy of being written down, and seemed to him quite remarkable.

53 Cambry, *Voyage*, p. 105, regarding the coastal 'peoples' of the Brest region: 'they throw themselves with the greed, the brutality of tigers on the prey that the sea brings them; one cannot tear it away from them.'

54 Regarding this source of concern, see Jean-Pierre Peter, 'Ogres d'archives', *Nouvelle revue de psychanalyse* (autumn 1972), and, under the direction of Michel Foucault, *Moi, Pierre Rivière, ayant égorgé ma mère, ma soeur et mon frère* ('L'Animal, le fou, la mort', by J. B. Peter and Jeanne Favret) (Paris, Gallimard-Juillard, 1973); see also our preface to Parent-Duchâtelet, *La Prostitution*.

55 This background is evident in Eugène Sue's work. The character of the sea-wrecker precedes that of the rag-picker, archetype of the

animality of the city. Cambry, notes Delouche, *Les Peintres de la Bretagne*, vol. 1, p. 19, tends to see sea pirates in all of coastal Brittany.

56 Corbière mentions the actions of the sea-wreckers in *La Guêpe* and in his poem 'Gens de mer'.
57 Chasles, *Scènes*, p. 93.
58 Hippolyte Bonnelier, *Vieilles femmes de l'île de Sein*, (Paris, Killian, 1826).
59 This is probably the most frenetic of the French *roman noir*, whose author associates himself, initially, with Walter Scott and Cambry.
60 Souvestre, 'La Cornouaille', pp. 691–2.
61 It is said that the painter saw such an event in 1831 while returning from Scotland. But it is important to take into account the interpretation given by the occasional witness who was sensitive to a recognized practice, codified even in the framework of grounding rights.
62 Souvestre, 'La Cornouaille', p. 691.
63 Girard, 'Le Mont-Saint-Michel', p. 165. We know of the impression made on Stendhal by the supposed energy of the people of Trastevere.
64 Souvestre, 'La Cornouaille', p. 691. Here is represented the tension which established itself between the piety of the pastor and the savageness of his parishioners; in the twentieth century, it inspired Henri Quéffelec to write *Le Recteur de l'île de Sein* and Jean Delannoy to create the film *Dieu a besoin des hommes*.
65 Bertrand, *Précis*, vol. 2, p. 168.
66 The picture of the sacrifice of the lamb responds to that of the beacon light.
67 Ducos, *Itinéraire*, vol. 2, p. 328.
68 Dauvin, 'Les Îles des Saints'.
69 C. Rouget de Kerguen, 'L'Île de Noirmoutier', *La France maritime*, 3 (1837), p. 312.
70 Anonymous article, 'Cayeux', *La France maritime*, 3 (1837), pp. 343 ff.
71 See T. S. R. Boase, 'Shipwrecks in English Romantic Painting', *Journal of the Warburg and Courtauld Institutes*, 22 (1959), p. 343.
72 Ducos, *Itinéraire*, vol. 3, p. 51.
73 Jouy, in Garneray, *Vues des côtes*, p. 22.
74 Very thoroughly studied, but which only touch the surface of our subject.
75 Brosse, 'Byron et la mer', pp. 71 and 76.
76 See Custine, *Mémoires*, and Lavallée, *Voyage pittoresque*, passim.
77 See Brosse, 'Byron et la mer', p. 71.
78 Ann Radcliffe, *The Italian, or: The Confessional of the Black Penitents* (London, 1797).

79 E.g. Amédée Gréhan, 'Les Côtes de Normandie', *La France maritime*, 1 (1852), p. 394.

80 On a later date, Jacqueline Lalouette (*Les Débits de boissons en France, 1879–1919* (thesis, Paris I, Oct. 1979) notes the persistent mixing in the cafés of Finistère.

81 De Kerguen, 'L'Île de Noirmoutier', pp. 313–4. It seems probable that this description is influenced by that of the Highlands Inn in Walter Scott's *Waverley*.

82 We find a long, precise description of it in the previously mentioned work of Audoin and Milne-Edwards, vol. 1, pp. 60–7. See also La Tocnaye, *Promenade d'un Français en Suède et en Norvège* (Brunswick, 1801), vol. 2, p. 174; the farmers of the shores of the fjords gave seaweed to their animals for feed. Or else Pictet, *Voyage de trois mois*, p. 112; the author describes the gathering of seaweed on an Irish beach.

83 'Men, women, children, carriages, all are already on their way towards the rough rocks whose soaked seaweed hangs, like the hair of a drowned cadaver and lends the rocks an aspect of desolation.' 'The tumultuous procession' moves on to the strand. 'There rises a boulder which has been mistaken, from far off, for an ancient cathedral; here is another one in the shape of a portico; further off lies a city in ruins or piles of broken armour' (Souvestre, *Voyage*, p. 212).

84 See above, p. 82.

85 Habasque, *Notions*, vol. 2, pp. 315 and 316. These beaches are situated in the district of the Côtes-du-Nord.

86 Except from 1816 to 1818.

87 Claude Berthou, farmer of Pleubian, won the race in 1807; dressed in the fashion of the Bas-Bretons, his spurs were actually nails stuck in his clogs. To increase the ardour of his mount, Canaris, people say he gave it a full bottle of brandy to drink (ibid., vol. 2, pp. 316 and 317).

88 See plate 24.

89 Zola sets the desperate fight against the menace of being devoured as a counterpoint to the fate of the heroes of *La Joie de vivre*, set by the sea-shore. At the time that he described them, walks like those of Pauline and Lazare or the strand had become common practice.

90 Delouche (*Les Peintres de la Bretagne*, vol. 1, p. 178) emphasizes the interest of his *Paysage breton, côte à marée basse*, from c. 1835.

91 Which does not stop it from being rehashed.

92 Jules Michelet (*La Mer*) complains of the invasion of tourists indifferent to the sea.

93 When tourists arrived on the island of Iona in 1826, children dressed in rags flocked to them; the local cicerone presented himself;

when they left, the tourists were harrassed by children selling aquamarine. The same was true in Wales, along the Giants' Causeway (Ducos, *Itinéraire*, vol. 3, p. 297).

94 Souvestre, 'Le Traîneur de grèves', p. 9.

Notes to chapter 10

1 On the metaphor of the ship of State, see S. Pressouyre, 'L'Emblême du Naufrage à la galerie François ler', in *L'Art de Fontainebleau* (Paris, 1975).

2 Lafont de Saint-Yenne, *Réflexions sur quelques causes de l'état présent de la peinture en France, avec un examen des principaux ouvrages exposés au Louvre, le mois d'août 1746* (The Hague, 1748), pp. 100 ff.

3 Cabantous, *La Mer et les hommes*, p. 148.

4 Goethe, *Italienische Reise*, pp. 318 ff.

5 Cabantous, *La Mer et les hommes*, p. 148.

6 On this subject, the previously mentioned account of La Platière's travel is enlightening.

7 Bernard Cousin, *Le Miracle et le quotidien. Les ex voto provençaux, images d'une société* (Aix-en-Provence, sociétés, mentalités, cultures, 1983), pp. 90–102, 131–5, 253–4, 282–4.

8 Used abundantly by Fénelon in his *Télémaque*.

9 There is a good example in the previously mentioned work of Cambry (1795), *Voyage*, pp. 110–11. 'It is on the Pointe Saint-Mathieu that friends, mothers, and lovers extend their arms, proffer their children, burst into tears when the ships leave ... It is there that we wait for them and greet them ..., we follow them from the shore ... impatience, screams of happiness, handkerchiefs waved in the air, hurried steps, anxiousness, heartbeats, convulsions; all sorts of feelings, of emotions, of love, friendship, and fear, all the movements determined by the heart are manifested on this arid rock.'

10 See Anne Vincent-Buffault, *History of Tears: Sensibility and Sentimentality in France, 1700–1900*, trans. T. Bridgeman (Macmillan, 1991), and, concerning the preceding century in particular, Sheila Page Bayne, *Tears and Weeping, An Aspect of Emotional Climate Reflected in Seventeenth-Century French Literature* (Tübingen, Narr, 1981).

11 It is because he is recognized as having the power to create reality while correcting it that Joseph Vernet disturbs and affects amateurs with such intensity. See Else-Marie Bukdahl, *Diderot critique d'art* (Copenhagen, Rosenkilde and Bugger, 1980), vol. 2, p. 262.

12 Quoted in ibid., p. 250.

13 See Valenciennes' indications, above, pp. 155–6.

14 Valenciennes, *Éléments*, p. 437.

15 A vast movement of which Philippe Ariès highlights only one aspect when he detects the rise of the interest shown in childhood.

16 Diderot, *Salons*, in *Oeuvres complètes*, vol. 7, p. 180: 'Salon de 1767'. Interesting introduction by Daniel Arasse.

17 E.g. regarding a seascape of Loutherbourg's (ibid., vol. 8, pp. 440 and 441): 'Ah! My friend, what a storm! ... in the midst of these agitated waters, one could see the two feet of an unfortunate man who was drowning tied to pieces of the vessel, and one shuddered; elsewhere the floating corpse of a woman wrapped in her clothes, and one shuddered; in another place, a man fighting against the waves that were throwing him against the rocks, and one shuddered ...'

18 Valenciennes, *Éléments*, p. 491.

19 See Cabantous, *La Mer et les hommes*, p. 152. The most famous of these duels are the following: on 17 June 1778, the fight of the frigate *La Belle Poule* against *L'Aréthuse*; on 11 Sept. 1778, that of the *Junon* against the *Fox* off the coast in front of Ouessant; in 1779, near that island, the fight between the *Surveillante* and the *Québec*; on 13 Prairial year II, the fight and shipwreck off the coast of Brest of the *Vengeur du peuple* (see Delouche, *Les Peintres de la Bretagne*, vol. 1, p. 103).

20 The forms of sociability themselves were often modified by this development which drew a host of young officers looking to get married towards the coastal regions of the Channel; these catches were all the more appealing because they stood good chances of becoming rich by capturing enemy vessels. See the novels of Jane Austen, *Pride and Prejudice* and *Persuasion*.

21 Off the coasts of Plymouth, Portsmouth, and Rochefort.

22 Which the prisoners milked for all it was worth after their liberation; see e.g. M. de Lezeverne, *Les Plaisirs d'un prisonnier en Écosse* (Paris, 1818), pp. 14–15. The author mentions the marine convict prison at Portsmouth. René-Martin Pillet, *L'Angleterre vue à Londres et dans ses provinces* (Paris, 1815), pp. 372 ff., evokes the suffering of the pontoon vessel laid up in Chatham Bay in 1813. Louis Garneray made this theme even more popular.

23 *Les Souvenirs d'émigration de Madame la Marquise de Lage de Volude, 1792–1794* (Évreux, 1869), accurately represents the anguish of travellers threatened by storms, fires, and pirates.

24 Geneviève Levallet-Hang, 'Philippe-Jacques Loutherbourg, 1740–1813', *Archives alsaciennes d'histoire de l'art*, 15th year, 1936, pp. 124 and 129. The waves of the *Eidophusikon* necessitated very painstaking work. A model of them was first made in earth, then cut out in soft wood. Then they were painted and heavily varnished to make them shine. Each one was manoeuvred around an independent axis, which created the overall aspect of a furious sea. Shortly after 1787, the Scot Robert Barker, creator of the first 'panorama', displayed in London a view of the English fleet at Portsmouth; a little later, there appeared *The Naval Battle* of 1 June 1795, *The Baths at Brighthelmstone*, *The Naval Battle of*

Aboukir, and above all, *A Storm at Sea with the Sinking of the Halsewell* (1786).

25 See the beautiful developments in the previously mentioned thesis by Delouche, *Les Peintres de la Bretagne,* vol. 1, pp. 117–18.

26 On this subject, Monique Brosse, 'Littérature marginale; les histoires des naufrages', *Romantisme,* 4 (1972), and Boase, 'Shipwrecks', pp. 332–46.

27 Later on, this fashion declined, at the same time as the taste for horror. The coasts were equipped; oceanography progressed at an amazing pace; and the steam-engine boats which took travellers over short distances were less dangerous than sailboats.

28 In particular in *Melmoth the Wanderer* (1820), a 'horror' novel; see *Romans terrifiants,* pp. 662–5. In Mary Shelley's *Frankenstein,* the beach scenes are also horrifying. See the episode of the island of the Orcades.

29 See William James, *The Naval History of Great Britain from 1793 to 1820* (London, Baldwin, 1822–4).

30 See Brosse, as well as for what follows, *La Littérature,* vol. 1, pp. 339–76.

31 Audoin and Milne-Edwards, *Recherches,* ch. 6: 'Recherches sur les naufrages'.

32 *Le Navigateur. Journal des Naufrages,* 1829. The editing team claimed that it had correspondents in the harbours. The other sections were dedicated to pirating, 'nautical tableaux', 'phenomena' (e.g. mirages at sea, Saint-Elmo's fire), the maritime legal code, steam navigation, and trips.

33 Cordingly, *Marine Painting,* pp. 96–113.

34 Thomas Luny (1759–1827), a disabled ex-sailor who retired in Devon in 1810, had people carry him each day to the sea; his clients were all naval officers. Thomas Whitcombe (1752–1827) displayed no less than fifty-six marine paintings at the Royal Academy.

35 His most famous seascape, *The Shipwreck, Fishing Boat Attempting to Save the Crew* (1805), was widely exhibited. Remember, Turner painted at least two works devoted to the battle of Trafalgar.

36 Jules Michelet, *Tableau de la France* (Paris, Les Belles Lettres, 1949), pp. 24–8, regarding the 'coast of Brest': 'There the two enemies are face to face: the land and the sea, man and nature. It is something to see when she is excited, the furious one, what monstrous waves she heaps up at Saint-Mathieu Point, up to fifty, sixty, eighty feet; the foam flies as far as the church where the mothers and daughters are praying. And even in the moments of truce, when the ocean is quiet, who has walked this funereal coast without thinking or feeling: *Tristis usque ad mortem?* ... nature is atrocious, man is atrocious and they seem to get along.'

37 See a special issue of the *Provence historique,* 1 (1983), and Cousin, *Le Miracle,* pp. 253 ff.

38 Brosse, *La Littérature*, vol. 1, p. 295; the author emphasizes the cathartic role of these thrillers.

39 Souvestre, 'La Cornouaille', p. 691.

40 Félix Pyat, 'Une tournée en Flandres', *Revue de Paris*, 33 (Sept. 1836).

41 Hartwig, *Guide médical*, pp. 20–1.

42 Bernardin de Saint-Pierre was the first, it seems, to pause with emotion before the prayer of the Cauchoise; see Pierre Trahard, *Les Maîtres de la sensibilité française au XVIII* siècle, 1715–1789* (Paris, Boivin, 1933), vol. 4, p. 105, concerning a text from the *Études de la Nature*, vol. 1, p. 510.

43 Brosse, *La Littérature*, vol. 1, pp. 368–9.

44 Water was then perceived as such by Hegel; see Yvon Belaval, 'L'*Encyclopédie* et la mer', introductory talk at the previously mentioned colloquium of Brest.

45 Girard, 'Le Mont-Saint-Michel', 1852 edn, vol. 4, p. 217.

46 Sir Arthur Wardour and his daughter enjoyed placing their feet on the damp, fresh, hard sand. Sir Walter Scott, *The Antiquary* (New York, Dutton, 1969).

47 Bousquet, *Les Thèmes*; see above, p. 168.

48 See J.-J. Baude, 'Les Côtes de la Manche', *Revue des deux mondes*, July 1851, p. 31.

49 Regarding this, consider the verse of Victor Hugo: 'Poor lost heads, you roll through the dark stretches' ('Oceano Nox'). At the bottom of the sea, the horror of the corpse is sometimes lessened by the poetic theme of marine metamorphosis, founded on the belief that, deep under water, the corpse takes on new shapes or becomes adorned with shells, marine plants, and even precious stones.

50 Habasque, *Notions*, vol. 1, p. 20.

51 Daniel Ligou, 'La Franc-maçonnerie des lumières et la mer', in *La Mer au siècle des Encyclopédies*, p. 77.

52 See the legends of trapped lovers (e.g. Souvestre, 'Le Traîneur de grèves'), the joint engulfment of Gwymplaine and Dea in *L'Homme qui rit,* and the slow disappearance of Gilliatt at the end of *Travailleurs de la mer.*

53 Bernardin de Saint-Pierre, *Paul et Virginie*, p. 60.

54 Brosse, *La Littérature*, vol. 1, p. 182.

55 Girard, 'Le Mont-Saint-Michel', p. 215. 'A thousand stories', says the author, 'stay in the memories of the inhabitants of the coast. Dismal stories, standard dramas with only two actors.' Here, 'the ocean does not kill for itself'.

56 Habasque, *Notions*, vol. 1, pp. 30–1. The fear of sinking into quicksand is visible in the region of Biarritz, in an even more fantastic way. See *Rapport sur les opérations et les résultats de l'association de bienfaisance pour la sûreté des bains de mer établie à Bayonne*, July 1834, p. 7.

In 1820, three ladies who were bathing together panicked upon feeling the sand moving under their feet, let themselves fall into the water, and were drowned, although they were not exposed to any real danger. On 11 Sept. 1833 two women who were bathing together off the coast of Bidart were seized by the same unjustified fear; one of them let herself fall into the water and was only barely saved.

57 De Kerguen, 'L'Île de Noirmoutier', pp. 314–15.

58 Habasque, *Notions*, vol. 1, p. 25. Bonneville is the town in which Zola's novel *La Joie de vivre* is set.

59 Delouche, *Les Peintres de la Bretagne*, vol. 1, pp. 58–9.

60 Nathaniel William Wraxall, *A Tour through the Western, Southern and Interior Provinces of France* (London, 1784).

61 See Audoin and Milne-Edwards, *Recherches*, vol. 1, p. 192.

Notes to chapter 11

1 Roche, *Le Siècle des Lumières*, vol. 1, p. 232.

2 Jean-Marie André, *L'Otium dans la vie morale et intellectuelle romaine de l'époque augustéenne* (Paris, P.U.F., 1966), and Suzanne Barthèlemy and Danielle Gourevitch, *Les Loisirs des Romains* (Paris, C.D.U.S.E.D.E.S., 1975), to which we owe much. Not to mention John H. D'Arms's classic book *Romans on the Bay of Naples* (Cambridge, Mass., 1970).

   In the middle of the nineteenth century, Ludwig Friedlander (*Moeurs romaines du règne d'Auguste à la fin des Antonins* (Paris, Reinwald), trans. C. Vogel (1867) vol. 2, pp. 333–401 and 480 ff., and vol. 3, pp. 110–3) shows that he pays attention, more than Victor de Laprade, Daniel Nisard, and Mommsen, to the seaside resort. On the other hand, the previously mentioned work of Eugène de Saint-Denis, published in 1935 (*Le Rôle de la mer dans la poésie latine*, esp. pp. 97 ff., 105 ff., and 467 ff.), unites erudition and anachronism: the description of the antique seaside resort is apparently influenced by the figure of the modern beach.

3 See above, p. 124.

4 See Barthèlemy and Gourevitch, *Les Loisirs des Romains*, p. 20.

5 E.g. 'Une excursion à Ostie', text of a third-century Christian mentioned in ibid., pp. 83–5.

6 See the already detailed analysis of Friedlander, *Moeurs*, vol. 2, pp. 391 ff.

7 After the Senate's session. Among the most prestigious of these maritime villas, we should mention that of Tiberius Sperlonga in the south of Latium and his cave adorned with statues of Ulysses and Polyphemus; see F. Coarelli, *Lazio, Guide archéologique de Laterza* (Rome, 1982), pp. 343–52.

8 See Président de Brosses, *Journal*, p. 64.

9 Esp. *Le Siècle des Lumières*, vol. 1, pp 157 ff.

10 Cambry, *Voyage*, pp 203–5 and ff.

11 La Reynière, *Lettre*, p. 12.

12 See below, p. 278.

13 Millin, *Voyage*, vol. 3, pp. 350–1. The existence of such *bastides*, used for the same purpose, is noted in the region around Nice in 1795 by Albanis Beaumont (*Travels through the Maritime Alps from Italy to Lyons across the 'col de Tende' by the Way of Nice, Provence, Languedoc* (London, 1795), pp. 89–90).

14 Daniel Lescallier (*Voyage en Angleterre, en Russie et en Suède fait en 1775* (Paris, Didot, year VII), pp. 37–8) emphasizes the rise of this fashion: 'The captains of ships, and the general officers who are financially comfortable, the lords and many rich individuals, even among those who do not care for navigation, find pleasure in building and rigging yachts, or leisure vessels, 80- or 100-tonners more or less, which they use during the nice weather to take small trips up and down their coasts, in France, in Holland, sometimes all the way to Lisbon or Cadiz ... These individuals created a club called the Navigating Club, and gave themselves a green uniform with white ornaments and gold bands.

The Naval Minister goes round the King's harbours in a yacht every year ... The King and the Queen also both have their own yachts, superbly gilded and decorated.'

In August 1785, Milady Craven exclaims in Genoa her regret at being on a felucca: 'I cannot imagine anything more pleasant than having an English yacht, clean and practical, and the company of four or five intelligent individuals to go to Italy, along the coast, as I am doing: the view is magnificent' (*Voyage de Milady Craven à Constantinople par la Crimée en 1786* (Paris, 1789), p. 49). We have seen the pleasure that Henry Fielding took in such things.

15 See Gilbert, *Brighton*, esp. pp. 89 ff. Regarding the details of the Brighton installations, we owe much to Dale, *History*, and *idem*, *Fashionable Brighton, 1820–1860* (London, Country Life, 1947).

16 Burney, *Diary and Letters*, vol. 1, pp. 217–228, 280–311, and vol. 2, pp. 158–83.

17 For what follows, *Torrington Diaries*, pp. 87–103.

18 Specialists in historical geography have analysed, in its variety, the growth process of seaside resorts, distinguishing the ancient fishing villages, the new residential zones developed near the harbours, and the *ex nihilo* creations along the beaches or at pleasant sites. See especially the totality of the work done on this subject by Edmund W. Gilbert. Often of very good quality, but on a subject different from ours, are the works of H. G. Stokes (*The Very First History of the English Seaside* (London, Sylvan Press, 1947)), of Ruth Manning Sanders (*Seaside England* (London, Batsford, 1951)), of Anthony Hern (*The Seaside Holiday: The History of the English Seaside Resort* (London, Cresset Press, 1967)), and of

Janice Anderson and Edmund Swinglehurst (*The Victorian and Edwardian Seaside* (London, 1978)).

19 Very recent bibliographies and chronologies concerning this process can be found in: *Saison am Strand: Badeleben an Nord und Ostsee, 200 Jahre*, (Herford, Koehler, 1986). Catalogue of an exhibition held in Hamburg at the Altonaer Museum, 16 Apr.–31 Aug. 1986. More recent contributions include two works by J. K. Walton, 'The Demand for Working-Class Seaside Holidays in Victorian England', *English Historical Review*, 1981, and *The English Seaside Resort: A Social History 1750–1914* (Leicester, 1983).

20 It must be said that in July 1783 a pastor had presented the King of Prussia, Frederick II, with a petition demanding the opening of a bathing establishment on the island of Juist, near Norderney. The text of this petition can be found in the above-mentioned catalogue, *Saison am Strand*, p. 12. See also J. D. W. Sachse, *Über die Wirkungen und den Gebrauch der Bäder, besonders der Seebäder zu Doberan* (Berlin, 1835). It is necessary, furthermore, to point out the importance of the *Histoire naturelle de la mer*, published by Otto in Berlin in 1792 (*Naturgeschichte des Meeres*).

In 1793, the *Annalen der Britischen Geschichte* (vol. 7) describes the English sea baths of Deal, Weymouth, and Harwich. That year, while Lichtenberg's preference leaned towards the North Sea, Dr Woltmann campaigned for the Baltic.

As of the first years of the nineteenth century, the literature on sea baths is considerable in Germany. In 1801, the first work appeared on Norderney; that of Friedrich Wilhelm von Halem, which was reprinted in 1815 and in 1822 under the title *Die Insel Norderney und ihr Seebad*. In 1818 Auguste Ruge's work on the sea baths of Cuxhaven was published (*Über Seebäder im allgemeinen und besonders über das Seebad Cuxhaven*).

At that time, the big resorts began to publish their own records; e.g. *Annalen des Travemünder Seebades* (1817). Then C. G. Hecker described the resort of Putbus (*Über das Seebad bei Putbus*); in 1821, Johann Ludwig Chemnitz that of Wangerooge (*Wangerooge und das Seebad*), which was also described by Lasius (*Beschreibung von Wangerooge*); Haffner wrote about Zoppot, von Colditz about the island of Föhr, Pfaff about Kiel, Van der Decken about Helgoland, and Formey about Doberan. A number of articles from the *Hufeland Journal* are dedicated to sea baths in Germany. The bibliography contains still more titles. We would like to thank Robert Beck on this subject for the documentation he gathered, which nicely complemented the above-mentioned catalogue.

21 The article ('Warum hat Deutschland noch kein grosses öffentliches Seebad?') was reprinted many times; see *Vermischte Schriften*, ed. Ludwig Christian Lichtenberg, vol. 5, (Göttingen, 1803), pp.

93–115, and Abendroth, *Ritzebüttel und das Seebad zu Cuxhaven* (Hamburg, 1818).

22 *Reise eines Gesunden in die Seebäder Swinemünde, Putbus und Doberan* (Berlin, 1823) (journey made in 1822).

23 De Montulé, *Voyage*, vol. 2, pp. 58–9.

24 The documents sometimes provide different dates, which is understandable; it is somewhat arbitrary to try at all costs to date the origins of a resort.

25 In 1828, Dr Kind describes the season at Swinemünde (*Das Seebad zu Swinemünde* (Stettin, 1828).

26 In 1823, Friedrich von Varnstedt, who attended the court of Denmark, described in detail the resort at Wyk, highlighting the distractions offered to the bather; see *Saison am Strand*, p. 22.

27 Stierling, *Über Seebäder in den Niederlanden*, published in *Hufeland's Journal*. The records of sea baths at Travemünde were translated into Dutch; it appears that this translation encouraged the foundation of bathing resorts at Zandvort.

28 On the history of sea bathing at Ostend, see Pasquini, *Histoire de la ville d'Ostende et du port ... suivie d'un vademecum du Voyageur à Ostende* (Brussels, 1843); Hartwig, *Guide médical*; Dr Louis Verhaeghe, *Traité pratique des bains de mer* (Ostend, 1855); Yvonne du Jacquier, *Ostende et Spa* (Ostend, 1965), as well as the secondary source by Robert Lanoye, *L'Épopée ostendaise* (1971).

29 Lescallier, *Voyage en Angletene*, pp. 12–13.

30 *Mémoires de Brissot*, vol. 1, pp. 271 ff.

31 Ireland, *Picturesque Tour*, vol. 2, pp. 191–3.

32 The 500-foot long façade could be entered through a wrought-iron gate supported by two marble columns and adorned with an obelisk. On top of the third vault of terraced bathes, a terrace and a flower-bed were provided for walks.

33 Bertrand, *Précis*, vol. 2, p. 165 ff. See also Souquet, *Essai*, pp. 53–9. According to him, the practitioner prescribed salt-water as early as 1756. The first, very richly documented history of Boulogne is that of Ernest Deseille, *L'Ancien Établissement des bains de mer de Boulogne. 1824–1863* (Boulogne, Aigre, 1866). Precise statistics concerning the prices, machines, baths, and bathers can be found in it. E.g. in 1835, 30,000 sea baths were taken at Boulogne.

34 See the description of it given by Dr Mourgué, a physician at the sea baths of Dieppe, in *Journal des bains*, as well as that of Ange Pihan Delaforest, *Premier Voyage de S. A. R. Madame la Duchesse de Berry en Normandie* (Paris, Pihan Delaforest, 1824). There are a few details in Gaudet, *Notice médicale*, followed by a dossier on this subject at the Medical Academy in 1837 by MM Guersent, Lisfranc, and Bousquet.

On the history of Dieppe, Simona Pakenham, *Sixty Miles from England. The English at Dieppe, 1814–1914* (London, 1967).

More generally, on the sea baths in France or in Belgium, apart from Gabriel Désert's work *La Vie quotidienne sur les plages normandes du Second Empire aux années folles* (Paris, Hachette, 1983), see a few anecdotal works: Paul Jarry, 'Bains de mer au temps passé', in John Grand Carteret, *L'Histoire, la vie, les moeurs et la curiosité* (Versailles, Société générale d'imprimerie, 1928), vol. 5; Georges Renoy, *Bains de mer au temps des maillots rayés* (Brussels, Rossel, 1976).

35 Delaforest, *Premier Voyage*, pp. 130–1.

36 Ibid., p. 129. The author was inspired by the Dr Mourgué, *Journal des bains*, pp. 17–18.

37 Ibid., p. 130

38 Ibid., pp. 131–2.

39 Jouy (in Garneray, *Vues des côtes*, part 2 (1832), p. 19) still mentions the Marie-Thérèse bathes at La Rochelle, on which we were unable to find any documentation, but which, according to this author, already attracted a number of foreigners, as of 1830. In the first part of the book can also be found a description of the baths at Boulogne (p. 45) and at Dieppe (p. 37).

40 On 27 Messidor, year III (1795), the records of the deliberations of the directorate of the district of La Manche show the request of one Sir Chavoy who asked to be allowed to 'settle at Grandville', where his wife 'was taking sea baths'. According to the mayor (letter sent on 24 May 1836 to the sub-prefect of Avranches), there were very few bathers before 1827. That year, a 'cabin' was built for them on the rocks on the beach. In 1830 the establishment 'was transformed into a big reception hall where the bathers could get together every day and dance twice a week' to the sound of a piano. Nine 'small portable cabins' were available to these bathers, who then numbered seven or eight hundred each year at the resort. As of 1837 Heinrich Heine spent the season there, accompanied by Mathilde. On all these points, see *La Vie balnéaire et les bains de mer à Granville*.

On the sea baths in the area around Avranches, see Marius Dujardin, 'Histoire des bains de mer à Carolles et à Jullouville', *Revue de l'Avranchin*, 1954, as well as the work of vulgarization by Elie Guené, *Deux siècles de bains de mer sur les plages de l'Avranchin et du Cotentin* (Manche-tourisme, 1985).

41 Viel, *Bains de mer*, pp. 7–10.

42 See Désert, *La Vie quotidienne*, pp. 15–18.

43 Alexandre Dumas, *Mes mémoires* (Poissy, Bouret, 1866), vol. 2, pp. 275 ff.

44 Jean Chennebenoist, *Trouville et Deauville vus par Charles Mozin, 1806–1862* (Deauville, 1962).

45 E. Ducéré, *Napoléon à Bayonne* (Bayonne, 1897).

46 Bouet, 'Les Bains de Biarritz', pp. 317 ff. Concerning a previous

period: C. Hennebutte, *Guide du voyageur de Bayonne à Saint-Sébastien* (Bayonne, 1850).

47 Laborde, 'Biarritz' p. 27.

48 The President François Habasque (*Notions*, vol. 2, p. 313) noticed that bathers came during the summer to the little beach of Cesson, *c.* 1830: 'No bathing establishment exists there, everyone bathes where and how he or she wants. During the Empire, one saw many students there, on vacation days.' Since then, sea baths have been forbidden to them 'in the interest of their morals and their personal safety' (p. 314).

49 The study of sea-shore architecture started in France; see the very interesting article by Nathalie Glou, 'Villas balnéaires de la Manche' (*Monuments Historiques*, no. 1 (1978), pp. 34 ff.) The author gives a good description of the specific features of the seaside villa as well as its regional and social diversity.

50 Once again, we owe much for what follows to the previously mentioned works by Dale and Gilbert.

51 This seaside architecture term was used in Bath.

52 Blanqui, *Voyage*, p. 387.

53 But not George IV.

54 Count A. L. C. La Garde Chambonas, *Brighton, scènes détachées d'un voyageur en Angleterre* (Paris, Aillaud, 1834), pp. 79–85 and 281–8.

55 Delaforest, *Premier Voyage*, p. 207.

56 Habasque *Notions*, vol. 1, p. 362. The author daydreams here while contemplating the sea for quarter of an hour, 30 Apr.

57 Abbé de Saint-Non, *Voyage pittoresque ou description des royaumes de Naples et de Sicile* (Paris, 1781), vol. 1, p. 226.

58 The following description is inspired by Vivant Denon's journal, which is in the appendix of the Didot edition of the above-mentioned *Voyages* by Henry Swinburne, vol. 5, pp. 74–6; by the account of La Platière, *Lettres*, vol. 2, pp. 349 ff.; and by the summary made by Tuzet (*La Sicile*, pp. 431–48). The author borrows from the accounts of Johann-Heinrich Bartels's journey in 1786, published in 1789–91.

59 A need emphasized by Bartels, according to Tuzet, *La Sicile*, p. 438. Bartels visited the marina in winter.

60 Schedule described by Vivant Denon, appendix to Swinburne, *Voyages*, vol. 5, pp. 74 ff.

61 La Platière, *Lettres*, vol. 2, p. 349.

62 Denon, appendix to Swinburne, *Voyages*, vol. 5, p. 74.

63 Ibid., pp. 74 and 76.

64 Dale, *History*, p. 54.

65 The final installation was not completed until 1840.

66 Quoted by Gilbert, *Brighton*, p. 101.

67 Blanqui, *Voyage*, p. 387.

68 Chambonas, *Brighton*, p. 125.

69 Ibid., p. 25.
70 A. B. Granville, *The Spas of England and Principal Sea-Bathing Places* (London, Henry Colburn, 1841), vol. 1; 'Northern Spas', p. 160.
71 Ibid., p. 183
72 Ibid., p. 350
73 This is also the case along the French Riviera; see J. Voisine, 'Les Anglais en Provence au XVIIIe siècle', *Revue de littérature comparée*, 1956, pp. 15–27. The discovery of the Riviera as a climatic resort started in 1760 – i.e. after the sick had become accustomed to going to Montpellier, whose vogue had been reinforced by the Jacobites' emigration after their defeat in 1715. Among the first to spend the winter in Nice, after Smollett, were the dukes of York and Gloucester, brothers of George III. The second hired Albanis de Beaumont as family tutor; his works contributed to promoting the popularity of this region in England. Note that houses built with views on to the sea, walks on the jetty, and sea excursions were common among the English settled between Nice and Villafranca; see Beaumont, *Travels*, pp. 90–106.
74 Granville, *Spas of England*, p. 151.
75 Ibid., pp. 182 and 188.
76 See Gilbert, *Brighton*, pp. 15–16, and Dale, *History*, p. 30.
77 Gilbert, *Brighton*, p. 15.
78 *Gentleman's Magazine*, 'Diary of their Majesties' Journey to Weymouth and Plymouth', 59 (1789), pp. 1046–7, 1142–4.
79 Interesting on this proximity is the analysis of the above-mentioned journey to Plymouth.
80 In 1769 Marlborough House was built; it was then sold to Duke George Spencer in 1771, who kept it until 1786. The Duke also bought Grove House, built in 1779.
81 See Dale, *History*, p. 32.
82 The Duke of Morny founded Deauville, a very popular French resort, during the reign of Napoleon III.
83 Léon de Buzonnière, *Voyage en Écosse. Visite à Holy-Rood* (Paris, Delaunay. 1832), pp. x–xi.
84 Chambonas, *Brighton*, p. 376.
85 Ibid., p. 261: the very detailed schedule of a lady at Brighton.
86 A pejorative term which for two centuries distorted the play of influences between France and England.
87 See on the following, Delaforest, *Premier Voyage*, Mourgué, *Journal des bains*, and Apponyi, *Journal*, vol. 1, 26 June–17 Aug. 1828, 4–6 Aug. 1829, 19–30 July 1830, pp. 120–47, 176–94 and 276–89.
88 Delaforest, *Premier Voyage*, p. 169.
89 Deseille, *L'Ancien Établissement*, p. 8. The date is 25 Aug. 1825.
90 Apponyi, *Journal*, 2 July 1828, p. 123.

91 In particular, that year, the Duke of Devonshire, Lady Hamilton, and Lady Dalrymple.

92 In this way, on 6 Aug. 1828, back from a trip to Brighton, Count Apponyi met at Dieppe 'The Duchess of Maillé, the Countess Albéric de Choiseul, the Duke of Coigny, the Prince and Princess of Beauffremont, Madame the Countess of Rougé with her daughter, ... Madame de Saint-Aldegonde with two very young girls, M. de Dreux-Brézé, son of the grand master of ceremonies, MM de Sainte-Maure, General de Bordesoulle, the Countess of Chastelux with her two daughters, and Mlle de Barante and her father, the Count of Barante, author of the *Histoire des ducs de Bourgogne*, the Count and Countess of Avaray, Madame de Fourdonnet ...', not to mention the Prince of Saxe-Cobourg.

93 Apponyi, mentioned in *Journal*, 15 Aug. 1828, vol. 1, p. 146.

94 Ibid., p. 284

95 Pakenham, *Sixty Miles from England*, p. 25.

96 A. Loève-Veimars, 'Souvenirs de la Normandie', *Revue des deux mondes*, Dec. 1834, pp. 94–5.

97 *Reise eines Gesunden.*

98 Kind, *Das Seebad zu Swinemünde.*

99 Dr W. Francken, *Scheveningen, sa plage, ses bains* (Paris, J. B. Baillière, 1899), p. 1.

100 Du Jacquier, *Ostende*, p. 23. Unfortunately, the house in which the royal couple was staying was not on the beach. Therefore, Queen Louise-Marie had herself hoisted by a hand elevator on to the terrace of the house, where she could knit while looking at the sea, protected by a veranda.

101 Remember, on this subject, that the Beau Brummel was present at the Brighton Pavilion. Thomas Creevey, who was invited to spend four months there in 1805–6, and again in 1811, evoked in detail the evenings with the Prince of Wales and the merry life at the villa. When he returned to Brighton in 1837, he found the resort detestable (*The Creevey Papers. A Selection from the Correspondence and Diaries of the late Thomas Creevey, 1768–1838*, ed. Sir Herbert Maxwell (London, John Murray, 1903), esp. vol. 1, pp. 47 and 146 ff. Regarding 1837, vol. 2, pp. 325 ff.). Madame de Boigne (*Mémoires* (Paris, Mercure de France, 1971), vol. 1, pp. 460 ff.), invited by the Regent in 1817, also described in detail her stay at the villa.

102 On this evolution, see Gilbert, *Brighton*, pp. 108–9 and 184, and Dale, *History*, p. 39

103 Pückler-Muskau, *Briefe*, vol. 1, p. 348.

104 Granville, *Spas of England*, vol. 1, p. 188.

105 See below, p. 280.

106 Granville, *Spas of England*, vol. 1, p. 269.

107 Details in ibid., pp. 346 ff.

108 Ibid., pp. 348–9.

109 See on this subject the documents concerning regulations on the beach at Boulogne in Deseille, *L'Ancien Établissement*. On the other hand, control seems to have been harder to enforce on the Ostend beach; hence the admittedly belated complaints of a Bruges newspaper reported by Yvonne du Jacquier (*Ostende*, p. 27): 'Before one's eyes is the disgusting spectacle of rascals whose obscene nudity and scandalous walks from the dune to the sea upset onlookers.' A few ladies, bathing near the lighthouse, were accosted by men who were hardly reassuring; 'they almost fainted from fear'. 'Already certain families have left this resort, where their morality and that of their children is not protected.' We suspect, in fact, that the norms of modesty and the fear of social or sexual contact became accentuated between 1811 and 1850. (See above, pp. 83–4.)

110 The following details come from Kind, *Das Seebad zu Swinemünde*.

111 Du Jacquier, *Ostende*, p. 33; but the testimony is rather old since it dates back to the 1850s. Regarding Blankenberghe: Hartwig, *Guide médical*, p. 63. People started bathing there only in 1840; in Blankenberghe, 'the maritime population ... does not hide in the harbour as at Ostend, but rather spreads itself out on the platform'.

112 Details from the previously mentioned anonymous work *Reise eines Gesunden*.

113 See above, p. 148.

114 On the subject of Brighton's prestige as viewed by Jane Austen, see *Pride and Prejudice*.

115 William Cobbett, *Rural Rides* (London, Reeves and Turner, 1893), vol. 1, 10 Jan. 1822, pp. 91–3. *Memoirs of the Life of John Constable, Composed Chiefly of his Letters*, by C. R. Leslie (London, Phaidon Press, 1951), p. 12 (letter written from Brighton, 29 May 1824).

116 Charles Dickens, 'The Tuggses at Ramsgate', ch. 4 of *Sketches by Boz*.

117 It was apparently at Margate that the fashion of travelling along the beach on donkeys originated. A woman offered this excursion around 1800 for a shilling an hour; the practice was introduced in Brighton in 1806 (see Gilbert, *Brighton*, p. 20).

Notes to Conclusion

1 K. D. Fines, 'Landscape Evaluation: A Research Project in East Sussex', *Regional Studies*, 2 (1968), pp. 41–55.

2 Charles Avocat, 'Essai de mise au point d'une méthode d'étude des paysages', in *Lire le paysage*, pp. 11–37.

3 See H. Flatrès-Mury, 'L'Évaluation des paysages bretons', in *Lire le paysage*, pp. 39–59.

4 On all these problems, see also Jacques Bethemont, 'Élément pour
   un dialogue: géographie et analyse du paysage', in *Lire le paysage*,
   pp. 101–11.

INDEX

leverage in commodities. It only takes a 20 percent of a portfolio in commodities to offset an 80 percent investment in stocks and bonds because in the latter, you have either no leverage or at best 2 to 1 leverage (if you borrow on margin to buy your stocks). With commodities, however, your leverage is better than 10 to 1.

Further, with commodities, you can get your profits out immediately while maintaining your position. Thus, as you lose on stocks and bonds in a recession or other down period, you can withdraw funds from commodities profits to compensate on a daily basis, if needed.

You must be part scientist, part mathematician, and part artist to be successful in commodities. None have to be big parts.

B. R.

Dr. Lintner's conclusion was that by including commodities as a part of an overall portfolio, the investor could reduce the risk of volatility. That is, he could reduce his losses.

MODERN PORTFOLIO THEORY (MPT)

I have built on the work of Dr. Lintner and have modified his theories to come up with my own investment model which I call Modern Portfolio Theory or MPT.

The basic question I raise is, if commodities help to balance volatility in an overall portfolio, why couldn't an investor hone in on just the profits that the commodities offer? In other words, why use commodities just to balance out losses on stocks and bonds? Why not use commodities to make profits on their own?

Another way to think of this is the sawtooth pattern of investing. A typical investment portfolio over A period of time has this sawtooth appearance.

Dr. Lintner's study showed how to mitigate the effects of the sawtooth. He showed how to reduce the volatility so that lows of the sawtooth weren't as low. My theories suggest that we can use a similar approach to cut off the lows entirely and just keep the highs.

The key to doing this, of course, is to be able to invest in commodities during those periods when they will do

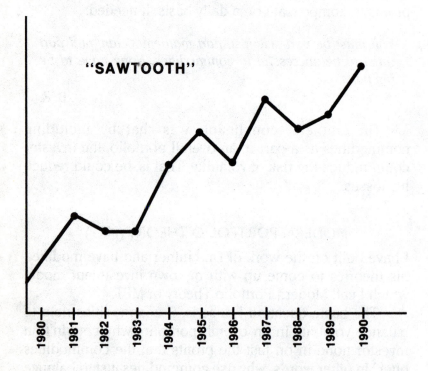

"SAWTOOTH"

The "sawtooth" appearance of a typical investment. (Few high-profit investments show steady profit or loss.)

well and to stay out of the market when commodities are flat and stocks and bonds are doing well. The trick is identifying those periods.

If only we had a yardstick or, better still, a barometer that would show us when to get into commodities and when to stay away. If such a barometer existed, then we could refine Dr. Lintner's theory to a point where commodities weren't just offsetting stocks and bonds, but were themselves a major source of profit.

The barometer that I found was gold.

GOLD—THE NEGATIVE CORRELATION

Gold is a commodity and is traded in several exchanges. (I'm not suggesting that you should necessarily buy gold— only that you should follow its price movements closely.) More importantly, gold is also a currency.

Since antiquity, gold has been a currency of established value. The Egyptians knew of gold, the Romans used it, and it was treasured during the Middle Ages. During the First and Second World Wars many French families survived occupation only because they were able to use gold coins, which they had hoarded, to buy food and protection at a time when their regular currency had no value.

There's no money that speaks louder than gold.

King Faisal

Even today, major countries—including the United States—produce legal tender gold coins. South Africa has the Krugerrand, Australia the Nugget, Canada the Maple Leaf, and the United States the Eagle.

When other currencies fail, gold has always retained its value. That's why in Brazil today (the economics of which we discussed in the last chapter) gold retains its buying power while the cruzero deteriorates.

Because of gold's status as the "currency of last resort," as the only money to retain its buying power, it has a unique position in the world of finance. Gold has become the ultimate negative correlation to the economy.

When there is a threat to our economy, whether it is from inflation or banking collapse or military adventure, where do people turn financially? Do they immediately try to save paper dollars? Do they buy stocks and bonds? Do they go into real estate?

Those in the know do only one thing. They buy gold. Knowledgeable investors convert their assets to gold at the threat of any economic upheaval. This is why each time there is even just the hint of inflation rising, according to "official" sources, gold moves upward. When the U.S. warships entered the Persian Gulf on their "peace keeping" mission, gold rose in price. When Brazil or Mexico or Argentina or some other Third World country threatens to default on its loans, gold goes up. When the stock market falls, gold rises.

Gold is negatively correlated with the economy. In good times, gold does badly. In bad times, gold does very well indeed.

That's why I suggest that gold can be that barometer which we use to gage our investments. We don't necessarily have to invest in gold. We just need to keep track of what it's doing.

On that day that gold moves up in price, that's the day we should begin investing more heavily in commodities. On the day that gold jumps $25 an ounce, that's the day we should consider realigning our investments so that we have less of a commitment to stocks and bonds and more of a commitment in commodities. On that week that gold goes up $25 or more an ounce per day *every day*, that's the week we should consider dumping all our other investments and doing nothing but investing in commodities.

MPT VS. DR. LINTNER

Dr. Lintner's study demonstrates that commodities can be a safe investment as part of an overall portfolio of 40 percent stocks, 40 percent bonds, and 20 percent commodities. The leverage, liquidity, and negative correlation of commodities would offset losses in stocks and bonds, while during good times, gains in stocks and bonds would offset losses in commodities. Dr. Lintner showed how to take some of the volatility, and hence the risk, out of investing.

MPT carries this a step further. In my newsletter I frequently advise my clients to change that ratio. When it appears that inflation, hard times, or some other problem is rearing its head and chomping down on the economy, I have suggested investing as much as 40 or 50 percent in commodities, with the remainder of the portfolio in stocks, bonds, special situations, and real estate.

He who hesitates is lost.

Anonymous

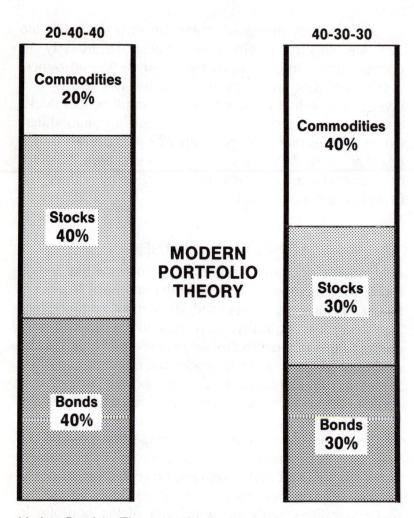

Modern Portfolio Theory juggles the ratios of commodities, stocks, and bonds to increase the profit potential.

I can easily foresee a time when the gold barometer shoots up when I might advise 90 to 100 percent investment in commodities. Because in that situation, the only place where you won't lose money will be the commodities future market.

THE BOTTOM LINE

Are commodities safe?

As part of a balanced portfolio in the manner described by Dr. Lintner, yes, I think they are safe. Their leverage and liquidity make them safe when balanced with other investments such as stocks and bonds. As part of the MPT which I advocate, yes, I think they are safe.

Does that mean that you won't lose money in commodities? No, it doesn't. Even in an inflationary period with gold giving clear signals by price increases, even following Dr. Lintner's theories or MPT, you could lose your shirt. You could go broke if you're in that 85 percent who are at the bottom of the learning curve. You could go broke if you try to invest in commodities on your own and don't know what you are doing. You could lose every penny you put in if you choose the wrong broker, the wrong money manager, or the wrong financial advisor.

But please keep in mind that what I am saying is that if you lose money, it's your fault, not the fault of commodities. Remember, for every dollar lost, there is a dollar won. Commodities is a zero sum game. Somebody is *always* winning. Often that person is winning big.

In the next chapters we'll see how to avoid losing and how to win big in commodities futures.

5

UNDERSTANDING
YOUR
WEAKNESSES

The weakest person (and, incidentally, also the most foolish) is the one who says he has no weaknesses. In commodities a person who believes he has no weaknesses will also very quickly become the poorest.

There's always the very real possibility that you could be wrong.

B. R.

At the outset, it's important to grasp the fundamental truth that commodities are an emotional investment. The volatility, with prices changing moment to moment requiring numerous quick decisions, makes it so. (This is the reason that when looking for a money manager, always ask to see an *actual* track record, not a *simulated* track record. Simulations show how a system would have operated in the market if it had been used. Usually these come out as shining examples of profit. An actual track record is how a system did perform in the market with real people managing it for real money. It's the only true measurement.)

In the emotionally charged climate of commodities, each of us has our own personal investment weaknesses— you have yours and I have mine—and I don't intend to get into those. However, we all have certain investment weaknesses in common and in this chapter I intend to expose two of them: greed and fear.

WHEN I GOT STARTED

The best way I can think of to illustrate these weaknesses is to describe my own experiences, wild though they might have been, when I got started in commodities. I probably made every mistake and hit every bonanza there was. I think this example should prove illustrative.

Back in 1969 I was a stock broker, having worked for McGraw-Hill (as the youngest managing director they had ever had). I quit McGraw-Hill because overnight I began making triple my former salary.

At the time, I wanted nothing to do with commodities. To me it was the proverbial crapshoot, and anybody who got involved with it was a nut. Besides, I just didn't understand commodities then. Rather than take the time to learn, it was a lot easier to say, "That's just like going to Las Vegas." (Reread the last chapter if you think that statement is correct!)

At the time—this was 1969 and 1970—I began to make big money, well into six figures, and as a result, I had big tax consequences. (If you'll remember the time, it was a bear market, but I was doing hedges and warrants for my accounts and myself and thereby virtually creating a new source of profits. But, that's another story for another book.) At the time it didn't escape my notice that my income resulted in my paying something over two-thirds of my earnings in taxes.

So I talked to the partners of the firm I worked for, and one of them explained that I could use the commodity market to convert ordinary income to long-term capital gains. In short, I would end up paying at a far lower tax rate. (Don't get too steamed up about this strategy—the I.R.S. managed to do away with it several years ago—I'm just explaining it to show how I literally "backed into" commodities.)

The minute the partner said commodities, I yawned and started to lose interest. But the tax problem wasn't going to go away, and the more the partner explained, the more I got interested.

Napoleon Hill said, "Work always for 'onemanship.'" He meant always have only one goal and wait until it's finished before you start the next.

B. R.

The basic strategy was to buy gold in the physical market and sell it short in the futures market. By borrowing on the physical purchase from a bank and by closing out the short position on December 31 and opening it again in January 1, you were able to deduct all the carrying charges on the physical gold, while offsetting them in the futures market. The net result was that you got a huge deduction for carrying charges which, if you played the strategy well, was enough to offset your entire income!

I was hooked onto commodities. I didn't get into the business to make a profit. I did it to legally reduce my taxes.

Shortly thereafter I ran into a gentleman who was a Cuban national. He had come over to this country with basically no money, but he still had his Cuban contacts and he came from an extremely wealthy family with sugar processing interests. This man was investing in sugar futures.

The only thing I knew about sugar was that you put it into your coffee. Anyhow, relying on his knowledge of sugar, I bought four contracts of my own which back in 1971 cost $2,000.

The importance of this investment was that it was not a tax move. It was my first for-profit venture into commodities. (Actually, I was making well into six figures, I had bought a big house and car and was living fairly well. As a result, even though I was making plenty of money, I was still broke! It's a common phenomenon amongst the newly rich.)

I actually used my MasterCard to borrow the $2,000 I put up for the sugar contracts! I bought sugar at two cents. Soon after, it went up to four cents.

During the time it went from two to four cents, I bought additional contracts. (This is called pyramiding, and I'll show you why you *shouldn't* do it in a later chapter.)

You have to understand that I wasn't acting as a commodity broker at the time. I was investing my money with another broker who made the commissions. I was doing this on my own.

If I had real savvy of the market, I could probably have made half a million dollars during this run-up. As it was, because of the *leverage* available in commodities, I made $45,000 in one year on a $2,000 investment.

I had never seen money come in like that before. All I knew was that this kid was getting into commodities. I got licensed as a commodities broker.

A man's reach should always exceed his grasp.

Of course, I had never "been to the well." In commodities lingo that means that I hadn't seen my account nor my clients' accounts lose all their equities in bad trades. I had only seen the sunny side of commodities, and I had no idea things could go wrong. I thought commodities was the easiest way to make money that I had ever run into, and I couldn't understand why I'd never found it before.

(Incidentally, my MBA is from Columbia University and Columbia's School of Business was one of the first to have an actual commodities department. You can get an education in commodities at Columbia, and I had been there earning an MBA and had been aware of these programs and still hadn't wanted to trade commodities!)

After making $45,000 on a $2,000 investment, however, I was a "done dog." With the advice of a Ph.D. in agro-economics, I went into grains. He laid out this fundamental case that wheat, soybeans, and corn were going to go up because of the fundamental supply and demand relationship.

I popped my whole $45,000 into the grains. As it turned out, the Russians had secretly bought up all our grain reserves. (It wasn't a secret from our government—they're the ones who hushed it up. It was a secret from the general public.)

Roughly eighteen months later I had transformed that original $2,000 borrowed on a credit card into $1.2 million. At that point, I quit the market for reasons I'll explain shortly.

GREED—THE GREAT MOTIVATOR

It's important to be perfectly clear about why I got into the commodities market, why I opened contracts, why I took the risks involved. The answer, plain and simple, was greed.

I wanted to make money, lots of money. I wanted to make it quickly and easily, and commodities was the avenue I found for making it.

Greed is usually considered an undesirable quality to have. The truth, however, if we're man (or woman) enough to admit it, is that we all have greed. If the opportunity to make big bucks with a minimum amount of risk and effort presents itself, we will take it. I have yet to meet a person who will give up the opportunity to make a quick profit.

Greed, the desire for profit, is what motivates people to get into the commodities market. Perhaps we see an opportunity in grains. We realize that for a few thousand dollars, we stand to make $20,000 or more on a single contract. So we begin to trade.

That is exactly what happened in my case. The typical story of someone who begins trading commodities is that they stumble onto it. Once they see the money to be made, they are hooked for life.

There's only one rule for success, only one way to express it—persist.

B. R.

THE OTHER SIDE OF THE MARKET

Thus far we have been discussing the up side of the market, the profit potential that tickles our greed and gets us started trading commodities. There is, however, a down side, as I noted earlier. I ran into this as well, albeit in a most strange manner.

On the way to making $1.2 million in 18 months, I took a very strange turn. I was establishing a heavy position in grain. I called up the trading department in the firm where I was working and gave them a complex order based on a technical signal. Essentially the order said this: "I want to buy wheat or corn or soybeans or soybean meal or soybean oil or a half dozen other grains. I want to buy *one* of these, and here's how I want you to determine which one it is that I buy. I am giving you a certain price. If they start moving up in price, the first one that hits that price, buy it. Then cancel all the rest." (This is technically called a "stop-buy" order.)

Be sure you understand what I was ordering. I wanted to end up with a contract in only *one* commodity. I was leaving the choice of which commodity to end up in to the broker. He was to make the decision based on which of the group of commodities I'd selected made the biggest move upward first.

Confident that I had placed my order, I left town to attend a business meeting in another city. Of course, the broker in the trading department had misunderstood my order!

The following Monday I was still out of town, and I called in to see if my order had been placed. I about had a heart attack when I heard what had happened. Yes, the prices had gone up. However, instead of placing a contract for *one* commodity, the broker had put me into all of them! I had open contracts in virtually every grain across the board. My exposure, in terms of loss, was potentially hundreds of thousands of dollars.

FEAR—THE OTHER SIDE OF THE COIN

If greed had motivated me to get into the market, fear was pushing me to get out. Here I was looking at a potential loss of hundreds of thousands of dollars, of my entire equity. For the first time, I was looking at going to the well . . . and I saw that the well could come up dry.

My mind went blank, and all that greed that had been pushing me to invest was instantly gone. "Out," I told myself. "You've got to get out before you're killed." I tasted fear.

I counted to ten and then counted again to try to get my pulse steady. Then, with my fingers shaking, I called up the senior partner of the firm. As soon as he got on the phone, I blurted out what happened and then said, "Our broker made a horrible mistake. He did not fill the order the way I wanted. I want out. I'm not accepting these trades, because it could kill me. I just want out." I must confess that looking back on it, my request must have sounded more like a plaintive demand. However, I was consumed with fear.

At this point the senior partner laughed and said, "That's perfectly fine. However, would you mind if I take them on . . . they are *all currently up the limit!*"

I did a quick calculation. All the contracts up the limit made me a fast $600,000!

I tried searching my mind for that fear that just a very few moments earlier had been devouring my being. It had suddenly become a very insignificant, very small thing. In fact, I couldn't even find it! Very quietly I said back into the telephone, "No, I believe I'll keep them."

A bull sometimes makes money. A bear sometimes makes money. But a pig never does.

Roy Longstreet

I hung up and thought about it. With a sappy, satisfied smile, I turned to a friend nearby and told him that I was the boy wonder of commodities. It was my great ability in the markets and my tremendous knowledge that had made me a fortune overnight. (Remember, this was back in the early 1970s when a hundred thousand was more like a million.)

Wasn't I smart, I told myself. And the more I told myself, the more I believed it. And the more I believed, the more I began to think that if I could make $600,000 over the weekend, why couldn't I make another $600,000 during the next week? Greed, which had been obliterated by fear, came roaring back.

I called the trading broker back and told him, "Leave the contracts on."

True to form, as soon as I said this, fear again crept up. Leaving the contracts on left me still exposed. If the price turned, I could still lose everything I had made and more. So I added, "On the first sign of weakness in the market, sell it all out . . . I'll take my profit."

Mind you, I've just made $600,000 on an error. All I have to do is to sell it all out immediately and I can pocket the cash. But greed tells me that I'm so smart, I can stay in the market and make another $600,000 by tomorrow. Fear tells me to protect myself, so I tell the broker to sell it out if the price starts to fall.

Now I've taken care of both fear and greed and I'm one self-satisfied slaphappy fellow.

THE DARK DAYS

I hung up and decided to spend the afternoon playing bridge. (In those days I was an avid bridge player, and not too bad at the game if I say so myself.)

I hung up at approximately 11:00 in the morning. The grain markets closed at 12:15. The bridge game wouldn't start until 12:00, so I went across the street and bought myself a new Cadillac. I just wrote out a check for the most expensive car in the dealership, intending to pay for it with equity from my commodities account. (Remember, you can withdraw your profits at any time.) This was really a thrill for me because it was the very first time I had ever paid for a car with cash.

> *I shall return.*
>
> General Douglas MacArthur

Off I go to play bridge. We had just started on the first rubber when I get a telephone call from my secretary. I go to the phone thinking she is about to tell me how much richer I am. Instead, she informs me that at ten minutes past 12:00, five minutes before the markets close, the markets all reversed and *went down the limit!*

I sat down rather abruptly. Down the limit meant I could have lost as much as I had made. However, I tried to calm the butterflies in my stomach. I had told the broker to "take me out at the first sign of weakness." Hopefully he had. I was praying, "Dear Lord, let me not be in the market."

I left the bridge tournament and went to my office where I called the broker. This is what he told me had basically happened. The markets had turned so suddenly that there were no traders on the other side. He had tried to get me out, but he couldn't do it before the closing bell. I was still in . . . and down the limit!

My heart sank. In just a few hours I had gone from congratulating myself on my ingenuity to kicking myself for my stupidity.

THE MORAL

I was finally able to get out. The next few days the markets were down the limit. But by borrowing a huge sum of money, I was able to do an offsetting trade in the cash markets (the markets closest to the spot or current month which have no trading limits) and trade out. (By the way, this is a strategy that you may want to keep in reserve if you ever get trapped in a limit move.)

Never continue only *because you have begun.*

B. R.

When it was all over my equity position was $480,000. Not bad you might say? Consider, had I taken $2,000 up to $480,000 in eighteen months? Or had I taken $1.2 million down to $480,000 in a few days?

My first wife told me I had taken $1.2 million down to $480,000 and shortly thereafter divorced me to make sure I didn't lose any more of her money. My own personal feeling is the other way round. I think I took that $2,000 up to nearly half a million.

Regardless of whether you think I was a hero or a bum because of this trading, you should see what this story teaches. It shows that the main motivating emotional forces in commodities are fear and greed. Greed got me in, fear got me out. It happened to me that way . . . and when you get into commodities (if you're not already trading, it will, I guarantee it will, happen to you.)

If you think about it, fear and greed motivate those who don't even get into commodities. The reason I stayed out of commodities initially was that I didn't really understand them. I couldn't really explain what trading was. Every time I invested in something, my wife would ask me, "Well, why did you do it?" I'd give her the supply and the demand fundamentals and she'd understand. But, what if I went home and said, "Dear, I invested in soybean oil."

She would surely ask, "What's soybean oil?" I would have to say, I haven't got the slightest idea!" And that made me afraid. So, initially I stayed out of the market. It was only when, virtually by accident, I made money in the market, that greed took over and nudged me in.

THE IMPORTANCE OF DISCIPLINE

A successful trader learns to have a system that in effect controls greed and fear. True discipline, if I'd had it, would have been for me to have sold out the minute that I found out that there had been a mistake made, because true discipline would have demanded that I stick to my predetermined trading program. (Remember, that program said I should be in *one* commodity, not all of them. If I had been in one and it was down the limit, I wouldn't have sustained anywhere near as great a loss.)

> *You must have a program. You must know your program.*
> *You must follow your program.*
>
> Roy W. Longstreet

The moment something happens that is different than the trading program, than the system, dictates, I should have liquidated. Had I done so, I would have been up $600,000—a gift from God.

In the next chapters we are going to look at the elements of trading systems. However, before doing that, let me make a point that I'll return to again and again. Very few people have the necessary discipline to stick to a trading program even once they get one. Far too often, greed or fear motivates us to forget our discipline.

HOW TO MAINTAIN PERFECT DISCIPLINE

To conclude my story, after I got out of the grain trades, I quit the commodity business for a time. I went to Hawaii, rented a room at a fine hotel, sat on the beach with a drink, and thought.

My conclusion was that I, personally, was not suited to trade my own account. I, personally, was too susceptible to the swings of fear and greed. In other words, I could not trust myself not to abandon my trading program when things either went unexpectedly well . . . or unexpectedly badly.

Did that mean that I should get out of commodities? Hardly. I had made over a million dollars (although I kept less than half of it) in less than 18 months of trading. Obviously, commodities was the avenue that was going to lead to my ultimate fortune. I just had to find a different vehicle to get me there.

The vehicle I selected was a manager who handled the trades for me.

It's important to understand that what I'm talking about here is significantly different than what I've discussed up to this point. Prior to this, I had been calling up the broker and telling him, "Trade me into this." Or, "Trade me out of that."

My trades, of course, had been predicted on both fundamental as well as technical information. Nevertheless, I was the one who essentially executed them. I was the one who picked up the phone and said what to buy or sell and when.

What I decided to do was to give up that responsibility. I would give it to someone else, someone I would hand pick.

That didn't meant that I was going to let that person have a field day with my money. Quite the contrary. I would work out a plan, a system. I would give that person very strict parameters. He would be able to trade this or that on his own. However, he had to follow the system that I dictated. Since it wasn't his money he was playing with, but rather was mine, he shouldn't have to contend with the greed and fear that threw me off.

It was a great plan, and back in the early 1970s, it set a precedent. I helped develop a system, later computerized, which a major brokerage house is still using to this day that allows them to trade commodities successfully.

Of course, I don't expect that you, the reader, will be much interested in developing a computerized trading system. In fact, you shouldn't. It would be like reinventing the wheel. Today, there are brokerage houses across the country that operate super-sophisticated systems that profitably trade their customers in and out of commodities.

What you need to do is to find the right person and give him or her the right parameters with regards to trading your account. Once you do that, you will have established a way of getting the discipline necessary to successfully trade commodities.

GREED AND FEAR

It's important to remember that greed and fear are always going to be with you when you trade commodities. The way to get around their influence is to have someone else handle the trades for you. In later chapters, we will see how to find the correct person to trade your account. But first, in the next chapter we will look at the ten basic strategies you need to know in order to formulate the instruction you will give to your ultimate commodities trader.

6

TEN STRATEGIES
FOR PROFITING
IN COMMODITIES

*There ain't no easy money lying around . . . and if there
is, ain't nobody trying to shove it into your pocket.*

Jesse Livermore

Jesse Livermore became famous in the financial world for
making $9 million shorting stocks going into the Great
Depression. Then he took his $9 million and ran it up
to $20 million in the commodities market. Mind you, this
was in 1929 and 1930 when most financial whiz kids were
jumping out of the windows of tall buildings.

Jesse became a sort of legend, a person who those
in the industry looked up to. During the dark days of the
Great Depression, Jesse was like a ray of hope.

In 1931 Jesse Livermore shot himself in the men's
room of the Roosevelt Hotel in New York City. The reason?
He had taken his $20 million down to $2 million in later
trades in commodities.

Keep in mind that during the Great Depression, even $2 million wasn't so bad. However, it was the thought of losing that $18 million that had been in his pocket that did Jesse in.

As it turned out, Jesse Livermore did most of those things that you shouldn't do in commodities. He pyramided his money. He got "married to his positions." He pulled out without letting his winners ride. There are some who say that the money he made at first was just a fluke . . . that Jesse, in reality, never really had a chance in the commodities market.

Maybe they're right. Most of the people I've seen who are the big losers were the ones who started out as the big winners. They, so to speak, fell into it. But once in, they didn't know how to keep their winnings coming.

A compulsive trader is too much in love with the market.

B. R.

In this chapter we're going to look at ten separate strategies. Each one is a lesson in itself. Each one tells a story. Each one is a "must." If you absorb all ten and follow them to the letter, there's no way you can lose in commodities—I guarantee it!

STRATEGY I—GO ONLY FOR THE TWO OR THREE TRADES *A YEAR* THAT CAN MAKE YOU MONEY

I have a friend who told me that his commodity broker wants him to put up enough cash so that he can make a minimum of five trades each time he goes into the market.

The broker tells him that you have to keep hitting at a position until you get a winner.

This friend ended up making 27 trades his first three months in commodities. The result, he was down $2,500, which I consider very lucky. He could have been down one heck of a lot more making that many trades.

The first rule of commodity trading is that every single year, at least on one or two commodities make a major price advance or a major price decline based on supply and demand factors alone. This has nothing to do with inflation or deflation. It happens whether the commodities market is hot or cold. (Of course, if it's hot, then there are many more than one or two opportunities.)

Any broker worth his salt should be able to find those one or two trades. If your broker can't identify those big price swings based on fundamentals, then he shouldn't be anywhere near the business.

If you're starting out in commodities, or even if you've been in them a while, you should put your $2,000 or $3,000 or whatever in the account and then patiently wait. Read the *Wall Street Journal* and *Investor's Daily* every day. Read about the weather in the Midwest, and the grain forecasts, and the supply and demand factors in copper. Read about the cattle feeding and frosts in the Florida orange juice belt. Don't trade . . . read!

Never aim lower than the stars.

Eventually you'll know what's going on in commodities and, if you're relying on a broker and he's any good, one day he'll call and say that there's an opportunity in copper . . . or in silver . . . or in orange juice . . . or whatever.

It might be three months or six months before he calls. But when he does, from all your reading, you'll know that he's right. And you'll make your trade.

If you've done your homework and if you've waited patiently and if this is in fact the big move for the year in a commodity, you'll make your $10,000 or $20,000 *per contract* and you'll be feeling mighty smug.

By the way, your total commission paid should be no more than a couple hundred dollars!

STRATEGY 2—USE A COMPUTER SYSTEM

The worst thing any investor can do is to try and reinvent the wheel. You could spend years trying to develop a technical system based on market signals and if, and that's a big IF, you were successful, you'd only end up duplicating some probably superior system that's already out there.
Of course, it's imperative that you fully understand what a computerized commodities investing system does.

Most computerized systems are based on technical signals. Technical signals are indications of what the market may do based solely on past price movements. While at first this may sound simple, it's actually quite complex.

Computerized systems not only watch the raw price performance of commodities; they also watch such things as moving averages. (Moving averages are typically based on 3, 5, 10, or 30 days. You add up the closing prices for each of the days in the period and then divide by the period. For example, on a 3-day moving average, you would add up the prices of the last three days and then divide by three. You would do this *every day*. We'll have a much more detailed explanation of this in a later chapter.)

In addition to keeping track of moving averages, the computers also watch patterns in price performance. We'll have a lot more to say about this in the chapter on charting. However, for now, let's just say that the computer compares current patterns in prices with historical patterns in prices.

A computer is like an "idiot savant"—it only knows a little bit. But it knows that better than anybody else.

B. R.

When the computer recognizes a pattern that indicates that the price is likely to go higher or lower in the immediate future, it sends a buy or sell signal to the broker. Brokers who follow these computer models then immediately buy or sell. (Sophisticated computer systems today actually generate the buy/sell orders themselves!)

Back in the early 1970s, with the assistance of a computer specialist, I created one of the first computer systems approved for commodity trading called TLA. This system used a complex formula for tracking moving averages and other pricing patterns. Today it is used by a major commodities firm to handle millions of dollars of transactions.

The important point to understand here, however, is that sharp commodity brokers and successful houses use sophisticated computer systems, and if you don't take advantage of them, you're missing the boat.

This is not to say that by arduous study and careful analysis, you can't do what the computers do. You can. But as I said at the outset, why reinvent the wheel? The computer systems are already in place. Take advantage of them.

Having thus touted the wonders of computers, let me end by noting the negative side. Today, so much trading is done by computer systems that the systems themselves influence the market. Often the various systems will all recognize a particular market fluctuation as a trend. When that happens, many of the systems simultaneously will send buy or sell signals out and an enormous amount of trading suddenly takes place. That's why for no apparent reason in any given day you may see the price of pork bellies jump up or the price of silver plummet. The computers are trying to get their masters in at the beginning of a trend.

What you should know is that *the computers are wrong 60 percent of the time*. That's correct, the best systems only work 40 percent of the time. Having said that, I'm sure many readers are wondering why I resort to computers that are so fallible? The answer is part of the next strategy.

STRATEGY 3—LET YOUR WINNERS RIDE

This is probably the first rule that most new traders learn about commodities. When you have a winner, let the profits ride. When you have a loser, dump it quickly.

It's simply said, and it would appear to be a simple rule to follow, but it's not. Let me give an example of why not.

Those who let their profits ride, last.

P. O.

Let's say you're trading currencies, for instance, the Japanese yen. You've gone long on the yen and the dollar is down, meaning the yen is up. Yesterday your account showed a $5,000 profit.

Today, the U.S. government intervenes. The dollar rises and the yen falls. Your account is down to a $3,000 profit. Your broker calls and tells you that the yen might fall even further. Then he says the seven most unholy words in commodity trading, "You can't go broke taking a profit."

Remember, greed brought you into the market and got you trading to begin with. You've made a profit; now you fear that you'll lose that profit. Those seven words your broker utters strike deep at your fear.

But, you ask, what if the market turns and the yen skyrockets? Won't I lose out on bigger profits?

To this your broker smugly replies, "You can always get back in."

So you close out your position and take your $3,000 profit. And at the end of trading that day the yen skyrockets . . . and you can't get back in! By tomorrow your profits would have been $6,000, the next day $9,000, and so on. But you've missed the train on this trip. You're out.

Of course, you say, I made $3,000 profit. Why should I worry?

You should worry because, as we saw in the last example, the very best computerized systems, which make trades far better than you alone can, are wrong 60 percent of the time. The only way you can make up for the 60 percent loss is to let those 40 percent winners ride. You may lose $1,000 or $2,000 each time on those 60 percent losers. Consequently, you better make $20,000 or more on those 40 percent winners!

The truth of this really came home to me a few years back when an account I had decided to really test the system. He reasoned that if the computers were wrong 60 percent of the time, let's do the *opposite* of what the computer says. When the computer says buy, we sell. When it says sell, we buy. As a result we'll be right 60 percent of the time instead of 40. We'll have to make money!

When everyone is bearish, it means that there's no one left to sell. When everyone is bullish, it means that there's no one left to buy.

The contrarian view

Sound like a good idea? Be sure you understand what's involved before you answer. The computer systems are constantly giving buy/sell signals. Now, instead of following those signals—because you know from past experience that they are wrong 60 percent of the time— you trade exactly opposite. That should make you a loser only 40 percent of the time. Winning more than half should be enough to let you ride to glory on your profits.

The person who did this used the old Commodity Research Bureau system, a competent computer model. The guy comes in and tells me that he has figured out a way to wealth. So we start doing it. After all, I was fascinated. What he said was logical. What if he was right? We could all make a fortune overnight.

The only thing was, he had to be quick in his trades. He had to dump the losers on the first sign of weakness. After all, he was betting against what the computers suggested.

In the initial days of using his system, it was fun. We made a heck of a lot of trades, and he was making a decent little return on his money, something like 30 or 40 percent annualized.

At the end of a year, however, he had made nothing. His account, in fact, was down from all the commission costs of all those trades. The computer system for that year, however, was up over 100 percent in profits.

Why? The computer was wrong 60 percent of the time. Yet, it made a 100 percent profit. The account, by going opposite to the computer, had to be right 60 percent of the time, but lost money. How could that be?

Consider how it works. The computer is only telling you what the price trend is. The computer says "buy" and we all jump in and buy. But six times out of ten the computer is wrong and the price moves against us, so we all jump out.

But those four out of ten times that the computer is right, it has pegged a *trend*. With a trend, once you're out, you can't get back in—the price movement is too far, too strong, too quick.

To put it another way, 40 percent of the time the computer picks the BIG winners. If you let those big winners ride, they more than compensate for the 60 percent of the time the computer is wrong.

On the other hand, if you go into those trades where the computer is wrong 60 percent of the time, they are small trades. There is no trend. There are no profits to let ride.

Thus, when we say that the computer is wrong 60 percent of the time, it means that the trend signals fail to materialize. Forty percent of the time, however, that trend does materialize and if you are in there and *if you let that winner ride*, you are on the road to riches.

The whole message of this strategy is to let your winners ride. You may go up, and then you may go all the way back down to zero. You may lament the money that you "lost" on the ride. But you're not trying to make money on the short rides. You're trying to identify the big roller coaster, the trend that will make you bundles. You'll never get rich off hiccups in the market. What you need is a full blown explosion.

STRATEGY 4—DIVERSIFY WITHIN COMMODITIES

There's a classic story told of an investor who made several hundred thousand dollars in a few weeks back in 1980. At that time we saw the biggest surge in silver prices in history. Silver went from around $6 an ounce to $50 in an incredibly short time. (Of course, from there it went back to $6, which provides another truism—a commodity always returns to its point [previous price]. But that's a story for another book.)

This investor who made the couple of hundred thousand in silver figured he had finally found the road paved with golden bricks. He quit his job and took up residence at his broker's office. His full-time career, now, would be investing in commodities. Not just any commodities, but silver, a market he felt he knew very well.

*You may just as easily be right for the wrong reasons,
as wrong for the right reasons.*

B. R.

The last time I saw him, he was applying for welfare. Silver went eight years without another dramatic move. Oh sure, it had its ups and downs, and he made money here and there. But the big moves that made him his hundreds of thousands just didn't happen again.

This is not to say that silver won't return to $50 an ounce. I absolutely believe it will. A commodity always returns to its point (previous price). However, it may not do it until the middle of the next century for all I know.

The point here is that this individual specialized in just one commodity, and that was his downfall. As noted in strategy one, there are one or two major moves every year based strictly on supply and demand. But those moves are almost never repeated in the same commodity. In the summer it may be soybeans; in the winter it may be copper. The next summer it might be pork bellies and so forth.

If you specialize in just one commodity or just one area of commodities such as grains or metals, you may go broke waiting for lightning to strike your field. On the other hand, it may be striking like crazy in your neighbor's yard.

When you enter commodities, you have to determine that you are in all commodities. You will trade wherever the big move is. Put simply, you need to be willing to diversify your trades.

If you're a big investor with $50,000 to $100,000, you can diversify all over the board. If you're a small investor with under $5,000, then you'll just have to be patient. You'll have to follow all the commodities and pick the winners (and hope that you've made the right choice).

Many computer systems handle diversification for you. They automatically put you into the commodity which offers the best chance of being caught in a major move.

Diversifying by investing in several computer systems works even better. Here you have your capital spread among several models, and if any of them hit, you're a big winner.

Of course, you need to set up limits. I'll discuss limits in a later chapter, but for now I just want to talk about one kind of limit that is absolutely necessary in diversification through computer systems.

Limiting Diversification by Computer System

Remember, the computer is not a thinking being. It's just a machine which follows the program that's been placed in it to the letter. In almost all cases, the program is designed to identify trends. The very success of that program, however, can be your undoing.

Consider currencies as an example. You can trade in Japanese yen, German deutsche marks, British pounds, Swiss francs, French francs and so forth. The point is that there are a lot of different individual currencies under the general heading of currency.

What you need to know is that very frequently all the currencies will react together. The reason, quite simply, is that they are used to short the U.S. dollar. How do you short (sell) dollars when you don't own dollars? Simple, you buy the currency of some other nation. What that means is that when dollars go down in value, the value of other currencies goes up.

The computer, however, doesn't necessarily know this. The computer program is designed to identify trends. So on Monday the dollar begins to fall and the Japanese yen is up. Suddenly the computer signals that you should buy yen, so you go long in yen.

Always strive to diversify. But always be sure you haven't diversified all into the same area.

B. R.

By mid-morning, the German mark is up and the computer sends another signal, so you go long in marks.

By noon, there's a general run on the dollar and the computer signals to go long francs, pounds, and every other currency. Do you do it?

Not if you have a half a brain in your head.

You might get a dozen different buy signals from a dozen different commodities. But in reality, they are all one signal. They are all signaling one thing only, that the dollar is falling. If you heed all those signals and buy all the commodities, you haven't limited your exposure and increased your chances of profit through diversification. Instead you've increased your exposure and limited your chances of profit through concentration. All of your contracts will depend on one thing—the dollar. That's not diversification in my book.

If the dollar reverses and moves upward, all those contracts will also reverse and you'll lose big bucks. You have to understand that many commodities are *linked*. The currencies are linked and often move in tandem. The same holds true for the grains, the precious metals, livestock, financial instruments, and other areas.

What I do is tell my broker, who can then plug it into the computer system, that if there's a signal to buy or sell in one commodity, such as yen, go ahead and do a trade. If there's another signal in a *linked* commodity, do another trade. But after that, *ignore any further signals from linked commodities*. In our example, I might be in yen and deutsche marks. But that's all. If there is a trend and the computer was correct, I'll make plenty. But, if the computer was wrong and it's not the start of a trend, only a market hiccup, I've limited my exposure.

Yes, you want to diversify. But you don't want to diversify all over Hell's half acre. You want to be sure you're ready to trade in any commodity that signals it's making a major move. You don't want to trade in all the linked commodities in that field.

STRATEGY 5—TRADE SMALL IF YOU'RE A SMALL INVESTOR

Quite frankly you need a minimum of $50,000 to trade effectively by computer. Most investors, however, don't have $50,000. Rather, they have closer to $5,000 or less. (It's a fact that in this country less than 8 percent of the population can write a check for $10,000 without first calling the bank and arranging for a loan.)

Never get into trades that are over your head.

B. R.

If you're a small investor starting out, there are two strategies that I advocate you consider—pooling and waiting for special situations. We'll consider them separately.

Pooling

A "pool" is a legal designation. It involves pooling a bunch of money from different investors. Instead of having one investor with $50,000, you pool ten investors with $5,000. The risk is the same and the return is proportionately the same. (If the single big investor makes 50 percent on his money, or loses 50 percent, the small investor does the same percentage.)

To run a pool, the operator must be registered as a pool broker with the Commodity Futures Authority and must meet a series of fairly strict requirements. In the previous examples I've given of people getting into commodities, their names were on the account in the brokerage house. In the case of a pool, the pool's name is on the account. The money is insured, and there is virtually no possibility of fraud or of getting the pooled money confused with other money. In other words, in a pool you are normally just as safe as you would be in an individual account.

When you go into a pool you're joining up with 10 or 30 or maybe 200 or 300 other people. The money is no longer in your name, but you get all the benefits.

And the benefits can be significant. Most importantly, a pool has enough clout to afford a good money manager with a good computer system. For a small investment, you're playing right up there with the big boys.

There are, however, some problems with pools. Basically, brokers don't like them. Why don't brokers like pools? It's simple. Consider, if you have 20 investors and they each make a trade, how many commissions do you get? Twenty, right? On the other hand, if 20 people pool their money, how many commissions are there to make? That's right, only one. A pool is not a very advantageous way to handle investors from a broker's perspective.

On the other hand, firms love pools. Brokerage houses want you to go into pools that they've organized so that they get control of the money and have captive commissions. They promote pools as commodity's version of a mutual fund or a limited partnership. Unfortunately, that's not always the case. These could end up being nothing more than churning pools where the profits are all eaten up by the commissions.

What you want, therefore, is what's hardest to find. You want a pool that's run in such a way that the parameters are to your liking—few trades, maximized computerized assistance. How do you find such pools?

No question, it takes a lot of searching. In later chapters we'll go into methods of finding brokers and money managers who can direct you to good pools.

Special Situations

The other method of investing in commodities if you don't have more than a few thousand dollars is to open your own account and place your own trades. Of course, you can't possibly diversify. So, instead you have to study the market very carefully and hope that you can identify those one or two major moves a year on your own.

Essentially what you are looking for is a "special situation." By that I mean a trade that's special, one that offers great opportunity. For example, back in 1986 the Chairman of the Federal Reserve board said, "The dollar is too high. We're going to let it drop in value."

Always keep most of your capital in reserve. That way you'll always be ready to take advantage when opportunity calls.

B. R.

Now, there aren't many times in history that the person running the central bank of the most powerful country in the world announces his intentions regarding his currency. If you had your ears open, you would have realized this was a "special situation." Forget supply and demand and technical signals. This is information leading to an almost certain winner. So, if you had your ears open, you went long in yen or deutsche marks or just about any other currency. And for three years you were sitting on the hundreds and hundreds of thousands of dollars you had made.

Special situations don't come up all the time. They don't even come up very often. If you have limited funds, what you have to do is offset that limitation with research, readiness and most important of all, patience.

STRATEGY 6—DON'T PYRAMID

Pyramids fail. Not some pyramids or a few pyramids, but for the average investor, all pyramids fail, period.

What's a pyramid? Let's say you go long in copper with one contract. Copper goes up a few cents a pound and you have a $1,500 profit. Now you have a choice. You can take your money out. Or you can use it to open a new contract in copper.

If you pyramid, you take your money out and use it to open a new contract at the higher price level. If the price goes up, you do it again opening a third contract at a yet higher level.

Each time copper goes up, you use the profits on previous contracts to open new ones. The progression of your investments look like an inverted pyramid.

PYRAMIDING

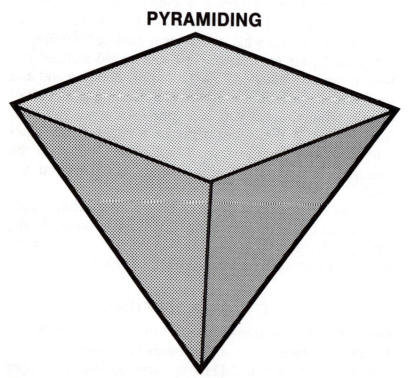

Pyramiding—note that the pyramid is always out of balance, constantly threatening your investment.

What's going to happen? In your mind you're projecting that as long as copper goes up, you'll have more and more contracts on and be making more and more money. Greed (remember the great motivator?) has taken firm hold of your senses.

The problem, of course, is one of timing. You can never know when copper is going to peak out and turn down. In your mind you are always projecting a higher price. But one day copper turns down.

Now you've got losses. Just as you took $1,500 out, you've now got to put $1,500 back in. Only now, you don't have the $1,500! Rather, you've spent it all on the other contracts. You're spread out thin as paper in contracts at ever higher price levels.

All that has to happen is for the price at the top to drop just a fraction, and your whole pyramid comes tumbling down. So now you've lost all the money you would have made, *even though you were right on copper*. If you had simply kept your one original contract, you would have done quite well.

Reverse Pyramids

Reverse pyramids, on the other hand, work quite well. In a reverse pyramid you believe that a commodity is going to make a move. You believe a drought is going to force grain prices up—or that U.S. economic pressures are going to force the dollar down. So you decide to get into the market. You buy soybeans or deutsche marks.

Only instead of opening just one contract, you put as much money into the market as you can. You open two or three or five contracts.

There are old pilots and there are bold pilots, but there are no old, bold pilots.

Air Force motto

On day one we open the contracts. On day two the market moves the way we hoped and we doubled our money. We now trade out one of the contracts. If we had five, we now have four. We've taken some of our profits and reduced our exposure.

On day three the market goes up and we double our money again. We close out another contract, taking our profits.

We keep going in this manner, pulling our money out so that in the worst case scenario, we leave with a profit. By day ten, we may have only one open contract. If the market now turns against us, we may lose a few bucks on that contract, but we will have made a bundle on the other four.

Our scenario looks like a reverse pyramid, which is to say, like a real pyramid. Instead of adding money to the market, we've taken it out. Reverse pyramids do work for the average investor and are a means of controlling exposure and increasing profits.

Pyramiding for the "Not So Average" Investor

Thus far I've been decrying the pyramid while extolling the virtues of the reverse pyramid. As I said, this is virtually always true for the average investor. But for the above average investor, it's a different story. Let me give two cases in point:

In the first case we have a dentist. Here's a fellow who buys gold and silver as part of his business. Because

94

he uses the metal to fill people's teeth, he's aware of the market in a kind of peripheral way.

It is 1979 and the dentist realizes that the silver market is going wild. There are opportunities to profit that may never repeat in history. Silver is on its way up from $6 to an incredible $50 an ounce.

He buys, and as the price goes up, he begins pyramiding his money in the fashion I've described above. As long as silver continues going up, he's making thousands, actually hundreds of thousands.

On the other hand, we have Omar Sharif. Omar Sharif, besides being a famous actor, is also a world class bridge player, a statistician, and an extremely intelligent man. On the other hand, his gambling is legendary and while most people think he's very wealthy, frequently he's quite the opposite.

Sharif is playing in bridge tournaments, and he's also keeping his ear to the wire listening to what's happening in the world of silver. He's making lots of money in the movies, and he decides to enter the silver market, just like my dentist, only with one big difference. Sharif leaves all investing to professionals. He finds a money manager. This is a person who directs the investment of Sharif's money. The money manager doesn't make the investments himself. Rather, he goes out into the market place and finds the very best specialists. Sharif's money manager picks an expert in commodities futures.

Sharif's money manager gives this commodities expert roughly $100,000. Now Sharif has just entered the commodities futures market.

The silver boom continued through 1980 when the rules on margins were changed and the Hunt brothers, who it turns out were later convicted of market manipulation, weren't able to continue buying silver. At that point the price collapsed.

What were the results of the biggest market boom in silver in history for my dentist friend and for Omar Sharif? The dentist came out without a penny! At one time, he had hundreds of thousands of dollars in profit. But because he pyramided, he gave it all back.

Wisdom is the ability to know yourself.

B. R.

What about Sharif? His expert commodities broker, under the watchful eye of his money manager, parlayed his original $100,000 into roughly $4 million. He got out before the market peaked and kept his profits (with which he bought a small casino so he would always have a place to gamble and could always win because he'd be the house).

Let's look at it more closely. Did Sharif's manager pyramid his money? Yes, but very modestly. When his original contracts were up sharply, a small portion of the profits were used to open new contracts. Unlike the dentist, however, his money manager didn't aggressively pyramid all the profits hoping for an even bigger grand slam.

In addition, Sharif didn't pay much attention to the minute-by-minute movements of the market. He generally kept abreast of what was happening and called in occasionally to see how his equity was doing, but kept his distance. He concentrated on playing bridge and making movies.

On the other hand, the dentist virtually gave up his dental practice. Never mind that this was a long-range play in a very unique and special situation. The dentist was on his phone first thing in the morning, before he bathed, talking to his broker to find out how the European markets were doing. Then he made his first daily market decisions.

After a shower, shave, and breakfast, he'd be back on the phone to find out how the New York markets were behaving, and he'd be making more decisions and trades.

He drove to work, saw a few patients, and called again to see what the markets looked like and made more decisions. All during the day he'd be taking breaks and buying and selling on the phone with his broker. At one point he was up over $600,000 in profits! But he stayed in.

Sharif, on the other hand, did exactly what his money manager told him to do. One day, when silver was still up there, his manager called and said they were getting out. Sharif traded out of all his contracts. A few weeks later the silver market collapsed.

Sharif generated several thousand dollars in commissions for the commodities broker and paid his money manager a modest fee, considering the total profits. My dentist friend generated close to a hundred thousand in commissions. And, as I said, when it was all said and done, the dentist had nothing and Sharif walked away with close to $4 million.

The moral here is that a professional money manager directing a commodities broker can be successful in moderate pyramiding. The reason, quite simply, is that he has the distance to be objective. He's working with your money, so you have the greed/fear motivations. He, on the other hand, has the distance to operate a plan/model.

This is not to say that the average person couldn't be successful with moderate pyramiding. It's only to say that I've never seen it happen. The average person very quickly deteriorates into a situation where he's driven by greed. He pyramids far too aggressively and then, when the market turns, he loses it all.

That's why I said at the beginning that all pyramids fail for the average investor. If you're going to be investing your own funds, use a reverse pyramid, as described, but do not try to pyramid. Your timing won't be perfect, you'll lose control, and the market will stomp on you.

On the other hand, if you're working with a professional and that professional at some point comes to you and says, "We're up substantially on your contract(s). Now, I think it's relatively safe to take a small portion of the profits and pyramid into a new contract," you're probably safe in taking his advice. His objectivity is going to make the difference.

STRATEGY 7—DON'T GET MARRIED TO A POSITION

No, this tip doesn't relate to love and sex (at least not directly). Rather, it refers to that expression that anyone who has ever invested in commodities has said at one time or another, "If only I had followed the system!"

Being married to a position is like marrying another person—it's for better or worse. Only with commodities, it's almost always for worse.

Let me illustrate what I mean when I talk about being married to a position in terms of my own personal life.

When I met my wife, my second wife, she, of course, knew that I was supposed to be this whiz kid with finances. However, she wouldn't invest a penny of her money into anything I suggested.

> *It's easier to change your attitude than it is to change the market.*
>
> B. R.

As an aside, she had been told by a fortune teller that she was going to be married five times and profit from the death of each husband. She'd already buried two husbands, each of whom had left her substantial sums of money.

I can't help but note that from her first marriage I did very well. I adopted her son, Tony, who has been a support in life and in business. From her second marriage I ultimately inherited the Roy McVicker Memorial desk and credenza set. It's a $50,000 (appraised value) piece of furniture, and since her second husband was a U.S. Congressman (Democrat), I can't help but wonder how he accumulated a $50,000 desk on $40,000 a year salary? But that's a story for a different book.

Her CPA (actually her dead husband's CPA) told her, "Mary, we don't want to invest with Bill. We don't want to invest with any money manager—it's too risky. We'll put all your money in CDs (certificates of deposit)."

My wife and I truly love each other; however, financially she was married to this position of sticking her money into CDs as advised by her CPA. It was her money, so naturally I didn't try to interfere. I did try to talk her out of it—to diversify, to try stocks, commodities, and real estate. But she was financially married to the CDs. She was convinced they were the only and the safest way to invest.

What I didn't know at the time was that the CPA, when he got her to sign the paperwork for the CDs, also put in there that he had the right to borrow against the CDs. He borrowed and borrowed and lost all the money! Ultimately, my wife got nothing, and the accountant went to prison.

My wife now believes in spreading her investments around. She is no longer married to CDs. (She also believes in investments that she can wear. She wears into the six figures in furs and in jewels because she likes to have things that no matter what, she knows she can get to them and they'll still be there.) By the way, if that fortune teller's prediction was correct, that she's going to marry five husbands and profit from the deaths of all of them, I've got the next one bamboozled—I've got trusts and corporations set up so that money goes where it belongs—to the kids. But that's another story.)

Marrying a position is like knowing—with the kind of certainty that comes only from God—that your investment is not only going to make you a lot of money, but also be safe. When you have that certainty, then you're married to a position. In commodities, it's also almost a sure bet that you're going to lose.

I once wrote an article in my newsletter on corn. It was a special situation, something that is unique and you can't count on, but when it happens, you better be there to take advantage of it.

Corn was around $1.80 a bushel. I had studied the fundamentals and the technicals, and I wrote, "I have never been so sure of anything in my life as I am that corn is going to go to $2.50." (That's a substantial move which could yield fantastic profits given the leverage of commodities.)

I remember what a good friend of mine, Joe, said. (I was carrying his account.) He called as soon as he read the newsletter and said, "I don't want anything to do with corn!"

Naturally I was surprised. I asked, "Joe, I'm dead sure. This is like shooting ducks in a barrel. Why don't you want to take advantage of it?"

I'll never forget his reply. He said, "Your certainty is why I don't want to get in. It's going to cost me too much money."

The only thing you can't learn quickly is experience.

B. R.

Too much money? At the time, I really didn't know what he meant. I was absolutely married to my position on corn. I put Joe aside and I went flying ahead with corn, and all my friends and accounts went along as well.

What happened? I was dead right. Corn went to $2.50. But I got killed along the way. My timing was wrong.

Corn moved to $1.90 and then took a twenty cent drop. It dropped to $1.70. A lot of people got out because of that. (A twenty cent drop was thousands of dollars per contract.) Those who got out weren't there when corn finally did get to $2.50.

That's just what Joe meant when he said it was going to be too expensive.

I was absolutely convinced that corn was going to make the big jump. So I stayed in there, married to my position. No matter what the market told me, no matter what people with common sense told me, I hung in there, losing a pile of money on that twenty cent drop.

If I hadn't been married to my position, I would have gotten out of corn as soon as any signs of weakness developed. But no, I was convinced.

The moral to this story is that the moment you marry a position, you give up your ability to make intelligent market decisions. And the moment that happens, you start losing money.

My solution for this problem is to hire a money manager. I currently do that. I pay him 6 percent a year plus 15 percent of the profits. (Believe me, there are plenty of profits.)

Yes, we talk all the time. Today, if I'm absolutely sure that corn is going to explode in price, if I'm convinced beyond a doubt, I'll suggest to him that he buy corn . . . and then I'll leave him alone. Will he buy corn? Maybe . . . and maybe not. *He's not married to the position.* He'll look at it objectively and make his decision. And I've learned enough through the bitter experience I've just recited to know that I'm better off living with his decision, even though I might be totally opposed to it.

This, of course, gets back to why you need a money manager or a broker to handle your commodities investing for you. If you can't afford these, in a later chapter I'll show you how you can set up a system so that you can be your own money manager, what I call your own "judiciary."

STRATEGY 8—FOLLOW GOLD

As I noted in an earlier chapter, gold is a barometer of economic health. More specifically, gold is a barometer of inflation. There has never been a time when inflation has gone up sharply that gold did not likewise increase in value.

It's important to understand precisely how this phenomenon works. In actuality, it is often the case that gold does not truly increase in value—rather, currencies decrease in value while gold retains its buying power.

Let's take the Mexican peso as an example. Between 1986 and 1988 the peso went from roughly 100 to the dollar to 1,000 to the dollar. That means that whereas before it took 100 pesos to buy a bottle of milk (worth roughly one American dollar), it later took 1,000 pesos to buy the same bottle of milk (which now cost roughly one and a quarter American dollars—the dollar was declining in value as well).

During that same period of time, the price of gold *in pesos* went from roughly 4,000 pesos an ounce to about 50,000 pesos an ounce. Do you see what's happening here?

From the perspective of pesos, the price of gold rose enormously. From the perspective of gold, the value of the peso declined enormously. Which perspective is correct? Both are!

If the mountain won't come to Mohammed, Mohammed will go to the mountain.

The Koran

The whole point is that gold retains its value while currencies, including both the Mexican peso and the American dollar, lose theirs.

Therefore, one way to tell when *significant* inflation is beginning is to watch the price of gold. Gold will signal the start of the next inflationary period.

Gold's Relevance to Commodities

As also noted earlier, commodities are likewise inflation sensitive. If we forget for a moment about fundamental (supply and demand) and technical (charting) price movements, there is another influence on the commodities market and that is, simply, inflation.

Commodities have roughly the same relationship to inflation as gold does. Consider that a bushel of soybeans may cost $5.00 in January. Now, let's just say that the economy of the United States really goes to pot and the United States has a 100 percent rate of inflation during the year. What could we reasonably expect that same bushel of soybeans to sell for *overlooking fundamental and technical* vagaries? The answer, quite predictably, is $10.

You see, soybeans are a commodity which has intrinsic value. Regardless of what's happening to currency, soybeans are going to be bought and sold and eaten. Assuming that the supply and demand factors remain constant and that there aren't any technical aberrations, the price of soybeans should fluctuate in inverse proportion to inflation, just as the price of gold should.

The problem with soybeans, however, is that we can never eliminate fundamental and technical factors. There may be a drought reducing the next crop or an especially wet period enhancing it. Thus, the price may be moving